Chemical Mechanical Polishing in Silicon Processing

SEMICONDUCTORS
AND SEMIMETALS
Volume 63

Semiconductors and Semimetals

A Treatise

Edited by R. K. Willardson
    CONSULTING PHYSICIST
    SPOKANE, WASHINGTON

Eicke R. Weber
DEPARTMENT OF MATERIALS SCIENCE
AND MINERAL ENGINEERING
UNIVERSITY OF CALIFORNIA
AT BERKELEY

Chemical Mechanical Polishing in Silicon Processing

SEMICONDUCTORS AND SEMIMETALS

Volume 63

Volume Editors

SHIN HWA LI

ROBERT O. MILLER

ACADEMIC PRESS

San Diego   San Francisco   New York   Boston
London   Sydney   Tokyo

This book is printed on acid-free paper.

COPYRIGHT © 2000 BY ACADEMIC PRESS

ALL RIGHTS RESERVED.

NO PART OF THIS PUBLICATION MAY BE REPRODUCED OR TRANSMITTED IN ANY FORM OR BY ANY MEANS, ELECTRONIC OR MECHANICAL, INCLUDING PHOTOCOPY, RECORDING, OR ANY INFORMATION STORAGE AND RETRIEVAL SYSTEM, WITHOUT PERMISSION IN WRITING FROM THE PUBLISHER.

Requests for permission to make copies of any part of the work should be mailed to: Permissions Department, Harcourt, Inc., 6277 Sea Harbor Drive, Orlando, Florida, 32887-6777

The appearance of the code at the bottom of the first page of a chapter in this book indicates the Publisher's consent that copies of the chapter may be made for personal or internal use of specific clients. This consent is given on the condition, however, that the copier pay the stated per-copy fee through the Copyright Clearance Center, Inc. (222 Rosewood Drive, Danvers, Massachusetts 01923), for copying beyond that permitted by Sections 107 or 108 of the U.S. Copyright Law. This consent does not extend to other kinds of copying, such as copying for general distribution, for advertising or promotional purposes, for creating new collective works, or for resale. Copy fees for pre-1999 chapters are as shown on the title pages; if no fee code appears on the title page, the copy fee is the same as for current chapters. 0080-8784/00 $30.00

ACADEMIC PRESS
A Harcourt Science Technology Company
525 B Street, Suite 1900, San Diego, CA 92101-4495, USA
http://www.academicpress.com

ACADEMIC PRESS
24–28 Oval Road, London NW1 7DX, UK
http://www.academicpress.com

International Standard Book Number: 0-12-752172-0
International Standard Serial Number: 0080-8784

PRINTED IN THE UNITED STATES OF AMERICA
99 00 01 02 03 04 EB 9 8 7 6 5 4 3 2 1

Contents

PREFACE xi
LIST OF CONTRIBUTORS xiii

Chapter 1 Introduction 1
Frank B. Kaufman

   I. CMP: A UNIQUE AND EVOLVING SEMICONDUCTOR FABRICATION
      TECHNOLOGY — PAST, PRESENT, AND FUTURE 1
      1. *The History and Development of CMP as a Unique Semiconductor
         Fabrication Technology* 2
      2. *CMP: The Current State of the Art* 3
      3. *The Evolution of CMP Technology into the New Millenium* . . . 3

Chapter 2 Equipment 5
Thomas Bibby and Karey Holland

   I. INTRODUCTION 5
  II. CMP EQUIPMENT DESIGN EVOLUTION 8
      1. *Throughput Improvement* 9
      2. *First-Generation Polishers* 10
      3. *Second-Generation Polishers* 11
      4. *Third-Generation CMP Equipment* 13
 III. CARRIERS 16
      1. *Gimbaled Carriers* 20
      2. *Floating Carriers — The Use of Hydrostatic Pressure* . . . 21
      3. *Linear Polisher and Pressure Control* 23
  IV. PLATENS 24
      1. *Temperature Control* 24
      2. *Orbital Polisher and Direct Slurry Delivery* 25
   V. PAD CONDITIONING 25
  VI. CMP EQUIPMENT INTEGRATION 30
      1. *Post-CMP Cleaning* 30
      2. *Integrated Metrology* 35
      3. *Slurry Reprocessing* 37

VII.	COPPER POLISHING AND CMP TOOL REQUIREMENTS	38
VIII.	300-MM CMP TOOLS	40
IX.	CONCLUSION	41
	REFERENCES	42

Chapter 3 Facilitization 47
John P. Bare

I.	INTRODUCTION	47
II.	OUTLINE	48
III.	SLURRY DISTRIBUTION SYSTEM OVERVIEW	48
IV.	SLURRY HANDLING	49
V.	SLURRY DISTRIBUTION SYSTEMS	52
VI.	SLURRY DISPENSE ENGINES	53
VII.	SLURRY BLENDING TECHNOLOGY	54
VIII.	SLURRY MEASURING TECHNIQUES	55
IX.	DAYTANK REPLENISHMENT	59
X.	MIX ORDER	61
XI.	PIPING SYSTEMS	62
XII.	PIPING SYSTEM VARIATIONS	63
XIII.	MATERIALS OF CONSTRUCTION	64
XIV.	SLURRY SETTLING	65
XV.	SLURRY ROOM LOCATION	66
XVI.	PRESSURE AND FLOW CONSISTENCY	66
XVII.	BACK-PRESSURE DEVICES	67
XVIII.	SLURRY CONSUMPTION RAMP	68
XIX.	SYSTEM REDUNDANCY	69
XX.	VALVE BOXES	69
XXI.	STORAGE TANKS	71
XXII.	AGITATION	74
XXIII.	METROLOGY	76
XXIV.	FILTRATION	78
XXV.	SLURRY SYSTEM MAINTENANCE	83
XXVI.	WASTE DISPOSAL	84
	REFERENCES	87

Chapter 4 Modeling and Simulation 89
Duane S. Boning and Okumu Ouma

I.	INTRODUCTION	89
II.	WAFER-SCALE MODELS	90
	1. *Macroscopic–Bulk Polish Models*	90
	2. *Sources of Wafer-Scale Nonuniformity*	92
	3. *Empirical Approaches to Wafer-Level Modeling*	97
	4. *Status of Wafer-Level Modeling*	97
III.	PATTERNED WAFER CMP MODELING	98
	1. *Feature-Scale Models*	100
IV.	DIE-LEVEL MODELING OF ILD CMP	104

1. Topography Modeling: Layout Density Dependence 105
2. Modeling of Thickness Evolution 106
3. Planarization Length and Response Function 108
4. Characterization—Determination of Planarization Length 113
5. STI CMP Modeling 118
6. Models for Step Height Reduction 120
7. Applications of Density Models 124
V. Models for Metal Polishing 125
1. Tungsten CMP Modeling 126
2. Tungsten CMP—Contact Wear Model 128
3. Copper CMP Modeling 131
VI. Summary and Status 132
References 133

Chapter 5 Consumables I: Slurry 139
Shin Hwa Li, Bruce Tredinnick, and Mel Hoffman

I. Introduction 139
II. Abrasives 140
1. For Oxide Slurry 141
2. For Metal Slurry 142
3. Agglomeration 143
4. Milling 145
III. Slurry Solution 146
1. For Oxide Slurry 146
2. For Metal Slurry 149
IV. Comparisons Among Slurries 150
1. Oxide Slurries 150
2. Tungsten Slurries 151
References 153

Chapter 6 CMP ConsumablesII: Pad 155
Lee M. Cook

I. Introduction 155
II. Classes of Pads and Their Manufacture 156
1. Classes of Pads 156
2. Primary Manufacturing Processes 156
III. Structure, Properties, and Their Relationship to the Polishing Process . 162
1. Local-Level Models for the Polishing Process 162
2. Die-Scale Models 166
3. Wafer-Scale Models 167
4. Impact of Structure and Properties on the Polishing Process 169
IV. Application to Semiconductor Processing 170
1. Dielectric CMP 170
2. Metal CMP 177
References 180

Chapter 7 Post-CMP Clean 183
François Tardif

    I. INTRODUCTION 183
    II. SURFACE CONFIGURATIONS AFTER CMP PROCESSES 184
    III. CLEANING REQUIREMENTS AFTER CMP PROCESSES 184
        1. *Particle Effects* 184
        2. *Metallic Contamination Effects* 185
        3. *Damaged-Layer Effects* 186
    IV. CORROSION EFFECTS 186
        1. *Electrochemical Corrosion* 186
        2. *Photoassisted Corrosion* 190
    V. SLURRY REMOVAL 193
        1. *Particle-Removal Mechanisms* 193
        2. *Scrubber Cleanings* 202
        3. *Wet Cleanings* 204
    VI. METALLIC CONTAMINATION REMOVAL 206
        1. *Metallic Contamination during the Full CMP Step* 206
        2. *Metallic Contamination Cleaning* 207
    VII. DAMAGED LAYER REMOVAL 208
        1. *Practical Determination of the Damaged Layer* 209
        2. *Elimination of the Damaged Layer* 209
    VIII. FINAL PASSIVATION 210
    IX. EXAMPLES OF PRACTICAL POST-CMP CLEANING PROCESSES . . . 210
        1. *STI and Silicon Oxide CMP* 211
        2. *Tungsten CMP* 212
        3. *Copper CMP* 212
    X. CONCLUSION 213
    REFERENCES 213

Chapter 8 CMP Metrology 215
Shin Hwa Li, Tara Chhatpar, and Frederic Robert

    I. INTRODUCTION 215
    II. REFLECTOMETRY 216
        1. *Substrate Modeling* 218
        2. *Measurement Patterns and Points* 220
        3. *Measurement of Patterned Wafers* 224
        4. *Integration Issues* 225
    III. DEFECTIVITY MONITORING 226
        1. *Laser Scanning Method* 228
        2. *Digital Image Comparison Method* 228
    IV. NONCONTACT CAPACITIVE MEASUREMENT 229
        1. *Flatness* 231
        2. *Bow and Warp* 232
    V. TOTAL X-RAY FLUORESCENCE 233
    VI. STYLUS PROFILOMETRY (FORCE MEASUREMENT) 236
    VII. ATOMIC FORCE MICROSCOPY 236
    VIII. FOUR-POINT PROBE 241
    REFERENCES 243

Chapter 9 Applications and CMP-Related Process Problems ... 245

Shin Hwa Li, Visun Bucha, and Kyle Woolridge

- I. Introduction ... 245
- II. Oxide CMP Within-Wafer Nonuniformity (WIWNU) ... 246
  1. Equipment and Consumables ... 246
  2. Recipe Parameters ... 251
  3. Effect of the End Effector ... 252
  4. Deposit Less, Polish Less vs Deposit More, Polish More (Uniformity vs Planarity) ... 257
  5. BPSG vs TEOS ... 261
- III. Post-CMP Oxide Thickness Control ... 262
  1. Lack of Endpoint Detection System ... 262
  2. Wafer-to-Wafer Thickness Nonuniformity (WTWNU) ... 263
- IV. Defectivity ... 265
  1. Scratches ... 266
  2. Other Defects ... 270
- V. Tungsten CMP Problems ... 273
  1. Metal Contamination ... 273
  2. Oxide Erosion ... 277
  3. Polish Rate and Uniformity ... 277
- VI. Other Problems ... 277
  1. Missing Alignment Mark ... 278
  2. Dishing ... 279
  References ... 280

Index ... 283
Contents of Volumes in This Series ... 289

Preface

In the engineering world, many ideas and plans are conceived in the information exchanges that occur over lunch. In the summer of 1997, one of the editors (Shin Hwa Li) was assigned to assist our CMP module at the Crolles facility of STMicroelectronics near Grenoble, France. One day, in the company cafeteria, while sitting near the window and taking in the gorgeous sub-Alpine scenery with a delicious French meal, he was joined by several site engineers. Having become aware that he was the CMP engineer responsible for both oxide and metal processes in another fab, they asked him questions regarding CMP fundamentals, equipment, consumables, application and the like. At that moment, the thought occurred to him that a CMP textbook would be very useful. To his knowledge, such a book did not yet exist. Ironically, while he was fielding and responding to the many questions, one of the engineers (Jimmy Huang) remarked somewhat jokingly that he should write a book about CMP.

The idea remained fixed in Li's mind, and after returning to the United States, he seriously considered the possibility. He discussed this with his manager (and coeditor, Bob Miller), who, being of the mindset that almost any idea is worth pursuing, immediately liked the idea. In performing a literature search, we found only one such book extant (*Chemical Mechanical Planarization of Microelectronic Materials*, by J. M. Steigerwald, S. P. Murarka, and R. J. Gutmann, John Wiley & Sons, New York, 1997). We read it and discovered that the book, although an excellent reference, is academically oriented. We felt that since the CMP technology was becoming a major and critical part of the semiconductor manufacturing environment, a practical, industrially oriented text, directed more toward the level of sustaining engineers, should be made available.

It is true that many industrial papers and articles have been published, but each one is usually focused on a single specific subject; furthermore, the literature is diffused into many different journals and/or conference proceedings. At some point, a textbook becomes an essential tool for imparting a coherent and comprehensive understanding. On the other hand, since CMP technology is currently still in its early stage of use, a book with only one or two authors would lack some of the insights that come from multiple sources, so we elected to edit a book that combined the inputs of a number of authors who are recognized experts in the field.

This book is designed primarily to help all engineers whose work impinges on CMP in the semiconductor industry: process, process development, and integration engineers; device and product engineers; and equipment, vendor applications, and vendor field engineers. At the same time, it is written at a level where students in colleges should find it a useful aid in bridging their knowledge between academic principles and actual practice. Perhaps more importantly, we intend this book to generate interest in the subject and to usher more people into joining and exploring the CMP world.

The book contains nine chapters. Chapter 1 is an introduction. Chapters 2 and 3 are devoted to equipment and facilitization issues; if it has been decided to introduce CMP to a fab's production line, these two chapters are useful to help the engineers to choose the tools and handle miscellaneous issues in a plant. Chapter 4 is designed to help engineers do advanced analysis and prediction of CMP interaction with devices or integrated circuits. Chapters 5 and 6 deal with the issues of CMP consumables: slurries and pads. Every experienced CMP engineer knows that CMP is a highly consumables-dominated process. The choices, correct use, and combination of consumables lie at the heart of the technology and are a fundamental determinant as to whether a CMP process will succeed. These chapters require a clear elucidation and understanding. Chapter 7 pertains to post-CMP cleaning. This is a new, and increasingly important topic in CMP. As design rules of integrated circuits become smaller, the sensitivity to unwanted residuals or contamination is greater. This chapter will help engineers properly use chemistry to achieve cleanness. Chapter 8 deals with CMP metrology; it is intended to help readers understand the importance and limitations of CMP metrology. Since most available metrological tools were not originally designed for CMP, their adaptation to it is encumbered by pitfalls and incompatibilities. Finally, Chapter 9 involves applications (where CMP is, or might be, utilized in industry) and deals with CMP-related process problems (what the open issues are, and how they might be resolved).

Shin Hwa Li would like to thank his previous graduate school advisor, Prof. G. B. Stringfellow, University of Utah. He provided invaluable teaching, not only in science but in interpersonal relationships as well. More recently, he gave guidance in presenting a book proposal to our publisher (Academic Press). Both editors would like to also thank Dr. Z. Ruder,

Academic Press, for allowing us the opportunity to produce a book, and for his encouragement. We are indebted to Dr. M. Fury, AlliedSignal, who initially helped us to select and organize the topics and the authors of each chapter. Many thanks go to C. R. Spinner, L. Jorgensen, and J. P. Rossome, all of STMicroelectronics, who encouraged our efforts at editing and allowed our use of data in the manuscripts of the two chapters we produced. Shin Hwa Li would also like to acknowledge F. McClung of EKC Technology. Upon his recently joining EKC, he gave Li freedom to finish this book. Most of all, the editors are very grateful to the outside contributors of each chapter; namely, Dr. F. Kaufman, Cabot, for Chapter 1, Dr. K. Holland and Dr. T. Bibby, Speedfam/IPEC, for Chapter 2, Dr. J. Bare, BOC Edwards, for Chapter 3, Dr. O. Ouma and Dr. D. Boning, MIT, for Chapter 4, Dr. L. Cook, Rodel, for Chapter 6, Dr. F. Tardif, LETI, for Chapter 7. All made sacrifices in their busy schedules to finish each chapter.

Finally, we acknowledge personal debts of gratitude: Shin Hwa Li to his parents, Weiwen and Moyuan, and his wife, Yina. They instilled in him faith, hope, and love, as have his daughter and son, Crystal and Jason, who give him endless joy. Bob Miller is grateful to his wife and son for time editing that would otherwise have been spent with them.

SHIN HWA LI
ROBERT O. MILLER

List of Contributors

Numbers in parentheses indicate the pages on which the authors' contribution begins.

JOHN P. BARE (47), *BOC Edwards, Santa Clara, California*

THOMAS BIBBY (5), *SpeedFam-IPEC, Chandler, Arizona*

DUANE S. BONING (89), *Massachusetts Institute of Technology, Cambridge, Massachusetts*

VISUN BUCHA (245), *STMicroelectronics, Phoenix, Arizona*

TARA CHHATPAR (215), *STMicroelectronics, Phoenix, Arizona*

LEE M. COOK (155), *Rodel Inc., Newark, Delaware*

MEL HOFFMAN (139), *EKC Technology, Hayward, California*

KAREY HOLLAND (5), *SpeedFam-IPEC, Chandler, Arizona*

FRANK B. KAUFMAN (1), *Cabot Corporation, Microelectronics Materials Division, Aurora, Illinois*

SHIN HWA LI (139, 215, 245), *EKC Technology, Hayward, California*

OKUMU OUMA (89), *Massachusetts Institute of Technology, Cambridge, Massachusetts*

FREDERIC ROBERT (215), *STMicroelectronics, Phoenix, Arizona*

FRANÇOIS TARDIF (183), *LETI, Grenoble, France*

BRUCE TREDINNICK (139), *EKC Technology, Hayward, California*

KYLE WOOLDRIDGE (245), *STMicroelectronics, Phoenix, Arizona*

CHAPTER 1

Introduction

Frank B. Kaufman

CABOT CORPORATION
MICROELECTRONICS MATERIALS DIVISION
AURORA, ILLINOIS

I. CMP: A UNIQUE AND EVOLVING SEMICONDUCTOR FABRICATION
   TECHNOLOGY—PAST, PRESENT, AND FUTURE 1
   1. *The History and Development of CMP as a Unique Semiconductor
      Fabrication Technology* 2
   2. *CMP: The Current State of the Art* 3
   3. *The Evolution of CMP Technology into the New Millenium* 3

I. CMP: A Unique and Evolving Semiconductor Fabrication Technology — Past, Present, and Future

Chemical mechanical planarization (CMP) has become, in a few short years, a required semiconductor processing module used in fabrication facilities worldwide. The history and development of CMP technology, the current state of the art, and how this novel processing technology will evolve are current topics of enormous interest presented and discussed in countless forums throughout the world. These forums typically involve scientific conferences and workshops, user groups, trade shows, and published articles and patents. There has been a noticeable lack of archival texts dedicated to these topics. The current work, conceived, organized, and edited by Drs. Shin Hwa Li and Robert Miller, with the significant effort of a distinguished set of authors, promises to add significantly to the archival record dedicated to CMP technology. Hopefully, the work discussed within will help make the next 10 years of planarization technology development as fascinating and interesting as the last.

1. THE HISTORY AND DEVELOPMENT OF CMP AS A UNIQUE SEMICONDUCTOR FABRICATION TECHNOLOGY

Arguably, the relentless and never failing capability to fabricate integrated circuitry on Silicon wafers in a manner that continually meets or exceeds Moore's law is the physical basis for the hugely successful microelectronics industry. To drive that processing engine, critical advancements in IC fabrication processing must be available in a timely fashion. Other requirements for processes to become mainstream and adopted worldwide are a viable infrastructure to support use of the technology, extendibility to future chip generations, and no obvious competitive process technologies that displace the initial invention.

CMP is a unique process technology for a variety of reasons. Perhaps the most far-reaching, however, has been that CMP is an enabling technology that allows chip makers to readily drive lithographic patterning steps to smaller dimensions. By virtue of the ability of CMP to provide flat surfaces, initially for interlevel-dielectric (ILD) layers and in W plug processing, it has enabled chip designers to make use of advanced lithographic patterning techniques, providing the continuous ability to shrink chips to smaller dimensions and seamlessly add additional levels of wiring. CMP came along just in time to provide an extendible processing technology that enables chip makers to stay ahead of depth-of-field limitations in evolving optical exposure tools. An ages old, "retro" technology related to glass polishing and metallographic finishing, thus enables optical lithography to work.

Additional uniqueness, during the development phase of this newest semiconductor fabrication technology, stems from three additional considerations: the fact that this is a fundamentally "dirty" and wet technology, the lack of systematic mechanistic fundamentals that existed during the development phase, and the rapidity by which the process went from development laboratory curiosity to mainstream manufacturing process. In all three aspects, adoption of CMP represents a situation that is a true paradigm shift from the typical way in which technological advancements become mainstream, high-technology, semiconductor manufacturing processes. One personal perspective on the successful and rapid adoption of the technology against many odds and conventional wisdom is that it succeeded due to the collective technical intuition and wisdom of process engineers and technical managers, that CMP represented a true advance that could be made to work despite enormous odds and technical difficulties early on.

2. CMP: THE CURRENT STATE OF THE ART

This topic is well covered by the contributions in this volume. CMP continues to be viewed as a surprisingly unique and flexible semiconductor fabrication technology by virtue of its ability to "make manufactureable" potential fabrication sequences that are either too cumbersome or too low in yield to be fabricated in any other manner. Using virtually any CMP polisher, a variety of materials of interest to IC fabricators can be planarized. These materials include insulators, semiconductors, interconnect metals, and barrier metallurgies. This means that once a user becomes adept in polishing one kind of material, typically oxide and W at first, other materials of interest and other semiconductor processing sequences become viable.

Currently, two examples of CMP-enabling-specific and novel processing steps to be made useable, from a manufacturability point of view, involve shallow trench isolation (STI) and Cu damascene. Historically, damascene architecture, which conceptually connects both STI and Cu BEOL processing, has been known since ancient times. It is a technique for producing jewelry, used for some time in Pacific Rim locations. In the late 1980s, the damascene or "inlaid metal" concept was first proposed by several different authors in journal publications as a novel approach for fabricating semiconductor devices. However, it was not until both advances were made in metal fill technology and unexpected breakthroughs were produced in the ability to use a polishing technique (to clear the metal high spots while leaving untouched the metal in the trenches), now referred to as CMP, that the concept was shown to be manufactureable. The result, a viable multilevel Cu damascene technology sequence, promises to significantly alter the way BEOL structures are fabricated in the near future.

3. THE EVOLUTION OF CMP TECHNOLOGY INTO THE NEW MILLENIUM

The year 2000 will mark 15 years since the initial CMP patents were filed by IBM. Opportunities for expanding use of CMP in existing chip technology continue to flourish. In addition, the challenges ahead for CMP technology to keep pace are formidable in this third wave of the evolution of the technology. Increasing concern about improved within-wafer nonuniformity, better planarity (flatter surfaces), and lower defectivity levels are all requirements for advanced, sub-0.25-micron devices. In addition, there is increasing attention to overall cost of ownership and logistics as the total number of CMP operations increases dramatically.

The chapters in this volume serve to provide a window on the evolution of CMP to meet these increasing challenges. Additional aspects of the demands to be placed on CMP technology will be provided by other features of the aggressive and relentless evolution of the semiconductor industry: the move to 300-mm wafers, increasing chip-level integration for system-on-a-chip in application-specific designs, and totally new uses for CMP processing (i.e., in fabricating metal gates) that will continue to force innovation and accelerate manufacturability.

It will be interesting to observe how CMP evolves and matures to meet these newest challenges. Increasing attention by university groups suggests that predictive modeling and the development of a fundamental understanding on some aspects of these processes will help significantly. On a final, more personal note, it certainly has been fascinating watching and being part of the growth of the technology. Further exciting developments are likely, as is obvious from the work presented in this volume.

CHAPTER 2

Equipment

Thomas Bibby and Karey Holland

SPEEDFAM-IPEC
CHANDLER, ARIZONA

I.	INTRODUCTION	5
II.	CMP EQUIPMENT DESIGN EVOLUTION	8
	1. Throughput Improvement	9
	2. First-Generation Polishers	10
	3. Second-Generation Polishers	11
	4. Third-Generation CMP Equipment	13
III.	CARRIERS	16
	1. Gimbaled Carriers	20
	2. Floating Carriers—The Use of Hydrostatic Pressure	21
	3. Linear Polisher and Pressure Control	23
IV.	PLATENS	24
	1. Temperature Control	24
	2. Orbital Polisher and Direct Slurry Delivery	25
V.	PAD CONDITIONING	25
VI.	CMP EQUIPMENT INTEGRATION	30
	1. Post-CMP Cleaning	30
	2. Integrated Metrology	35
	3. Slurry Reprocessing	37
VII.	COPPER POLISHING AND CMP TOOL REQUIREMENTS	38
VIII.	300-MM CMP TOOLS	40
IX.	CONCLUSION	41
	REFERENCES	42

I. Introduction

The acceptance of chemical mechanical planarization (CMP) as a manufacturable process for state-of-the-art interconnect technology has made it possible to rely on CMP technology for numerous semiconductor manufacturing process applications. These applications include shallow trench isolation (STI), deep trench capacitors, local tungsten interconnects, inter-level-dielectric (ILD) planarization, and copper damascene. In this chapter, we discuss the design evolution of CMP tools with this trend in mind and how the various approaches seek to maintain or improve high levels of performance and customer satisfaction. We review the key components that are required for CMP and discuss the benefits and disadvantages of both evolutionary and revolutionary approaches. We conclude with a summary of some of the equipment requirements of CMP tools in the next few years. The growth of CMP has been explosive, and with this growth has come a tremendous amount of published work ranging from trade journals and web site publications to refereed journals and to the patent literature. It is not possible to cite all of these sources, and the references listed in this chapter are intended to reflect the broad growth in the industry, not that of a particular subset of that group.

CMP began as a local and global planarization technology to enhance optical lithography process windows and metal interconnect reliability [1, 2]. The initially unanticipated yield enhancement brought by CMP over other planarization technologies such as etch-back has lead to the widespread acceptance of CMP for state-of-the-art interconnect applications, namely, interlayer dielectric and tungsten plugs [3]. By the beginning of 1998, more than 40 companies were using CMP in production, and CMP has emerged as a critical enabling technology as minimum device geometries shrink to below $0.35\,\mu m$ [4]. The reduction of minimum dimensions is paving the way for new applications: shallow trench isolation and <0.18-μm-generation copper dual-damascene interconnect technology [5–7]. CMP is required for both of these processes.

The rapid acceptance of CMP for advanced processing has stimulated dramatic growth in the supporting capital equipment market. For several years prior to the downturn in the semiconductor industry in 1998, the CMP market had demonstrated an average annual growth rate of slightly over 30%. Such dramatic growth has attracted many companies, and as of this writing more than 25 companies were either selling or trying to sell CMP tools. Many of these companies have a history of silicon wafer substrate polishing. In this chapter, however, we consider only those that are involved in planarization. Table I shows the market share of the top 10 in 1997. It is unlikely that all 25 companies will succeed in the extremely competitive CMP market [8]. Many will fail. Others will follow the example of IPEC and SpeedFam and merge.

The key issues affecting industry use of CMP during semiconductor chip manufacturing are the high cost of ownership (CoO), the lack of industry-wide CMP technology and integration knowledge, and the less than thorough understanding of the underlying science behind CMP. (Process integration issues and detailed discussion of the science of CMP processes are discussed in the other chapters.) Although leading semiconductor manu-

TABLE I

MAJOR SUPPLIERS OF CMP EQUIPMENT [9]

Company	1997 Market share (%)
IPEC	26.2
Ebara	22.2
Speedfam	21.5
AMAT	14.5
Strasbaugh	7.7
Sumitomo	2.0
Cybeq	1.4
Okamoto	1.4
Lapmaster	1.3
Toshiba	0.7
All others	1

facturers have world-class CMP development capability, the bulk of manufacturers lack the resources to compete with the leaders in the industry in CMP process development. The growing need in the market for CMP has led CMP tool suppliers to help fill that gap. Thus, the leading CMP equipment suppliers not only sell CMP tools, but they also offer CMP process implementation and integration support as well.

The cost per CMP wafer pass is high when compared to other semiconductor manufacturing processes. This aspect of CMP reflects the rapid transfer of CMP into production, as well as the large amount of slurry and pads that are consumed during CMP. CMP CoO has improved in the last few years due to several factors.

1. Raw throughput is higher (second- and third-generation polishers produce 25 to 60 wfr/hr in production).
2. Film thickness metrology has been integrated into CMP tools, which reduces or eliminates lost productivity due to test wafer queuing delays.
3. CMP tools are being built in a dry-in-dry-out configuration, which enables CMP processes to be optimized for minimum defectivity in addition to flatness. CMP tools with integrated cleaners also enable CMP tools to be operated in cleanrooms, which reduces wafer transport times. These tools cost more, which would otherwise increase the CoO. However, by being able to merge the polisher and the cleaner into a single tool, the overall cost of ownership of obtaining a clean, flat wafer declines.

The cost of consumables remains the largest component of CMP CoO, and in some geographic regions, this factor must now take into consideration the large amount of deionized (DI) water that is used to clean wafers.

CMP processes for oxide planarization (ILD and STI) rely on slurry chemistry to hydrolyze and soften the SiO_2 surface. Mechanical abrasion then controls the actual material removal. Thus, the key process output control variables (i.e., removal rate and nonuniformity) are strong functions of the mechanical properties of the system, namely, the down force and the relative velocity between the pad and the wafer. Metal CMP processes such as copper CMP rely more on chemical oxidation and dissolution of the metal than mechanical abrasion to remove the metal overburden. Consequently, careful control of the chemistry of the CMP process is more important for these CMP processes than it is for oxide CMP. Thus, CMP tools and processes optimized for ILD may not be optimal for metal CMP and vice versa.

In the following, we discuss the evolution of tool designs from first-generation modified silicon polishing tools to third-generation tools designed specifically for CMP and based on revolutionary concepts. This development path has been driven by the need for higher throughput and lower cost of ownership. CoO improvements result from many factors. One of the most important is due to the tool design and its effect on throughput. Then, for any given tool, process improvements contribute substantially to minimizing the CoO. Thus, we follow the discussion of increased throughput methods with discussions of carrier designs, platens, pad conditioning, and integrated CMP equipment. These changes set the stage for the implementation of copper metallization, which we discuss next. Copper is emerging as the preferred conductor for 0.15-μm technology. There will be some applications such as dynamic random access memories (DRAMs) where aluminum CMP is likely to play a role, but in terms of the number of CMP steps, copper CMP will vastly outnumber aluminum CMP. Efforts to shrink chip designs using 0.25-μm technology are enabling current fabs to meet the demand for performance. However, it is estimated that by the year 2002, fabs built for 300-mm tools will begin to see significant production usage. Thus, we close this chapter with a discussion of 300-mm CMP tools, the challenges they present, and the prognosis for their arrival by 2002.

II. CMP Equipment Design Evolution

The first CMP tools, based on rotational platen silicon wafer polishing tools, had low throughput values (10 to 18 wfr/hr). Since the implementation of these first-generation tools for polishing, CMP tool design has experienced two additional generations of development and innovation. The second generation emphasized evolutionary improvements, while the third generation has involved revolutionary polishing designs. Both second- and third-generation equipment designs are likely to remain in production for many years. Further innovations in planarization technology are likely to be blended into the existing planarization options, but such a digression is beyond the scope of this discussion.

1. THROUGHPUT IMPROVEMENT

Raw throughput improvement can be achieved by reducing the amount of material that must be polished away, increasing film removal rates, and increasing the wafer-handling efficiency. Wafer-handling efficiency is addressed in the following sections.

The ideal deposition process would leave a perfectly flat surface. That does not happen, so a planarization process is required to maintain depth of field requirements. For a dielectric planarization process, the ideal planarization process would remove only material in the "up" areas and remove no material in the "down" areas. Metal CMP involves the removal of metal overburden, leaving filled plugs or vias (single damascene) or filled vias and inlaid metal lines (dual damascene) with no removal of metal in the inlaid region and no removal of dielectric.

The most efficient of CMP processes requires the removal of the least amount of material, especially material in "down" areas. Throughput for efficient CMP processes are often robot-limited, although sometimes the polish time is increased deliberately to allow the use of lower down force to obtain better planarization results. The throughput of less efficient processes is limited by the time it takes to remove the excess material. For these processes, the throughput can be improved either by increasing the removal rate or by polishing more wafers at a given time. Since the removal rate increases with increasing relative velocity and with increasing down force, an increase in either of these parameters can increase throughput. However, steps to increase removal rates can be used only if there is no unacceptable degradation of the CMP process window (i.e., increase in the within-wafer nonuniformity, increase in the planarization or remaining topography, etc.).

An alternative approach to increasing throughput is to increase the number of wafers being polished. Polishing multiple wafers on one platen concurrently has been demonstrated to substantially increase overall tool throughput, but also increases the risk of damage to all the wafers on the pad in the event one of the wafers breaks. Significant improvements have been achieved in minimizing this risk. Alternatively, some second- and third-generation CMP tools polish several wafers simultaneously, but on multiple pads (usually one wafer per pad).

Third-generation CMP tools use revolutionary polish mechanisms to achieve increased raw throughput values. In addition to improved throughput, third-generation polishers add other process flexibility or planarization quality enhancements. Advantages and disadvantages of these designs are compared in the following.

2. First-Generation Polishers

First-generation tools typically use a single robot to move wafers through the tool, and this robot also holds the wafer on the carrier head. These tools typically have two large rotating platens. A relatively hard polish pad covers one head while a relatively soft buff pad covers the second platen. The carrier presses the device side of wafers against the polish pad on the platen at a point offset from the polish platen center of rotation. Slurry is deposited close to the center of the pad and centrifugal force spreads the slurry across the pad. The pad condition (i.e., structure, absorbency, and nap) strongly affects slurry transport to the pad–wafer interface. The polishing platen on these tools is large, typically 55.9 cm (22 in) in diameter, or more than 7.5 times the area of a 200-mm wafer.

The wafer is polished by the interaction of the wafer surface with the slurry chemistry, abrasives, and the pad. During wafer polishing, the pad surface structure is also planarized; however, this process is not well understood. The protruding pad material is compressed by contact with the wafer and is abraded by interaction with the slurry particles and the wafer surface. Residual pad material, abrasive particles, and redeposited material from the wafer surface are deposited in pad pores, thus filling up the pores, causing a glazed appearance and diminished polish performance. In conventional CMP tools, the pads are planarized and thinned more in proportion to the wafer–pad contact. Pad wear by these processes results in a bowl-shaped trench being worn into the pad. Thus, pads must be conditioned to (1) bring the pad back to flat, (2) remove material from pores, and (3) rebuild the nap.

Conventional CMP systems dispense slurry onto the polishing pad upstream from the wafer carrier, as depicted in Fig. 1. Centrifugal force causes slurry to move radially outward, where most of it encounters the wafer carrier. The slurry builds up on the leading edge of the carrier, but most of it falls off the pad and is discharged from the system. Actual slurry utilization is poor. In the absence of slurry reprocessing, the pH of the slurry used in CMP changes during usage [10]. The amount of slurry that actually

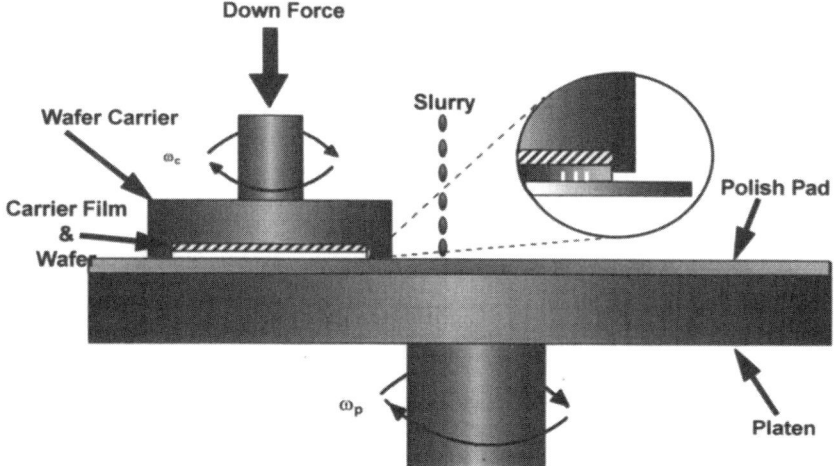

FIG. 1. Schematic diagram of a first-generation rotational CMP tool.

encounters the face of the wafer is a function primarily of the pad conditioner and the pad topography.

Early generation tools kept wafers sitting in water until the cassettes were completed and the whole boat of wafers could be placed in a batch post-CMP wafer cleaner. Keeping the wafers wet until they can be cleaned is critical to achieving the expected post-CMP defectivity and related yield improvements.

The removal rate on first-generation polishers can be improved by simply increasing the rotation rate. However, nonuniformity can suffer if the platen run-out is not controlled. Platen run-out is a measure of the wobble a platen undergoes during rotation and must be kept to a minimum to ensure good performance.

3. SECOND-GENERATION POLISHERS

Second-generation CMP tools are based on the classical rotating platen and rotating wafer design and include numerous evolutionary changes to increase raw throughput. There are generally two types: the single large platen [22 in (55.9 cm) or larger] that polishes several wafers concurrently on the same pad, and the single wafer per platen multiplaten systems. Large platens have the added benefit of increasing the relative velocity simply by moving the carrier toward the edge of the platen without increasing the platen rotation rate. Many of these systems have the option of being linked to post-CMP cleaners for dry-in-dry-out capability. The resulting second-generation tools featured significantly increased throughput, typically quoted in the range of 30 to 60 wfr/hr and from 0.5 to >1 wfr/hr per square foot of floor space.

a. Multiwafer per Platen Polishers

One approach to increasing the throughput is to increase the number of polish heads per platen. This approach results in a dramatic increase in raw throughput. However, there are several challenges with this approach. The most severe issue is the quantity of wafers put at risk. If a wafer breaks, the shards become embedded in the pad and immediately scratch and gouge the other wafers. When multiwafer per platen CMP tools were introduced and were shown to provide a greater than a factor of 2 increase in throughput, many semiconductor manufacturers were willing to take the risk of occasionally breaking a wafer. Since the introduction of these tools, significant improvements have been attained with multiwafer polishing systems, although the concerns linger.

A more subtle issue with this approach is load balancing. As long as there is a consistent number (and distribution) of wafers loaded into the carriers—for example, 5 wafers at a time (from a cassette containing perhaps 24 wafers)—polishing results could be relatively consistent. However, whenever there are fewer than 5 wafers, the pad wears differently (more slowly), and processing results may change. Also, should there be only a fraction of a cassette of wafers to polish, a scenario relatively common to application-specific integrated circuit (ASIC) manufacturers, polishing cycles in which 1 or 2 wafers to be polished would cause even greater challenges in load balancing for process engineers. Thus, there has been a trend toward single wafer per pad processing, for example, systems from SpeedFam-IPEC, Applied Materials, Lam, Obsidian, and others.

b. Sequential Rotational Systems

Another approach is to sequentially process wafers through a number of platens on a single system. This approach lends itself to multistep processing. For example, the first platen could be used for bulk removal without particular regard to scratching. The next platen could be operated under conditions optimized for planarization, while the third platen could be used for a final buff to eliminate scratches. It should be noted that to optimize the overall tool throughput, processing times on all three platens must be approximately the same to avoid bottlenecks. In other words, the throughput is limited by the slowest polish step. If the material being polished is copper, then there is a potential issue of corrosion occurring while wafers being polished on one of the other platens completes its portion of the polish cycle.

Sequential processing tools also increase the risk of multiple wafer damage. If a stray particle appears on one of the three pads, all wafers will be damaged by this particle and all three pads must be replaced. Also, in the event of a defective pad, it may be very difficult to identify quickly which pad needs to be replaced so all three pads may have to be changed. Tool utilization is also limited by tool inflexibility; for example, should a failure occur to any one component in this sequential tool, the entire system must be stopped while repairs are done.

4. Third-Generation CMP Equipment

Several suppliers have developed alternative approaches to traditional tools. These revolutionary approaches are designed to deliver improved CMP process results though capabilities such as high relative velocity for improved planarization and throughput, generalized pad and wafer motion, and unique methods of delivering slurry to the wafer–pad interface. Like the second-generation polishers, these systems also have the option of being linked to post-CMP cleaners for dry-in-dry-out capability.

a. Sequential Linear Polishers

As shown in Fig. 2, linear polishers feature a continuous belt kept in tension by rollers, and they operate much like a belt sander [11]. In use, the wafer is pressed device side down against the pad and slowly rotated against the rapidly moving belt. The current wafer processing sequence is similar to the second-generation-type sequential polisher utilizing two linear polishing stations and a single buffing station. Wafers that have completed their processing step must wait for the longest polish to be completed before the wafer can move to the next station. This design has been used to demonstrate very high relative velocities. These speeds have resulted in high film removal rates and, in conjunction with low down force and new polish-pad structures (e.g., a single polyurethane pad without a foam or felt subpad), has shown initial feasibility for improved planarization for STI [12] and for ILD. Publication of the planarization and removal rate improvements achieved by using such processing conditions is leading to the development of similar high-speed, low-down-force capability for most state-of-the-art CMP equipment.

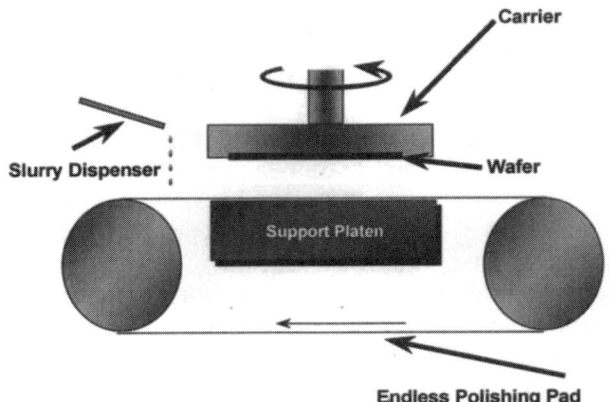

FIG. 2. The linear planarizer uses and endless polish pad moving at high speed to achieve high relative velocity values. The carrier rotates slowly to ensure uniformity.

b. Orbital Polishers

Orbital motion offers the capability of achieving high relative velocities without sacrificing tool footprint. This point is especially important as the semiconductor industry prepares to make the transition to 300-mm wafers. Several CMP tool concepts have been developed based on orbital motion. Some orbit the carrier while rotating the platen [13]. Others orbit the polishing pad while rotating the carrier [14]. Another design involves orbital (as well as arbitrary nonrotational) motion on a fixed polish pad [15].

Orbital CMP uses relative velocity values that are similar to the speeds of other CMP tools, but with a small tool footprint. Generally speaking, the processes are similar. Additionally, high relative velocity values can be achieved at high orbital speeds with no adverse effect on removal rate uniformity. Thus, with systems such as the one shown in Fig. 3, high pad speeds coupled with a low-down-force process and using single polyurethane polish pads can lead to significantly improved planarization. These trends are shown in Fig. 4.

Orbital pad motion also enables slurry to be delivered directly to the surface of the wafer, which improves slurry distribution [16]. However, grooves must be formed in the pads to ensure uniform slurry distribution. For production convenience, the grooves are Cartesian. The ensuing motion of the grooves, especially of the groove intersections, against the wafer causes a spirographic pattern to form if compound motion is not added. Pressure is applied by inflating two bladders, one behind the wafer backing

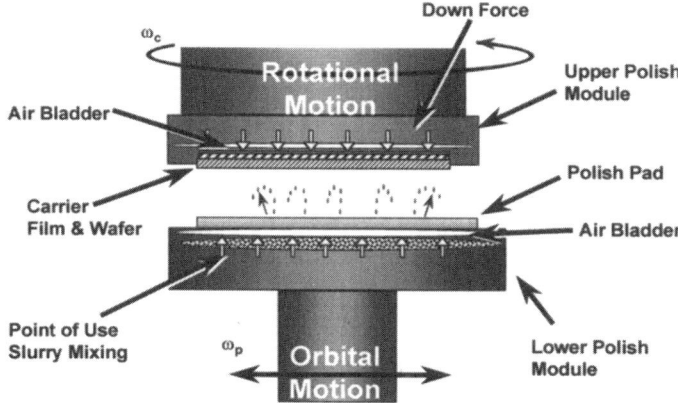

FIG. 3. Schematic diagram of a third-generation, orbital CMP head.

plate and the other under the flexible pad-backer assembly. A wide retaining ring is used to ensure the pad remains compressed during the entire polishing cycle. Controlling the pressure difference between the pad and the retaining ring optimizes uniformity in the edge exclusion region. The pad-backer assembly is held in place by a quick release buckle; pad assemblies

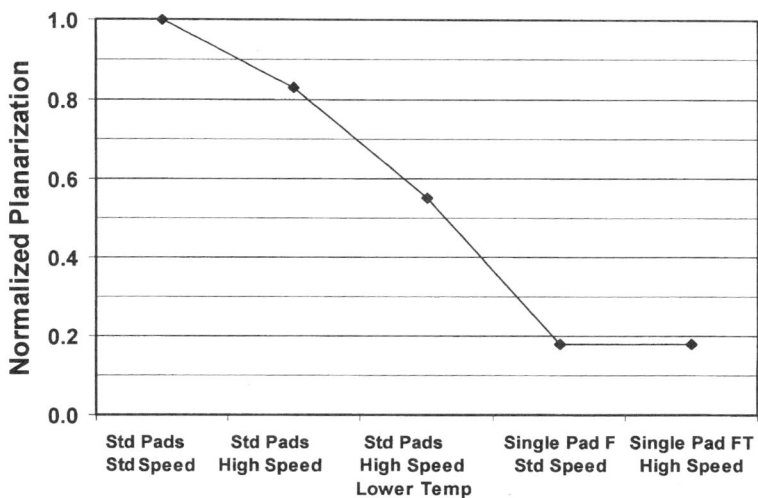

FIG. 4. Normalized planarization as a function of process.

can be changed in seconds. Production orbital polishers have four independent polish heads that can be randomly accessed for either single pad processing or multiple processing steps. Planarization capability on these tools has been demonstrated to be significantly better than that on first- and second-generation polishers. The rapid change capability of individual polish head pads minimizes the down time associated with conventional tools during pad change.

c. Pad Feed (Web) Polishers

A polish technology that is just emerging from development by several CMP capital equipment suppliers is the pad feed polisher. This equipment is based on some fairly recently developed polish pad rolls. These polish pads are in a roll similar to 35-mm camera film. The pad is fed out to the wafer polish table, a wafer is polished, the pad is conditioned, the pad is incremented forward, and then the next wafer is polished.

Successful polishing with such a methodology strongly relies on the pad characteristics being consistent from beginning to end. This method is particularly useful for pads with very consistent first polishes, but whose characteristics degrade rapidly with subsequent wafer polishes.

While this tool design has no significant polish performance or polish mechanism improvements over non-pad-feed polishers, it does have a strong advantage: tool utilization. With almost all polishers being used today except this one, the polisher must be turned off once a day to remove the worn polish pads, place new polish pads on the tools, and qualify the tool before restarting production. With a continuous-feed pad of adequate length, fabs can continue to process wafers without pad changes for up to a week. This could significantly improve CMP tool utilization by the fabs.

III. Carriers

Aside from the basic approach to polishing, the most critical component of a CMP tool is the wafer carrier. As with CMP tools, wafer carriers have evolved from roots in lens grinding and in the silicon wafer polishing industries to meet requirements specific to polishing silicon-based integrated circuits.

The basic function of the wafer carrier is to hold the wafer in place while the wafer is polished. A more detailed description of the basic function of a carrier is given following a summary of the processing issues being addressed by carrier designers. A carrier includes a means, such as vacuum, to hold the wafer in place while loading and/or unloading, and a retaining ring to keep the wafer from becoming dislodged from the carrier during the polish cycle. On many carriers, the wafer is placed against a carrier film that is used to compensate for small amounts of wafer bow, tilt, or warp. Finally, at the completion of the polish cycle, the carrier must release the wafer upon command.

In CMP, carriers must meet two additional requirements. First, polished wafers must be flat to within a predetermined specification across the wafer, but excluding the so-called edge exclusion region. The edge exclusion region is an annular region of the wafer at the wafer edge where the removal rate deviates significantly from that of the bulk of the wafer. Ideally, the width of this region is zero, and carrier design engineers continue to improve carriers to reduce the edge exclusion region to zero. Figure 5 shows the effect of reducing the edge exclusion and the clear need for minimizing wasted real estate on a wafer. In 1995–1996, typical processes were quoted with measurements made at edge exclusions of 5 to 7 mm. In 1997–1998 specifications including 3-mm edge exclusion became the standard. It is anticipated that by the year 2000 near zero edge exclusion will be the norm.

The edge effect arises from the interaction of the carrier with the pad. The dynamic action during polishing causes the pad to be compressed as the platen rotates. As the pad is exposed to a lack of pressure in the gap between the retaining ring and the wafer, the pad "rebounds," causing increased pressure against the wafer for several millimeters at the edge of the wafer.

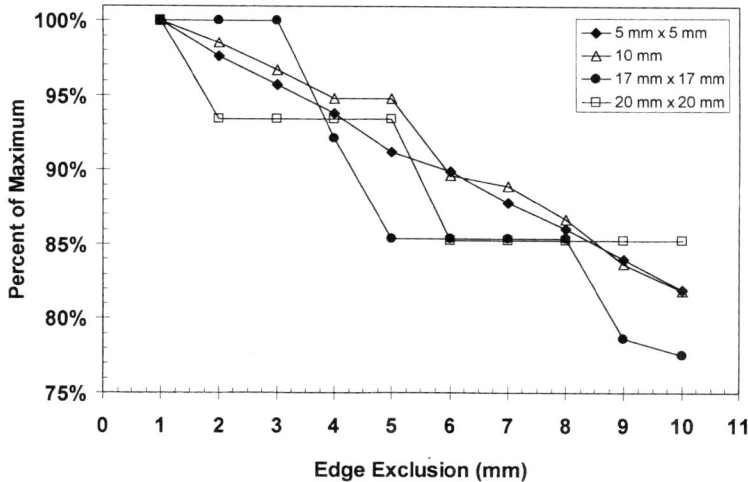

FIG. 5. Plot of the percentage of the maximum number of chips that can be obtained from a 200-mm wafer as a function of edge exclusion for several representative chip sizes. For small chips, the sensitivity is essentially linear, while for large chips there are plateaus in which further decreasing the edge exclusion has no impact on the percentage of possible yield.

This pressure induces significantly greater removal of material at the edge of the wafer. The pad then "bounces" off of the edge of the wafer, causing the adjacent region to experience lower pressure. Physical contact is maintained, but the pressure on the wafer in this region is much less than it is in the adjacent regions. Consequently, this narrow, annular region experiences a correspondingly lower removal rate. The net effect of this process is known as the so-called edge. This effect is shown in the diameter scan in Fig. 6 that reveals the rabbit-ear shape that is the signature of the edge effect. Figure 7 shows an expanded view of the edge effect.

Baker [17] derived an elegant description of the edge effect based on the mechanics of the pad and of the wafer. He argued that the peak in the edge exclusion region is proportional to $C^{1/4}t^{1/4}E^{1/4}h^{3/4}$, where C is the slope of the compressibility, t is the thickness of the base pad, E is Young's modulus for the pad, and h is the thickness of the top pad. This relationship indicates that there is a weak dependence of the peak on the material parameters of the pad. Wang *et al.* pursued a finite element analysis of the pad–wafer–carrier system, including the effects of both the polish pad and the carrier film, and determined that the edge effect was due to Von Mises stresses [18]. Both approaches agree that hard pads give better profiles in the edge exclusion region, but neither model fully addresses the overall interaction of all key process parameters. Wang and coworkers went on to show that the carrier film plays an important role with both the magnitude and location of the edge effect. They showed that thin, hard pads give the best edge exclusion [19–21].

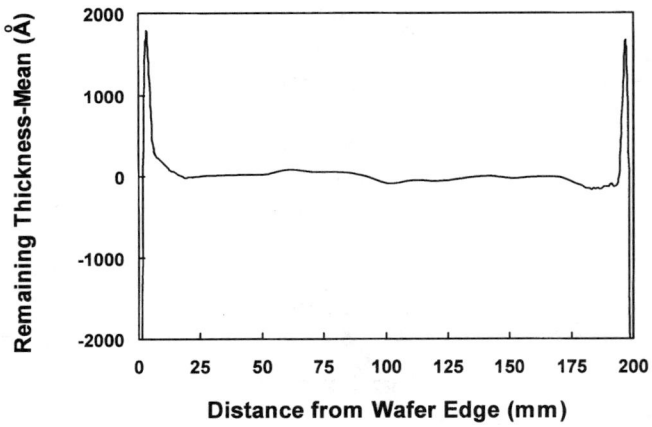

FIG. 6. Diameter scan of thickness (minus the mean) of a wafer that was polished using a first-generation rotational polisher. (Courtesy of SpeedFam-IPEC.)

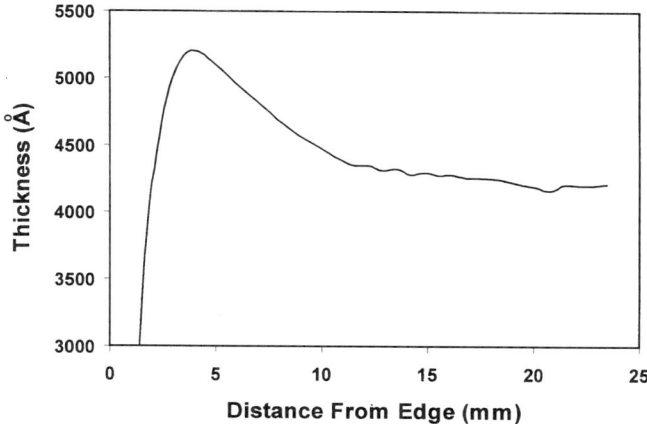

FIG. 7. Radial scan of film thickness (at the edge of a wafer) after removing 4000 Å of PETEOS on a wafer that was polished using a first-generation rotational polisher. (Courtesy of SpeedFam-IPEC.)

With respect to the edge exclusion, orbital tools differ from other CMP tools because during use virtually the entire pad is in compression. In contrast, rotational tools periodically compress and then release the pad as it passes under the carrier and rotates around the tool. Very little data is publicly available regarding the fundamental properties of pads, so the significance of this difference between tool types is not well understood.

The second additional requirement of CMP carriers is that they must allow the tool to polish a broad range of films with varying amounts of film stress. Film stress causes the wafer to deform, altering the pressure distribution across the wafer during CMP. These pressure variations cause characteristically fast or slow polishing across the wafer. This picture is further complicated by the time dependence of this stress during the course of the polish cycle.

Depending on deposition tool parameters, the deposited films may have lattice constants that differ from those of the films on which they are deposited. This mismatch induces stress in the wafer. This stress can be either tensile or compressive, and can have a significant impact on CMP [22]. A tensile film causes the edges of the wafer to bow up. When pressed against a polish pad such a film would exhibit a center slow polish profile (i.e., the center would polish at a slower rate than the outer region). A compressive film causes a center fast process. Although deposition tools afford significant latitude in adjusting the intrinsic film stress, process engineers often deliberately select parameters to induce a large amount of film stress to compensate for film stress from earlier processing or to reduce material cracking over underlying topography. In other cases, stress arises because the preceding processing steps were poorly optimized from an overall integration perspective, and sometimes the particular stress distribution is simply a result of a prior process being locked in for manufacturing purposes.

A number of approaches have been developed to accommodate film stress. The use of back pressure enables one to deal with modest bowing and for dealing with variable process conditions (e.g., incoming film nonuniformities or layer-to-layer nonuniformities) [23]. For more severe situations, the carrier face is built with a small amount of curvature. Typically, the center of the carrier extends out from the plane of the periphery by 5 to 20 μm. Once the appropriate amount of curvature has been identified, small amounts of back pressure are then used to fine-tune the process.

A variety of other methods have also been published, including several from the patent literature. These other methods involve introducing a spatially variable amount of back pressure to the wafer, for example, by including concentric rings within the carrier head that can be pressurized to different pressures [24].

1. Gimbaled Carriers

The simplicity of early carrier designs was complicated by their use as part of the robot used to move wafers from the load cassette to the unload cassette. On a so-called gimbaled carrier, the down force was applied to a central point on a plate behind the wafer, and it was assumed that the applied force was transferred through the wafer backing plate to be distributed uniformly across the wafer. Lateral motion of the pad then caused a torque to be applied to the carrier. To compensate for this rotation, a gimbal was built into the carrier at the point where the down force was applied.

These carriers were generally capable of providing 5–7% nonuniformity at 6- to 7-mm edge exclusion. The nonuniformity is more precisely called the within-wafer nonuniformity (WIWNU) and is defined by the standard deviation of a set of film thickness measurements on a wafer divided by the mean of that set. Smaller numbers denote better process control. Better performance was limited in part by the manner in which the carrier held the wafer during polish.

Even in the absence of the edge effect, several other issues limited the ability of these carriers to produce very flat wafers. First, the application of pressure to the center of the wafer, in spite of the rigid structure of the carrier backing plate, caused wafers to polish center fast. Second, poor slurry flow to the center of the wafer caused wafers to polish center slow. These two factors partially offset one another with slurry distribution believed to be the stronger of the two. Only in the late 1990s has the impact of slurry flow on CMP received any attention, and it is likely to play a much greater role in CMP as it becomes better understood [25, 26]. Process parameters, of course, also play a role as well as the film stress in the wafers being polished.

There are two basic methods used to address these effects: back pressure and the use of curved carriers. Back pressure consists of increasing the pressure applied to the wafer via a port behind the wafer plate. The use of back pressure is preferred since it is readily adjusted to compensate for a broad spectrum of issues. Adequate performance is attainable in most cases. However, in cases of extreme film stress, it is often necessary to use a curved carrier. Such carriers have a wafer plate with a small amount of curvature built into them. The maximum deflection is in the center of the carrier, and is typically about 5 to 15 μm.

2. Floating Carriers — The Use of Hydrostatic Pressure

Experience with the gimbal-styled carriers led designers to rethink the purpose of a carrier as well as the approach to designing better carriers. Two definitions emerged. First, the purpose of a carrier was to hold a wafer as flat as possible so that a uniform thickness of material could be removed from the wafer surface to eliminate surface roughness. This definition implies that removal is referenced to the pre-CMP surface (i.e., the front surface is referenced). This definition is essentially the one used by the raw wafer polishing industry. Any imperfections in the surface of the wafer must be accommodated by the flexibility in the polish pad. This definition does not take into consideration the potato chip shape of partially processed wafers with the topology of integrated circuit designs on them.

The second definition of the purpose of a carrier was to remove the overburden of material above a surface above the device plane. For present purposes, we define the device level as the boundary between the material one wishes to remove and the material one wants to keep. It is not necessarily planar, and it moves up with each layer. For oxide CMP, this layer lies within the topmost film layer. For metal CMP, this surface is defined by the topmost surface of the dielectric into which lines and vias are etched for a damascene process. This definition must accommodate a wafer with a modest amount of bow, tilt, warp, and total thickness variation. Furthermore, it must accommodate very modest amounts of bow, warp, tilt, and other factors that produce nonuniformity in the film being polished. The demands placed on carrier designers of such a definition are much greater than they are for designers addressing the first definition.

Thus, the goal in initial carrier improvement was to develop a means to apply more uniform pressure between the wafer and the pad. Significant improvement in carrier performance resulted from the use of pneumatic or hydraulic pressure. Various forms of design implementation have taken place and are described with terms such as "floating carrier," "bellows carrier," and "bladder carrier." Perhaps the best-known design is one patented by Cybeq [27]. The basic concept underlying the design improvement is the presence of a cavity into which pneumatic or hydraulic pressure is applied, and a flexible membrane that couples the pressure to the wafer. This concept is shown in a simplified manner in Fig. 8. Although the flexible membrane confines the pressure, its primary purpose is to respond homogeneously to the pneumatic pressure. In so doing it applies pressure uniformly across the wafer. The flexible membrane is connected to the bulk of the carrier through a wafer backing plate, but the action of the retaining ring is independent of the wafer; that is the wafer "floats." A key process parameter in this design is the differential pressure between the cavity and the ambient. There are other similar inventions involving the use of membranes to assist in applying pressure uniformly dating from the same time [28–31].

This approach has proved to be so successful that it has spawned a cottage industry of carrier designs based on the use of hydrostatic pressure. With some of these inventions a backing plate is present, with others [32]

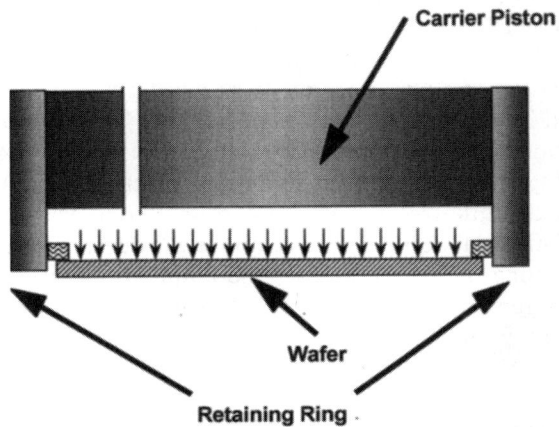

FIG. 8. Wafer carrier with a floating carrier plate. Pneumatic pressure is applied through an inlet resulting in hydrostatic pressure directly against the wafer. There is not necessarily any independent retaining ring control.

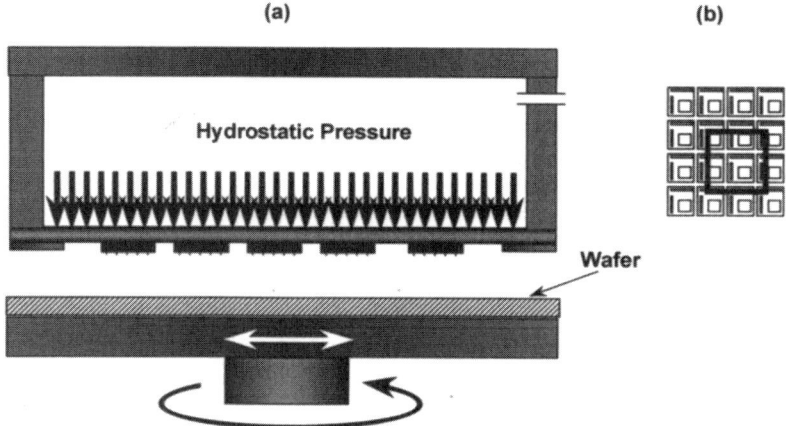

FIG. 9. (a) The distributed polish head extends the concept of hydrostatic pressure by integrating polish blocks in the carrier design to ensure that locally the spatial variation in the polish pressure is low. (b) Schematic representation of the region on a wafer that a polish block may contact.

it is absent. In one invention, the pressure applied to the retaining ring is adjusted separately from that of the carrier [33]. Another carrier design uses concentric pistons with independent pressure control [34]. A subsequent invention involves the use of a wear ring with independent pressure control to minimize the edge effect, and the use of a lip seal to allow limited vertical wafer motion independently of the retaining ring [35]. Another invention using multiple bellows achieves the same end: independent control of the pressure on the retaining ring as on the wafer [36].

All of these approaches are intended to address global planarization issues. We close this section with a brief note about an alternative approach that focuses on achieving planarization at the die level. The distributed polish head [37], extends the concept of hydrostatic pressure by integrating polish blocks in the carrier design to minimize the locally the spatial variation in the polish. As shown in Fig. 9, each polish block makes contact with several die, each of which may contain regions of high and low pattern density.

3. LINEAR POLISHER AND PRESSURE CONTROL

OnTrak invented an intriguing pressure controller in which fluid pressure is used to fine-tune the pressure presented by the force applied to the carrier against the polish pad [38–40]. Beneath a continuous polish pad is a support in which a collection of fluid channels are formed. Control of the fluid pressure in these bearings regulates the pressure distribution that the wafer experiences. These channels are controlled in groups. A central fluid bearing is larger than the others, and contributes most to controlling the removal rate in the center of the wafer.

IV. Platens

The platens on which wafers are polished have also evolved over time. Traditional polishing, as is done on first- and second-generation CMP tools, is done on a hard platen. The reason for a hard platen, of course, is to present as close to absolutely flat a surface as possible against which the wafer is pressed. Ideally, platens rotate perfectly. In practice, however, there is a small amount of run-out, or wobble, which limits the ability of the tool to polish films uniformly, especially at high rotation speeds.

Although typically not thought of as a platform for more than presenting a flat surface, platens are being exploited in several ways to enhance overall CMP performance. Included in these approaches are temperature control, slurry delivery routing, and pressure control.

1. Temperature Control

Temperature control is important for several reasons, and some platens are designed to allow either heating or cooling of slurry. First, the chemical aspect of CMP is temperature-dependent. This is especially the case for most metal CMP processes and for exothermic processes such as silicon polishing. Increasing the temperature increases the chemical reaction rate and hence the removal rate. However, other effects temper the benefits of increasing the temperature. Since the glass transition temperature for polish pad materials is approximately 65°C, increasing the temperature causes the material in currently used polish pads to soften. This causes the pads to deform more, so uniformity improves and planarity deteriorates. On the other hand, decreasing the pad temperature causes the pad to stiffen, which can be used advantageously to improve oxide CMP planarization performance. A stiffer pad deforms less, so planarization improves, especially on a feature scale [41]. On a wafer scale, however, excessive pad stiffness can lead to very good planarization, but with excessive material removal in some areas.

Second, friction during CMP generates heat that can affect the reaction kinetics as well as soften the polish pad. On most tools, heat transfer away from the center of the pad takes place only through the motion of the pad out from underneath the wafer. The rate of heat removal increases with increasing slurry flow, so one reason there is a limit on reducing the slurry flow rate is the effect it has on polish temperature. In addition, some processes such as silicon polishing are exothermic, so there is even more heat to remove than with either metal or oxide CMP.

Temperature control is accomplished in one of three general ways. One method is by controlling the temperature of the platen, usually by means of an integral channel in the platen through which temperature-controlled heat transfer fluid flows. Second, the temperature of the slurry itself can be regulated prior to being dispensed onto the platen. Finally, a means of heating the backside of the wafer can be built into the carrier [42, 43].

2. Orbital Polisher and Direct Slurry Delivery

Although several orbital designs exist, only those designs in which the platen orbits allow slurry delivery directly to the surface of the wafer without the expense of complex rotary union fittings are in widespread use. Slurry injection through the pad is particularly important with high-speed processes, where the removal rate is limited by pad absorbency and slurry transportation. The orbital platform sold by SpeedFam-IPEC is unique in that the pad is not rigid. The platen consists of a thin disk of stainless steel on which a 4.75-mm-thick polyurethane pad backer is glued. The pad backer has grooves in it to allow greater flexibility. It also contains holes between the grooves that are aligned with holes in the underlying backing plate through which slurry is delivered.

By delivering slurry directly to the pad–wafer interface, process engineers have a great deal of latitude in controlling slurry distribution across the wafer during polish. In other words, they can design processes that do not suffer the limitations of pH or oxidizer concentration gradients across the wafer. Oxide and metal CMP processes are very different, so it is useful not only to be able to inject slurry directly to the wafer surface, but also to control where on the wafer the slurry is delivered.

V. Pad Conditioning

Pad conditioning is the process of revitalizing a polish pad to produce reproducible, consistent results [44, 45]. During CMP, several processes take place that generate material that can be deposited on the pad and in the pores in the pad.

1. Polymer is abraded from the surface of the pad.
2. Inorganic material is removed from the topmost film on the wafer via chemical and/or mechanical processes.
3. Slurry particle agglomeration can take place in the slurry in which abrasive particles and colloids coalesce to form extended particles.

Any and all of these particles can become attached to the pad. As the surface of the pad accumulates more and more particles, the surface glazes and becomes smoother and less abrasive. Consequently, the removal rate declines. Also, as the pad becomes glazed it becomes smoother, which causes a decline in the ability of the polish pad to distribute slurry under the wafer. At present there is considerable debate as to whether CMP takes place in a contact regime or in a lubrication regime. It is likely that resolution of this issue will be necessary to establish what fundamental limits there are on slurry flow, as well as how these limits affect pad conditioning.

Some work has been done to correlate oxide CMP performance with pad properties [46]. This work indicated that the specific gravity of the pads and the cross-linking densities affect polish performance. Other work has been done to correlate CMP performance with slurry composition [47]. This work suggests that the friction during polish is proportional to the removal rate when the abrasive content is greater than 10%, and inversely proportional to the removal rate when it is less than 10%.

The porous structure of the pad shown [48] in Fig. 10 resembles a filter. If one considers the action of pad conditioning to clear the pores, then some study of polishing, that is, clogging of the pores may shed light on the mechanisms of pad conditioning. Figure 11 shows [49] a log-log plot of removal rate as a function of total polish time on a pad. The straight line indicates a power law governs the variation in the removal rate, that is,

$$\text{RR}(t) = \text{RR}_0 t^{-\gamma} \qquad (1)$$

where RR_0 is the initial removal rate and γ is a measure of how rapidly the removal rate diminishes in the absence of pad conditioning. For the data shown in Fig. 11, $\text{RR}_0 = 3301.6$ Å/min, and $\gamma = 0.0859$. An indication of the improvement in performance that this value of γ represents is that a log-log plot of the data shown in Fig. 12, which was obtained in 1993 on a first-generation CMP tool, reveals that $\gamma \sim 0.21$ to 0.31.

Pad conditioning serves to reverse the pore-clogging process, and there are two general approaches that seek to apply any of several pad-conditioning methods. One approach consists of concurrent pad conditioning. In this

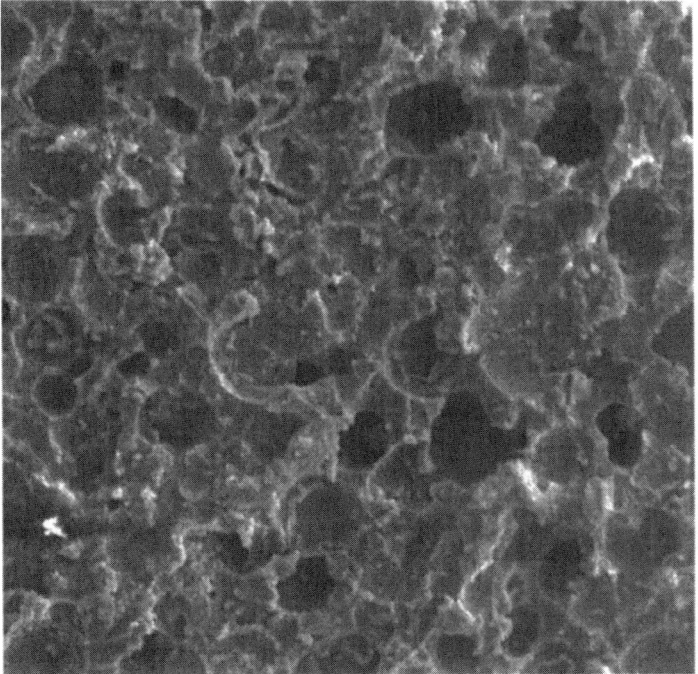

FIG. 10. Scanning electron micrograph (SEM) of a polish pad.

case, the pad is conditioned while the wafer is being polished. The second approach involves sequential pad conditioning in which the pad is conditioned prior to polishing each wafer.

Several methods have been developed for conditioning pads. The most common uses a diamond-coated disk that is pressed into the pad and rotated about its center axis as it moves radially across the pad. Conditioning disks are made of diamond-impregnated nickel or of a nonmetallic substrate on which a diamond grit has been epitaxially bonded [50]. The diamond grit scratches the surface of the pad, causing some particles to be dislodged. Once dislodged, they can be swept from the surface of the pad. Even if the particles are not removed from the pad (a likely scenario given the gluelike nature of silica), the grooves in the pad stemming from the action of the conditioner enable slurry transport to take place. Additionally, unless the pressure is exceptionally light, pieces of the pad itself are abraded as well. Depending on the pressure applied, the application of excessive pressure can lead to premature pad failure. An example of the significance

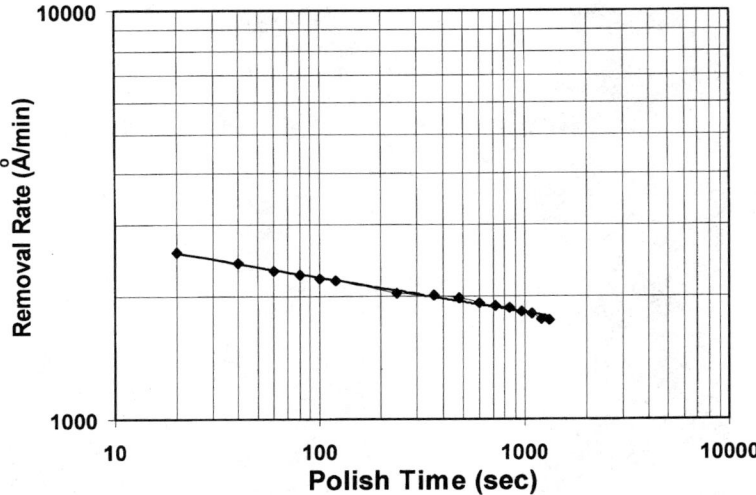

FIG. 11. Plot of instantaneous removal rate without pad conditioning vs polish time using Klebosol 1501 slurry and a Freudenberg FX9 polish pad.

of pad conditioning is shown in Fig. 12. This figure shows the removal rate as a function of wafer number. During the polishing of the first 43 wafers there was no pad conditioning, and the removal rate declined rapidly. Once concurrent pad conditioning was initiated, the removal rate exceeded its initial rate and remained steady.

Other methods of pad conditioning, such as the use of high-pressure deionized water [51] and acoustic energy [52] have been proposed for applications requiring aggressive pad conditioning. These methods show promise, but to date have not seen widespread use. Other CMP applications do not require such aggressive pad conditioning. For example, pad conditioning for tungsten plug CMP is done with a conditioner.

Pad conditioning is also used to shape the pad during polishing to improve uniformity. Figure 13 shows a plot of the pad thickness after polishing 481 wafers on a rotational platform. The bar in the figure indicates the location of the wafer during polish. The effect of pad conditioning on shaping the pad is evident. For the process shown in the figure, the pad was curved to enhance the removal rate in the center of the wafer.

Pad shaping is accomplished by adjusting the amount of time the pad conditioner spends at each radius [53]. Other control parameters process engineers may use include the down force exerted by the pad conditioner and the rotation speed of the conditioning disk. Typically, however, the

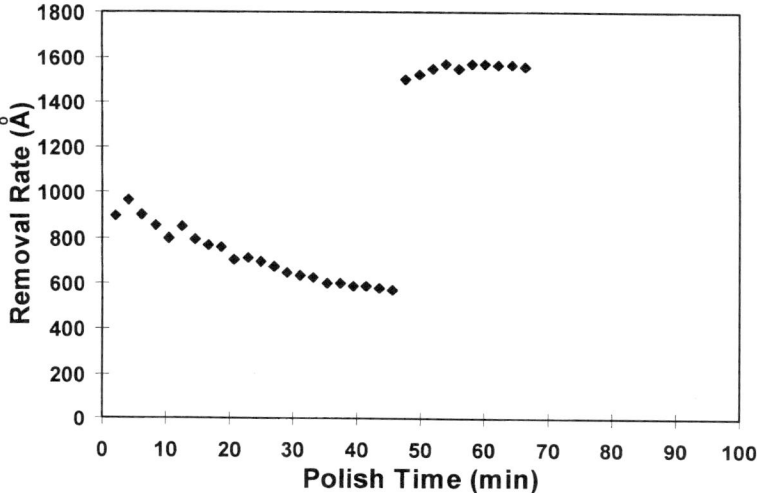

Fig. 12. Plot of removal rate as a function of wafer number showing the effect of pad conditioning. (Data c. 1993.)

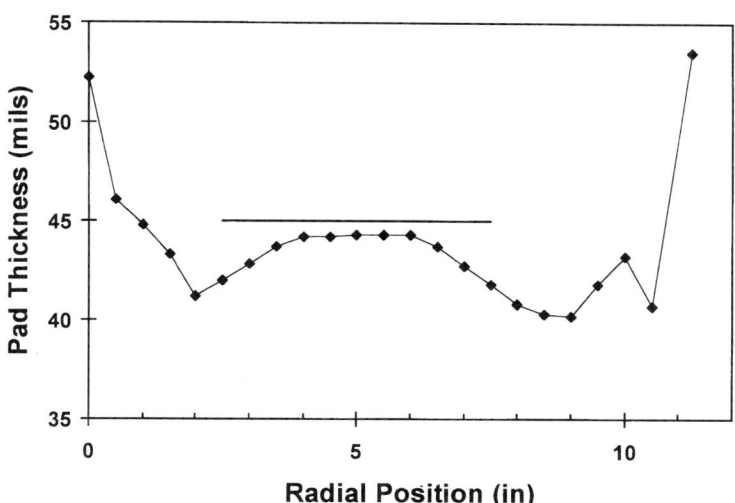

Fig. 13. Pad profile after polishing 481 wafers. The bar denotes the normal position of the wafer during polish.

pressure is adjusted to not only optimize the pad life and removal rate, but to cause the wear pattern in the pad to enhance the uniformity of the polished wafer. This result is achieved by increasing the pressure in regions where a lower removal rate is desired, or by increasing the dwell time in these regions. Care must be taken to condition the pad uniformly. Poorly chosen conditioning parameters can cause portions of the pad to be highly overconditioned while other portions are not conditioned at all. Alternative means of addressing the issue of nonuniformity due to nonuniform pad wear include the use of a platen with a carefully designed raised region under the path of the carrier [54], and the use of polish pads with regions of less compressibility under the wafer [55].

VI. CMP Equipment Integration

CMP tools have become much more sophisticated than first-generation CMP tools. The original tools met functionality requirements (most of the time!), but left much room for improvement.

The need for a reliable source of slurry during CMP stimulated the growth of a whole industry of bulk chemical distribution (BCD) systems. An extended discussion of BCD systems is beyond the scope of this work, but a few words are in order. Slurry distribution systems vary in sophistication from the simple laboratory system consisting of little more than a barrel of slurry and a pump to sophisticated delivery systems designed to supply slurry to tens of CMP tools in a high-volume production environment [56].

Suppliers of second- and third-generation tools are integrating cleaners and metrology equipment into CMP tools to improve efficiency, and to ease the integration into the main fab. In addition, such integration enhances the overall quality of the polished wafers because the CMP and cleaning processes can be optimized together in a single dry-wafer-out (DWO) process.

1. Post-CMP Cleaning

The semiconductor industry has evolved from treating CMP and post-CMP cleaning (PCMPC) as distinct processes to treating them as part of the overall CMP process. Early on it was recognized that cleaning processes (most commonly, brush cleaners) were essential for enabling CMP to be used. Even then, CMP processing was typically done in an isolated area well away from the rest of the fab. Post-CMP cleaning specifications are becoming ever more stringent.

To understand integration of post-CMP cleaners into CMP machines it is important to understand the seven basic rules that apply to post-CMP cleaning [57]. For an extended discussion of post-CMP cleaning, refer to Chapter 7. The Gotkis rules are:

1. *Never allow slurry to dry on wafers.* Strong bonds are formed when slurry dries, making the removal of the slurry nearly impossible. The bonds are formed between the abrasive, the abraded pad, the abraded material, and the wafer. The bond strength is such that even supplementary cleaning steps are insufficient to remove the slurry.
2. *Include a precleaning and/or buffing step as an integral part of the polish recipe.*
3. *Rinse and buff immediately after the bulk CMP step.* Rinsing removes the residual slurry and the polishing by-products including abraded film from the wafer, abraded polish pad material, and agglomerated slurry. Buffing removes most of the particles that are adhering to the surface as well as many mechanically embedded particles.
4. *Decreasing particle size means increased effort required to remove it.* This rule arises from the smaller interaction cross sections for collision and momentum transfer. Furthermore, the electrostatic forces are stronger for smaller particles, and they diminish in proportion to $1/r$ as opposed to $1/r^3$. Both of these factors lead to redeposition being a major source of small particles. Since mechanical action requires increasing amounts of work, chemical dissolution is more effective at removing small particles than is mechanical action.
5. *Minimize surface roughness to limit sites at which particles can adhere.* This point is especially true for small particles and removal methods exploiting mechanical means. A rough surface can shield small particles from momentum transferal from brushes.
6. *Avoid CMP chemistry that involves multicharged cations.* Such chemicals compress the double charge layer and activate slurry agglomeration and process defectivity. Ions such as Al^{3+} and Fe^{3+} may initiate agglomeration and scratching at concentrations as low as 10^{-4} to 10^{-7} M.
7. *Use any and all options for post-CMP cleaning.* Keep particles from adhering more strongly than they already are, and remove them as quickly as possible. Once they are removed they must be transported away from the wafer surface as quickly and efficiently as possible to minimize redeposition.

These rules basically state that post-CMP cleaning must be done as part of the CMP process, and it must be done as quickly as possible after beginning the CMP process. Rule 7 is driving the design of cleaners comprised of multiple methods of particle removal. The drawback to the use of multiple mechanisms of cleaning to produce cleaner wafers is a decrease in the reliability of the tool due to increased system complexity. A related factor is an increase in the overall CoO due to an increased footprint. Furthermore, cleaning methods that use a lot of consumables (e.g., DI water), also add to the overall cost of producing good, clean, flat wafers.

Although performance is the dominant factor driving post-CMP cleaning, the efficiency of any given method of removing particles is also very important. If the cleaning process for a given post-CMP cleaner takes longer than 90 seconds, a 40 wfr/hr cleaner will be cleaner-limited. Further improvements to the CMP tool that do not also reduce the time to clean wafers will not increase the overall throughput.

Most CMP tools now feature integrated post-CMP cleaners. However, of these the vast majority are double-sided brush scrubbers. An example is the OnTrak cleaner shown in Fig. 14 and two brush types are shown in Fig. 15. Brush configurations consist of either a series of rollers [58] or opposing pancake scrubbers. Typically, there are two brush stations, each of which

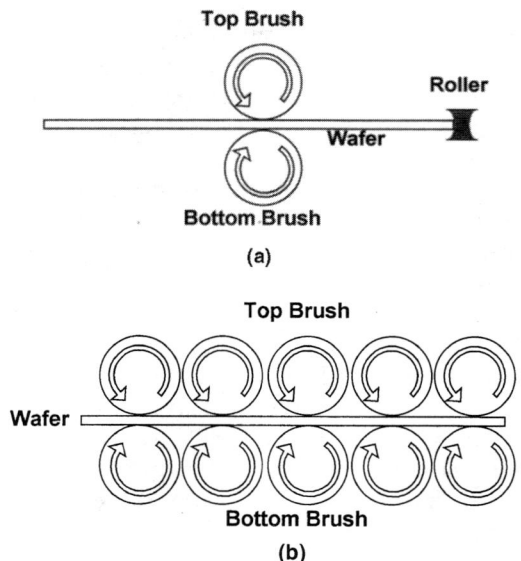

FIG. 14. Cross-sectional views of post-CMP cleaning using double-sided roller brushes. The two brushes roll in opposite directions to keep the wafer pressed against two rollers. The rollers are used to rotate the wafer and are withdrawn to allow the wafer to pass once the wafer is clean: (a) a single-brush configuration and (b) a multiple-brush configuration.

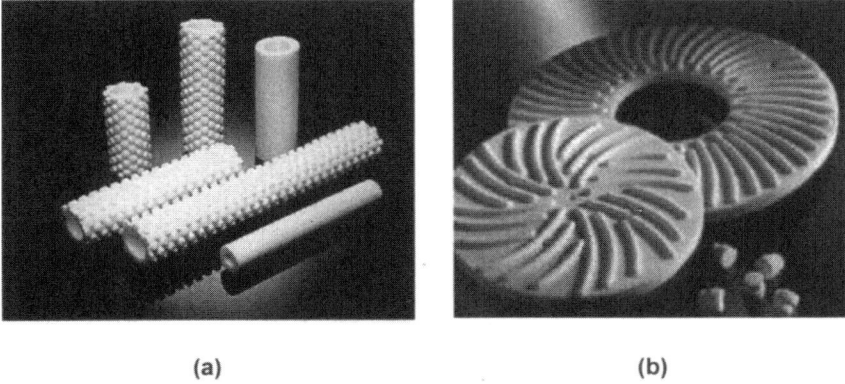

(a)   (b)

FIG. 15. Brushes used for post-CMP cleaning: (a) roller brushes (Merocel Corp. web site, reprinted with permission) and (b) puck brushes (Rippey Corp. web site, reprinted with permission).

can be used with active chemistry such as dilute HF to enhance cleaning performance. However, HF is well known for its safety issues, so there is considerable impetus to develop alternate PCMP cleaning solutions that provide comparable or better performance than dilute HF.

Double-sided brush scrubbing followed by a spin–rinse–dry step dominates the industry. An example of a double-sided brush scrubber is shown schematically in Fig. 16. The rollers keep the wafer positioned and rotating while the brushes clean debris from both the front and back sides of the wafer. The cleaning solution is delivered through the bristles themselves. Typically, the bristles are porous so that fresh solution can be delivered directly to the wafer and minimal amounts of debris accumulate on the brush. Increasing the down force increases the particle removal efficiency up

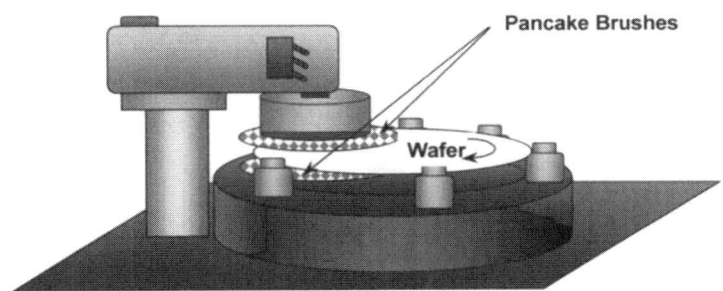

FIG. 16. Schematic of double-sided brush cleaner using pancake brushes.

to a point after which the pressure causes the bristles to bend too much and they break, causing bristle particles to deposit onto the wafer. Figure 14 shows a roller-based post-CMP cleaning method. With this technique wafers are transported through the cleaning region while the brush scrubs the top surface of the wafer.

There is concern about the impact that brushes may have on the wafer's surface, both because of the potential for scratching and shedding as well as the potential of cross-contamination and subsequent redeposition onto otherwise clean wafers [59]. Alternative methods, such as megasonic cleaning, have been proposed as noncontact cleaning methods. Megasonics is a term used loosely to describe all acoustic based methods of particle removal. Ultrasonic excitation in the 20 to 40-kHz regime or megasonic agitation in the 0.7–1-MHz regime provide cavitation and particle removal via streaming effects [60, 61] and has received growing interest for post-CMP cleaning, especially for wafers on which there are 0.35-μm critical dimensions and below.

Some preliminary work has also been done involving nanosecond pulses of laser light to very rapidly heat localized regions of the wafer causing an abrupt expansion wave to expel particles from the surface of the wafer. This method is intriguing from the point of view of using very small amounts of consumable materials. However, from a CMP tool point of view there are concerns about the size of a cleaner based on pulses of laser light using existing laser technology, as well as about throughput issues.

An alternative approach is the use of megasonic energy [62]. Figure 17 depicts a schematic representation of Verteq's Goldfinger. This cleaner uses a quartz arm to which a megasonic transducer has been affixed. The arm is swept across the wafer close enough that DI water (and, if needed, chemicals such as HF) sprayed at the wafer form a meniscus between the wafer and

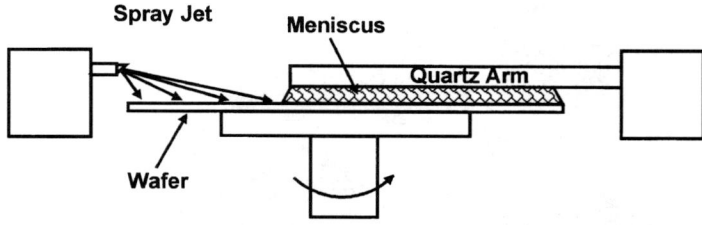

FIG. 17. Schematic representation of the Verteq Goldfinger that uses acoustic energy. The quartz arm contains a megasonic transducer that sweeps acoustic energy across the wafer while it rotates.

the quartz arm. This meniscus acts as an acoustic coupler and waveguide between the wafer and the transducer. Limited results have been reported to date, but those that have been reported have been encouraging.

2. INTEGRATED METROLOGY

Yield improvement is being addressed by integrating metrology using *in situ* methods to determine when to stop polishing, or in-line methods to monitor the quality of the process control with measurements immediately before and after CMP, but before the wafers have been removed from the CMP tool. By incorporating these measurement systems into run-to-run control systems, closed-loop control is being used to tighten process specifications [63, 64]. These systems also address the impact on throughput due to off-line measurements during process qualification following a pad change or other PM action. It can take as long as 2 hr to confirm that the tool is running properly. Table II shows a "nightmare" scenario in which a supposed 40 wfr/hr high-volume CMP tool is reduced to an effective peak throughput of 25 wfr/hr primarily because of the time currently required for metrology. In addition, while waiting for the results from the monitor wafers, the tool is idle, and the pad properties can change. Integrated metrology in CMP is therefore increasing in importance from a strong desire to an essential component of CMP tools.

a. In-line CMP Metrology

In-line metrology is a process control method suitable for regulating dielectric CMP. There are two major methodologies in use for in-line metrology: wet and dry. Wet measurements involve immersing each wafer in DI water immediately after polishing it and measuring the amount

TABLE II
EXAMPLE "NIGHTMARE" SCENARIO

40	Wfr/hr	Throughput at capacity
8	hr/shift	Shift time
1	hr	Lunchbreak and the like
2	hr	Metrology
200	Wfr/shift	Number of wafers per shift
25	Wfr/hr	Effective throughput

remaining at specified sites using pattern recognition software. Dry measurements are done in a similar manner, except that the wafers are measured after going through an integrated cleaner, and the measurements are done in air.

The advantages of measuring wet wafers are that the delay in obtaining measurements is minimized, and essentially no contamination issues arise. Since dry measurements are done following post-CMP cleaning, there is a delay of a couple of minutes between the time a wafer is polished and the time it is measured. In principle, with wet in-line metrology, the next wafer to be polished can benefit from a process modification due to the wafer just being measured. However, this argument is not very strong since a tool and/or process that requires immediate correction would be better off with an *in situ* endpoint sensor. In-line measurements are ideal for use as statistical process control tools and controlling relatively slow and predictable changes in the process, such as the effects of pad wear.

The advantage of dry in-line metrology is that the index of refraction difference between the top oxide layer and the air is much greater than that between water and the oxide, and consequently the signal-to-noise ratio is larger than is the case with wet measurements. Figure 18 shows [65] a plot of the spectral reflectance of a SiO_2–TiN–Al film stack in air and in water. It is clear that the spectral signature in both cases lends itself to determining the oxide thickness, but that the amplitude of the oscillations is considerably less in the case of measurements made in water than those made in air. As

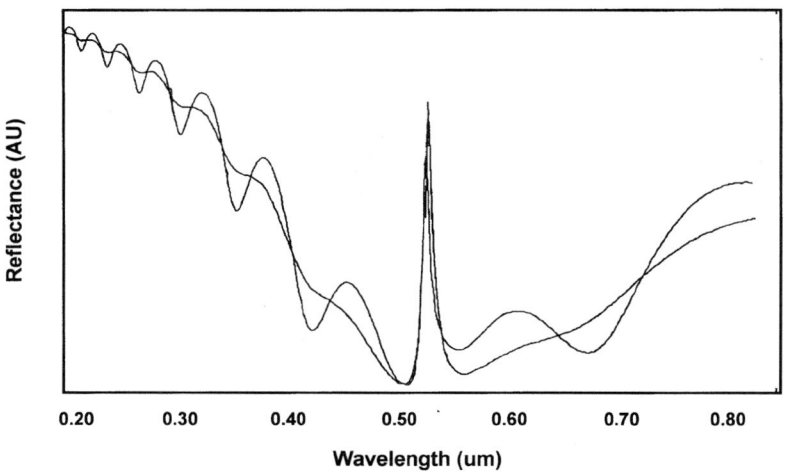

FIG. 18. Spectral response of a SiO_2–TiN–Al film stack in air and in water.

for contamination, if properly designed, dry systems pose minimal risks. Also, they use no DI water, so there is no additional ongoing cost.

The point is that both approaches can work. Nova manufactures a wet system [66] and is responsible for creating the market for in-line metrology tools for CMP. Nanometri produces a dry wafer measurement system. Run-to-run control systems using integrated metrology are beginning to exploit the potential of in-line measurements [67]. For proper control, both pre- and postmeasurements must be made. With rising throughput requirements on CMP tools, the necessity of measuring wafers twice per wafer pass imposes stringent requirements on in-line metrology tools.

b. In Situ Metrology

The process variations that have been intrinsic to CMP since the beginning of its use in semiconductor manufacturing have led to endpoint detection (EPD) being viewed as the holy grail of CMP.

Numerous approaches have been proposed for use in CMP for *in situ* EPD. They include optical, electrical, and acoustic sensing. Given the benefits of EPD, it is no surprise that many of these methods have been awarded patents. Some of these methods, most notably current sensing, have been developed to become commercially viable products while others remain laboratory curiosities. For a review of *in situ* endpoint detection methods up to early 1998, see the work of Bibby and Holland [68].

3. Slurry Reprocessing

Slurry usage in CMP falls into three categories: silicon polishing, dielectric CMP, and metal CMP. Within the latter two categories fall a variety of slurries depending on the specific material being polished. The chemistry and related issues are discussed elsewhere in this book. There are, however, a couple of points related to slurry that bear on CMP tool performance.

Unlike other parts of semiconductor wafer processing, the cost of consumables used in CMP is a significant cost to the end user. Consider the following example in which one wishes to remove 12,000 Å of oxide. An average blanket removal rate of 3000 Å/min translates to an effective removal rate of 6000 Å/min (due to pattern density) and results in a 2-min polish time. A slurry flow rate of 200 ml/min and a slurry cost of $10 per gallon (assuming volume pricing) imply a cost of $1.21 per wafer. A CMP machine with a throughput at capacity of 40 wfr/hr and operating at 65% of capacity would then consume nearly $275,000 per year in slurry. It would also produce approximately 27,500 gal of waste slurry that requires treatment at additional cost.

The cost of consumables is proportional to the amount of slurry required for each wafer polished. Reducing slurry consumption has been achieved by increasing the efficiency with which slurry is used. This has been achieved using direct slurry deliver systems in which slurry use is cut in half. Alternatively, reusing (or reprocessing) slurry has been proposed [69]. Such reprocessing systems actually improve the quality of the process because the slurry pH is adjusted to maintain peak performance. Although such systems offer potentially huge savings, they have not experienced market acceptance in spite of pleas from manufacturers for further development [70].

VII. Copper Polishing and CMP Tool Requirements

There are areas in which equipment advances are being made that will enable greater flexibility in processing. Such advances are particularly important as advanced interconnection concepts are implemented. In particular, copper CMP is evolving rapidly. Although tantalum is widely used as a barrier metal for Cu damascene, TaN, WN, and CoWP have also been reported for use in copper interconnection technologies. High barrier-layer-removal rates are essential to minimize metal dishing and dielectric erosion. It is also necessary to obtain high selectivity to underlying dielectrics to reduce dielectric erosion values and to minimize dielectric loss in field areas. To achieve high barrier-removal-rates without increasing Cu dishing it may be necessary to rapidly change the slurry when the bulk copper is cleared. Thus reliable and consistent endpoint technology will be necessary. Optical endpoint technology, such as the broadband optical endpoint system that produced the endpoint trace shown in Fig. 19, delivers this capability.

Depending on the ability of slurry suppliers to develop slurries capable of removing both copper and barrier layers without excessive dishing, it may be necessary to use multiple slurries, and possibly to polish wafers on more than one head. The point here is that flexibility is key. It is important that customers have the capability of delivering different slurries to a given head within a polish cycle and to different heads within a tool if desired.

In addition to controlling the standard process parameters such as down force and the relative velocity, it is also important to have random access capability to route wafers through a CMP tool to optimize both performance and throughput. Low-down-force processes and special CMP pads are likely to be necessary to reduce copper dishing just as they improve oxide planarization. Furthermore, a balance between high relative velocity to reduce copper dishing and moderate relative velocity to minimize the sheering of small oxide feature may be necessary.

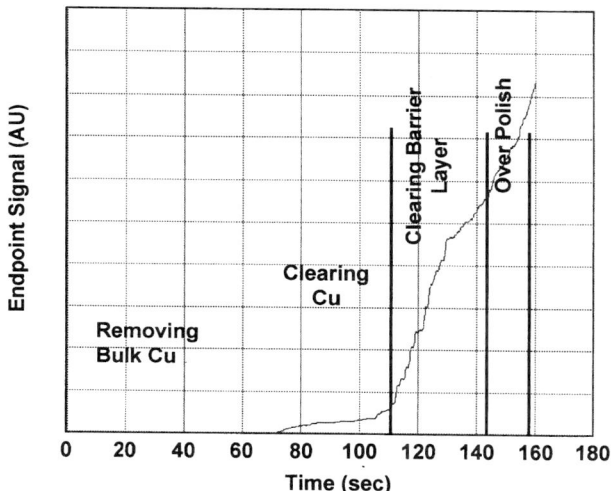

FIG. 19. Endpoint signal from a copper CMP process.

With the growing level of control of copper CMP process comes the requirement for temperature control. Copper CMP is highly temperature sensitive, so hot slurry will enhance the removal of bulk copper. However, barrier metals are significantly less reactive (with the slurries currently available), so there is a significantly greater mechanical contribution to the removal of the barrier layer(s). Consequently, hot slurry is of little benefit for barrier layer removal and may soften the pad and aggravate metal dishing and dielectric erosion.

Key performance issues include metal thinning and dielectric erosion. To minimize metal thinning and dielectric erosion it is advantageous to use a hard pad. An option on some tools is the capability of switching from one pad to another, and in some circumstances this is the best route. However, doing so can lead to a reduction in throughput, even in the presence of random wafer access capability. By being able to quickly cool the slurry, the process can be converted to a largely mechanical one using a relatively stiff pad. Figure 20 shows the benefit of multiple process steps on metal thinning and dielectric erosion.

In the future, pH control will also be a desirable capability. This capability already exists with tools on which slurries can be changed quickly and may be particularly useful with secondary slurries designed for specific barrier layers or for buffing.

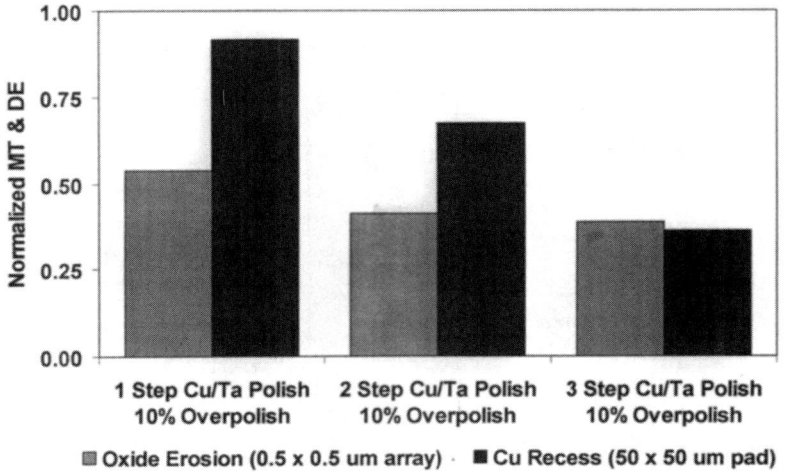

FIG. 20. Benefit of using a multistep CMP processes on an orbital tool using slurry delivered directly to the surface of the wafer.

Some work has already been done on reprocessing slurry, but to date the focus has been on oxide slurry reprocessing. Given the remarkably high contribution of slurry cost to the overall cost of ownership, one can reasonably expect that slurry reprocessing will play a greater role in the future.

VIII. 300-mm CMP Tools

The technology for developing 300-mm CMP processes has been significantly affected by technical factors and economic delays. Since IC manufacturers and capital equipment suppliers began to develop 300-mm-based fabs and equipment, the startup date of 300-mm-wafer-based IC manufacturing set out in the Semiconductor Industry Association (SIA) road map [71] of 2001 has varied by several months to several years. These variations in estimated startup dates are due to fluctuations in the economic climate coupled with very successful die shrinkage, and are due in part to the successful implementation of CMP. The economic slowdown of 1998 has cut into the available funds for process development, both of 300-mm-capable CMP tools and also of the other semiconductor processing tools. In spite of the challenges posed by the downturn within the semiconductor industry, most of the CMP equipment suppliers have 300-mm-capable tools either in development or ready for production.

The availability of thin film deposition equipment capable of processing 300-mm wafers has also been limited, but this improved significantly since 300-mm development work began. Furthermore, the means of evaluating either deposited or planarized thin films has been affected by the limited availability of 300-mm-capable metrology tools. At the beginning of 1996, there were only lab-based metrology tools; in 1998, several manufacturers are offering 300-mm-capable tools. The result has been process development that consists of wafers being sent from lab to lab at each step of the process. Thus, process development has been slow. An additional factor contributing to the modest pace of development are the high cost of wafers. While costs dropped from $\sim$ \$US 2500 in 1996 to a 1998 price of $\sim$ \$US 1000, they remain much too high for less than critical, strategically important development experiments.

One factor that is different from earlier transitions is the increased reliance IC manufacturers are placing on the capital equipment vendors to share some of the burden of the development costs associated with the transition to 300-mm wafers. Although this shift is complementary to the maturing capabilities of the equipment vendors, it is placing enormous strain on them. Few vendors are in a position to provide new tools, engineering support, and process support in multiple sites for upward of a year with only the promise of remaining in contention for the substantial capital equipment purchases that can justify such significant investment of resources.

IX. Conclusion

There has been dramatic improvement in the capabilities of CMP tools in the 1990s. Early limitations of low throughput and the high cost of consumables has driven suppliers to make significant improvements, especially in the area of throughput and process quality. Even with the dramatic increase in activity in developing copper CMP processes, the anticipated work in low k dielectric CMP has not changed the focus of effort. Ironically, the ongoing drive for ever improving processes has caused the focus of effort in tool design to remain essentially the same.

Emphasis in CMP technology in 1998–1999 has evolved from a nearly exclusive focus on traditional polish systems to the use in manufacturing of advanced second- and third-generation polishers. This advance in technology has translated into standard CMP processes for oxide, polysilicon, and tungsten CMP, and expanded interest in aluminum CMP, copper CMP, and low k CMP. The development of slurry reprocessing for oxide CMP, and the introduction to the market of integrated metrology for CMP by nearly all of the equipment suppliers is expected to have a significant impact on CMP cost of ownership. This broad range of developments indicates that CMP is becoming an integrated process in semiconductor manufacturing facilities with a diverse arsenal of supporting tools. CMP is an enabling technology for the less than 0.5 μm generation technologies used by integrated circuit manufactures, and is critical to further advances. The future of CMP is very bright, and the technology is poised for dramatic developments in the months and years to come.

REFERENCES

1. C. Kaanta, W. Cote, J. Cronin, K. Holland, P.-I. Lee, T. Wright, "Submicron Wiring Technology with Tungsten and Planarization," *Int. Dev. Meeting Technical Digest*, pp. 209–212, 1987.
2. S. Pennington, S. Luce, "Improved Process Latitude with Chemical-Mechanical Polishing," *Proceedings of the VMIC Conference*, pp. 168–172 Santa Clara, CA, June 9–10, 1992.
3. J. M. Steigerwald, S. P. Murarka, R. J. Gutmann, *Chemical Mechanical Planarization of Microelectronic Materials*, John Wiley & Sons, Inc., 1997.
4. K. Holland, T. Bibby, "State of the Art in CMP Integration for 0.35 μm and Beyond," *SEMI Technical Symposium*, Makuhari Messe, Chiba, Japan, Dec. 4–6, 1996.
5. D. T. Price, R. J. Gutmann, S. P. Murarka, "Damascene copper interconnects with polymer ILDs," *J. Thin Solid Films*, 308–309, pp. 523–528, 1997.
6. Y.-L. Wang, C. Liu, M.-S. Feng, J. Dun, K.-S. Chou, "Effects of underlying films on the chemical-mechanical polishing for shallow trench isolation technology," *J. Thin Solid Films*, 308–309, pp. 543–549, 1997.
7. P. C. Andricacos, C. Uzoh, J. O. Dukovic, J. Horkans, H. Deligianni, "Damascene copper electroplating for chip interconnects," *Meeting Abstracts, 193rd Meeting of the Electrochemical Society*, Inc., San Diego, CA, Abstract No. 254, May 3–8, 1998.
8. W. O'Mara, "The inevitable shakeout in the CMP equipment market," *Solid State Technol.*, pp. 54–58, May 1998.
9. Data Quest, May 1998 (entire CMP tool set for each company).
10. T. Bibby, J. A. Adams, K. Holland, G. A. Krulik, P. Parikh, "CMP CoO reduction: slurry reprocessing," *Thin Solid Films* 308–309, pp. 538–542, 1997.
11. R. Jairath, A. Pant, T. Mallon, B. Withers, W. Krusell, "Linear planarization for CMP," *Solid State Technol.*, pp. 107–114, Oct. 1996.
12. B. Withers, E. Zhoa, R. Jairath, "A Wide Margin CMP and Clean Process for Shallow Trench Isolation Applications," *1998 Proceedings of the Third International Chemical-Mechanical Planarization for ULSI Multilevel Interconnection Conference* (CMP-MIC), Santa Clara, CA, pp. 319–327, Feb. 19–20, 1998.
13. N. Shendon, D. R. Smith, U.S. Patent #5,582,534, Dec. 10, 1996.
14. T. Bibby, A. Zutshi, Y. Gotkis, J. D. Lee, J. Yang, C. Barns, K. Holland, "Advantages of Orbital CMP Technology for Oxide and Metal Planarization," presented at Semicon Taiwan, Hsinchu, Taiwan, Sept. 22–24, 1997.

15. N. A. Hoshizaki, R. O. Williams, J. D. Buhler, C. A. Reichel, W. K. Hollywood, R. de Geuss, L. L. Lee, U.S. Patent #5,759,918, Jun. 2, 1998.
16. J. R. Breivogel, S. F. Louke, M. R. Oliver, L. D. Yau, C. E. Barns, U.S. Patent #5,554,064, issued Sept. 10, 1996.
17. A. R. Baker, "The Origin of the Edge Effect in Chemical Mechanical Planarization," presented at the ECS Fall 1996 meeting, San Antonio, TX.
18. D. Wang, J. Lee, K. Holland, T. Bibby, S. Beaudoin, T. Cale, "Von Mises stress in chemical-mechanical polishing processes," *J. Electrochem. Soc.*, Vol. 144, No. 3, pp. 1121–1127, March 1997.
19. C. Srinivasa-Murthy, D. Wang, S. P. Beaudoin, T. Bibby, K. Holland, T. S. Cale, "Stress distribution in chemical mechanical polishing," *Thin Solid Films* 308–309, pp. 533-537, 1997.
20. D. Wang, A. Zutshi, T. Bibby, T. S. Cale, S. P. Beaudoin, "Effects of carrier film physical properties on W CMP," Submitted for publication in *Thin Solid Films*.
21. Y. Zhang, P. Parikh, P. Golobtsov, B. Stephenson, M. Bonsaver, J. Lee, M. Hoffman, "Wafer Shape Measurement and Its Influence on Chemical Mechanical Planarization," *Meeting Abstracts of the Fall Meeting of the Electrochemical Society*, Abstract No. 250, pp. 629–630, San Antonio, TX, October 6–11, 1996.
22. G. Morsch, J. Lohmuller, E. Hartmannsgruber, *Chemical-Mechanical Polishing Process Optimization: A Study Focused on Local Back Surface Pressure*, Semiconductor Fabtech Edition.
23. M. Nakashiba, N. Kimura, I. Watanabe, K. Yoshida, U.S. Patent #5,762,539, issued Jun. 9, 1998.
24. J. Coppeta, J. Rogers, A. Philipossian, F. Kaufman, "Characterizing Slurry Flow During CMP Using Laser Induced Fluorescence," *Second International CMP for ULSI Multilevel Interconnection Conference*, Santa Clara, CA, pp. 307–314, Feb. 13–14, 1997.
25. J. Coppeta, J. Rogers, L. Racz, A. Philipossian, F. Kaufman, "Pad Effects on Slurry Transport Beneath a Wafer During Polishing," *Third International CMP for ULSI Multilevel Interconnection Conference*, Santa Clara, CA, pp. 36–43, Feb. 19–20, 1998.
26. N. Shendon, K. Struven, R. Kolenkow, U.S. Patent #5,205,082, Apr. 27, 1993.
27. S. Yamada, Japanese Patent #4-13567, issued Jan. 1, 1992.
28. Y. Nakamura, Japanese Patent #4-171170, issued Jun. 18, 1992.
29. D. L. Olmstead, U.S. Patent #5,193,316, issued Mar. 16, 1993.
30. H. Yui, Japanese Patent #6-91522, issued May 4, 1994.
31. A. Strasbaugh, U.S. Patent #5,449,316, issued Sept. 12, 1995.
32. H. Kobayashi, H. Miyairi, U.S. Patent #5,584,751, issued Dec. 17, 1996.
33. P. D. Jackson, S. C. Schultz, U.S. Patent #5,643,061, issued July 1, 1997.
34. J. R. Breivogel, M. J. Prince, C. E. Barns, U.S. Patent #5,635,083, issued Jun. 3, 1997.
35. M. T. Sherwood, H. Q. Lee, N. Shendon, S. Spektor, U.S. Patent #5,681,215, issued Oct. 28, 1997.
36. M. Leach, *Semiconductor International*, pp. 137–144, Oct. 1996; Y. Bukhman, U.S. Patent #5,230,184 (1993).
37. D. E. Weldon, B. A. Nagorski, H. Talieh, U.S. Patent #5,593,344, Jan. 14, 1997.
38. A. S. Meyer, T. G. Mallon, B. Withers, D. W. Young, U.S. Patent #5,722,877, Mar. 3, 1998.
39. A. K. Pant, D. W. Young, A. S. Meyer, K. Volodarsky, D. E. Weldon, U.S. Patent #5,800,248, issued Sep. 1, 1998.
40. S. Morimoto, R. J. Patterson, U.S. Patent #5,127,196, issued July 7, 1992.
41. R. J. Walsh, U.S. Patent #4,450,652, issued May 29, 1984.
42. G. S. Sandhu, T. T. Doan, U.S. Patent #5,762,537, issued Jun. 9, 1998.
43. I. Ali, S. R. Roy, G. Shinn, C. Tipton, "The Effect of Pad Conditioning on Chemical Mechanical Planarization (CMP) of Interlayer Dielectric (ILD)," *1995 Proceedings of the*

First International Dielectrics for VLSI/ULSI Multilevel Interconnection Conference (DUMIC), pp. 311–317, Santa Clara, CA, Feb. 21–22, 1995.
44. I. Ali, S. R. Roy, "Pad conditioning in interlayer dielectric CMP," *Solid State Technol.*, pp. 185–191, June 1997.
45. K. Achuthan, D. Hetherington, M. Oliver, "Pad Characteristics and Polish Performance," presented at the 3rd Annual Workshop on Chemical Mechanical Polishing, Lake Placid, NY, Aug. 16–19, 1998.
46. S. Inaba, T. Katsuyama, M. Tanaka, "Study of CMP Polishing Pad Control Method," *1998 Proceedings of the Third International Chemical-Mechanical Planarization for ULSI Multilevel Interconnection Conference (CMP-MIC)*, Santa Clara, CA, pp. 44–51, Feb. 19–20, 1998.
47. S. Beaudoin, D. Castillomejia, S. Schultz, J. Herb, J. Lee, T. Bibby, private communication, 1999.
48. J. Adams, T. Bibby, private communication, 1999.
49. A. Inamdar, M. A. Fury, D. Towery, A. B. Stubbman, J. W. Zimmer, "Cost of Ownership Implications of a Novel CMP Pad Conditioning Device," *Third International Chemical-Mechanical Polish (CMP) for ULSI Multilevel Interconnection Conference*, pp. 169–171, Feb. 19–20, 1998.
50. M. A. Walker, K. M. Robinson., U.S. Patent #5,779,522, Jul. 14, 1998; M. A. Walker, K. M. Robinson, U.S. Patent #5,616,069, issued April 1, 1997; J. M. Mullins, U.S. Patent #5,578,529, issued Nov. 26, 1996.
51. S. H. Lee, H. D. Jeong, "The New Concept of Conditioner for CMP," *Third International Chemical-Mechanical Polish (CMP) for ULSI Multilevel Interconnection Conference*, pp. 209–215, Feb. 19–20, 1998.
52. P. Troung, L. R. Blanchard, "Utilizing Pad Shaping as a Method to Stabilize Removal Rate, Improve Non-uniformity, and Increase Pad Life for Oxide CMP," *Third International Chemical-Mechanical Polish (CMP) for ULSI Multilevel Interconnection Conference*, pp. 351–356, Feb. 19–20, 1998.
53. P. E. Marmillion, A. M. Palagonia, U.S. Patent #5,785,584, issued Jul. 28, 1998.
54. C. C. Yu, U.S. Patent #5,769,699, issued Jun. 23, 1998.
55. H. Mizuno, "Slurry Delivery System Design for Tungsten Chemical Mechanical planarization Manufacturing Site," presented at ISSM, Oct. 10, 1997.
56. Y. Gotkis, "Post-CMP Surface Defectivity: Classification, Origin, and Defect Reduction," IPEC Planar PACRIM Chemical Mechanical Planarization Symposium, Hsin Chu, Taiwan Sept. 30–Oct. 1, 1997.
57. A. Jha, A. Gupta, S. Zhu, A. Blazev, M. Kason, S. Basak, "Post-CMP Clean on integrated Dry-in/Dry-out Auriga C Tool," presented at the 3rd Annual Workshop on Chemical Mechanical Polishing, Lake Placid, NY, Aug. 16–19, 1998.
58. S. Jankovsky, "Key Technology Challenges for Post CMP Cleaning," presented at the 3rd Annual Workshop on Chemical Mechanical Polishing, Lake Placid, NY, Aug. 16–19, 1998.
59. G. Gale, A. Busnaina, F. Dai, I. Kashkoush, "How to accomplish effective megasonic particle removal," *Semiconductor International*, pp. 133–138, Aug. 1996.
60. A. Busnaina, I. Kashkoush, G. Cale, "An experimental study of megasonic cleaning of silicon wafers," *J. Electrochem. Soc.*, Vol. 142, No. 8, pp. 2812–2817, Aug. 1995.
61. M. Olesen, C. Franklin, "A Single Wafer 'Non-Contact' Post CMP Cleaning Technology," *Third International Chemical-Mechanical Polish (CMP) for ULSI Multilevel Interconnection Conference*, pp. 375–378, Feb. 19–20, 1998.
62. A. Hu, X. Zhang, E. Sachs, P. Renteln, "Application of Run by Run Controller to the Chemical-Mechanical Planarization Process," *IEEE Proceeding of the 15th International*

Electronics Manufacturing Technology Symposium, pp. 235–240, Santa Clara, CA, Oct. 1993.
63. A. Hu, R. Chan, S. Chiao, "Real Time Control of Chemical Mechanical Planarization (CMP) Process," *First International Chemical-Mechanical Polish (CMP) for ULSI Multilevel Interconnection Conference,* pp. 235–240, Feb. 22–23, 1996.
64. Courtesy of IPEC Precision.
65. G. Dishon, D. Eylon, M. Finarov, A. Shulman, "Dielectric CMP Advanced Process Control Based on Integrated Thickness Monitoring," *Third International CMP for ULSI Multilevel Interconnection Conference,* Santa Clara, CA, pp. 267–274, Feb. 19–20, 1998.
66. S. Runnels, I. Kim, "Algorithms for Reliability-Based CMP Optimization," *Third International Chemical-Mechanical Polish (CMP) for ULSI Multilevel Interconnection Conference*, pp. 95–102, Feb. 19–20, 1998.
67. T. Bibby, K. Holland, "Endpoint detection for CMP," *J. Electron. Mater.* Vol. 27, No. 10, pp. 1073–1081, 1998, and references therein.
68. T. Bibby, J. A. Adams, K. Holland, G. A. Krulik, P. Parikh, "CMP CoO reduction: slurry reprocessing," *Thin Solid Films* 308–309, pp. 538–542, 1997.
69. A. Philipossian, F. Sanaulla, K. Lopez, "CMP Consumables Pricing and the First and Second Laws of Thermodynamics," presented at the 3rd Annual Workshop on Chemical Mechanical Polishing, Lake Placid, NY, Aug. 16–19, 1998.
70. *The National Technology Roadmap for Semiconductors*, Semiconductor Industry Association, 1994.

Chapter 3

Facilitization

John P. Bare

BOC EDWARDS
SANTA CLARA, CALIFORNIA

I. INTRODUCTION	47
II. OUTLINE	48
III. SLURRY DISTRIBUTION SYSTEM OVERVIEW	48
IV. SLURRY HANDLING	49
V. SLURRY DISTRIBUTION SYSTEMS	52
VI. SLURRY DISPENSE ENGINES	53
VII. SLURRY BLENDING TECHNOLOGY	54
VIII. SLURRY MEASURING TECHNIQUES	55
IX. DAYTANK REPLENISHMENT	59
X. MIX ORDER	61
XI. PIPING SYSTEMS	62
XII. PIPING SYSTEM VARIATIONS	63
XIII. MATERIALS OF CONSTRUCTION	64
XIV. SLURRY SETTLING	65
XV. SLURRY ROOM LOCATION	66
XVI. PRESSURE AND FLOW CONSISTENCY	66
XVII. BACK-PRESSURE DEVICES	67
XVIII. SLURRY CONSUMPTION RAMP	68
XIX. SYSTEM REDUNDANCY	69
XX. VALVE BOXES	69
XXI. STORAGE TANKS	71
XXII. AGITATION	74
XXIII. METROLOGY	76
XXIV. FILTRATION	78
XXV. SLURRY SYSTEM MAINTENANCE	83
XXVI. WASTE DISPOSAL	84
REFERENCES	87

I. Introduction

Facilitization for chemical mechanical polishing (CMP) evokes many different images, depending on the perspective of the user. For a researcher in an R&D lab with a single polisher, the image may at first seem to be simply a 20-liter (5-gal) pail with a piece of flexible tubing connecting directly to the peristaltic pump of the polisher. For a fab operations manager, the picture is far more complex, with many drums of different slurries arriving and requiring storage and handling, mixing and delivering slurry to the polishers, and ultimately disposing of the waste. Regardless of the size of the installation, many of the issues and principles are the same, only the scale is different. Furthermore, the regulatory climate may be very different for R&D versus manufacturing. As a fab follows the typical progression from R&D to pilot line to full-scale manufacturing, changes must be continually made to automate the "art" of the lab scientist to the practical needs of the operators, equipment and facilities engineers, and environmental and safety authorities. Compromises must often be made, as some things that "work great in the lab" often behave differently in manufacturing—and rarely better. We follow, more or less chronologically, the issues encountered by CMP slurry and its users as the slurry makes its way from the loading dock to a distribution system to the polisher and on to early retirement, even though it has plenty of life remaining.

II. Outline

- Slurry handling
- Dispense engines
- Blending
- Piping systems
- Storage tanks
- Agitation
- Metrology
- Filtration
- System maintenance and cleanup
- Waste disposal

III. Slurry Distribution System Overview

A slurry distribution system is a part of the infrastructure of a semiconductor fab, rather than a stand-alone piece of equipment. Even so, it can be thought of as a collection of pieces that comprise the whole. Many of these individual components can be sized for the specific system or left out entirely, depending on the application. Figure 1 shows a generalized slurry distribution system layout.

Key items for CMP slurry distribution include

- Slurry supply drum or tote
- Slurry dispense module
- Daytank with agitation system
- Piping system
- Filtration station
- Metrology
- Assay monitoring and control
- Data logging–interface with factory monitoring system (FMS or CMS)

Support facilities and services associated with the slurry distribution system may include

- Slurry drum storage area
- Slurry drum preparation area
- Post-use drum cleanout area
- Slurry waste treatment or reclaim
- Chemicals, equipment, and parts for preventive maintenance.

IV. Slurry Handling

CMP slurry starts out in fairly healthy condition from the manufacturer; or at least that is the hopeful assumption we make here. From then on, there are many opportunities for the slurry health to deteriorate in the course of handling by people or equipment; fortunately, there are also some opportunities to monitor, control, or improve the slurry quality.

CMP slurry must be shipped and stored without exposure to freezing, or even near-freezing, temperatures. The commonly accepted lower limit for slurry temperature is $5°C$ ($40°F$). Below this temperature, irreversible damage to the electrically charged layer surrounding the abrasive particles can occur, resulting in changes to the zeta potential, reduced suspension, and other unpredictable changes in polishing performance. While this is a seasonal problem, loading docks and other uncontrolled areas should be avoided as storage areas whenever possible.

Although CMP slurry is supposed to be primarily abrasive solids, only those abrasives chosen by the slurry manufacturer should be used for polishing. The two largest sources of particulate contamination during

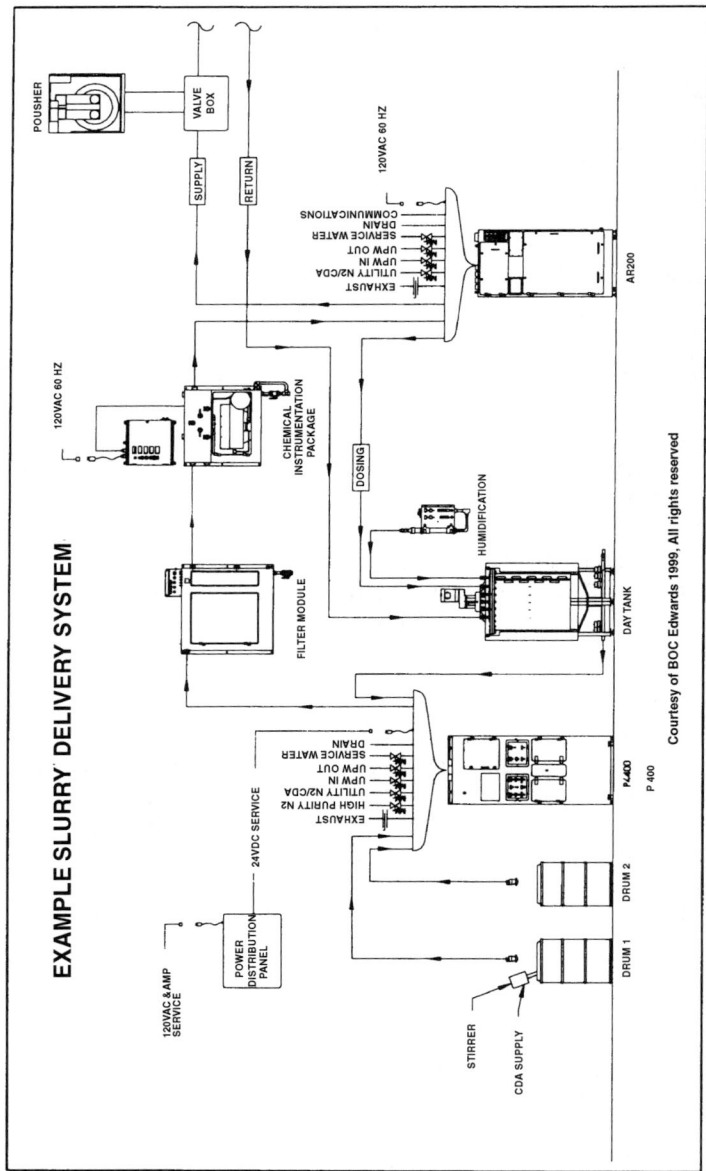

FIG. 1. Generalized CMP slurry distribution system with metrology, assay control, and filtration.

shipping, storage, and handling are foreign particles or dust, which is always present in loading dock type environments, and dried slurry. Of the two, dried slurry is more likely to be a real manufacturing problem. Dried slurry typically accumulates around the drum openings. Wiping caps with a damp cloth or even washing the cap area with water and a brush will minimize the number of dried slurry particles falling into the good slurry. While these dried slurry particles will not affect polish rates, they can cause scratches and defects. Dried slurry particles can also accumulate on drum connection heads and supply drum dip tubes, so care should be exercised to clean all equipment as thoroughly as is practical. The introduction of a little water from cleaning is much better than dried slurry particles from not cleaning.

CMP slurry is typically shipped to user sites in 20-liter (5-gal) pails for R&D evaluation or 200-liter (55-gal) drums or 1000-liter (265-gal) totes for manufacturing. The difference in handling ease of each container varies due to the size, shape, and weight of the container; the number, size, and location of openings in the container; and the need to insert stirrers, diptubes, or other devices into the container. The variety and nonstandardization of containers present a challenge for slurry distribution equipment manufacturers and additional cost for users: every different container potentially requires a different type of connection to the dispense equipment.

The 20-liter (5-gal) pails do not connect well to bulk slurry distribution systems. The small volume does not match well with volume required to prime even a small distribution system plus circulation piping and a small daytank. Thus these pails are most commonly used for R&D on a single polisher. The pails typically have two openings that can be used for withdrawing slurry and inserting a mechanical stirrer. Despite the drawback of an open container, which permits foreign particles to fall in and volatile components to escape, connection is typically made simply by inserting flexible tubing connected to a peristaltic pump and then to the polisher; a stirrer may be inserted through the second opening.

The 200-liter (55-gal) drums are the most common and work very well with bulk distribution equipment. Drums have two 2-in. openings that mate via threaded fittings with sealed dispense heads and sealed stirrer connections that minimize loss of water or other volatile components and minimize entry of foreign contaminants. All connections, including the drum hole caps, should be thoroughly cleaned to prevent dried slurry from falling into the drum. The need for stirring is very dependent on the type of slurry and will be covered later.

The 1000-liter (265-gal) totes are common for some well-suspended slurries and also work very well with bulk distribution equipment. Slurries that settle hard and require vigorous agitation for redispersion are not suitable for packaging in totes, since it is very difficult to reliably redisperse slurry that has settled into the corners of the cube-shaped totes. Dip tubes and stirrers can be inserted through the openings and sealed to minimize loss of water or other volatile components and minimize entry of foreign contaminants. Dimensional stability and repeatability of dip tube insertion depth is not as good with totes as with drums since totes are slightly flexible on all sides. If dip tubes are inserted from the top, it may be desirable to have an extralong dip tube to guarantee that it will reach the bottom with a special standoff on the bottom of the dip tube to raise the intake above the bottom and prevent drawing slurry from the very bottom of the tote. The bottom layer may contain large slurry agglomerates, which can scratch wafers or clog filters and slurry distribution piping systems. Recommendations from slurry manufacturers vary about using the last 2–3 cm (~ 1 in.) of slurry in the bottom of the tote. User practice varies about this same challenge and is usually unrelated to the recommendations by the slurry manufacturers or other industry experts; after all, the slurry is expensive and most users do not like to discard the last 20 or 30 liters. Some totes are equipped with bottom openings that can be connected to the supply line to the distribution equipment. The bottom openings, which are primarily intended for rinsing and cleaning the tote after use, will clearly draw the bottom layer of the slurry. Users must balance the benefit of using all of the slurry against the possibility of drawing agglomerated particles into the distribution system.

V. Slurry Distribution Systems

Slurry distribution systems have followed the same path of evolution that UPW (ultrapure water) and high-purity chemicals have followed, progressing from pail feed to fabwide delivery from a single source. A single R&D CMP polisher can certainly be conveniently fed from a small pail, typically 20 liters (5 gal), with a peristaltic pump. At this stage, process development is more important than yield improvement, which results from continuous improvement of a largely automated process. A single R&D polisher, used and cared for intermittently, probably presents a greater challenge to slurry distribution equipment than one operating full-time with proper maintenance. However, as the number of polishers increases and manufacturing output and yield predominate, manual slurry distribution yields to integrated bulk distribution.

Key differences between manual and bulk distribution systems are:

Manual distribution

- High traffic into polish areas
- High packaging costs

- Increased contamination risks
- Safety risks

Integrated bulk distribution

- Pump from remote location
- Purchase chemical in more economical bulk containers
- Uniform quality slurry to all points of use
- On-line slurry monitoring and control
- Slurry distribution loop with dual containment
- System interface to fab and process tool

VI. Slurry Dispense Engines

The term "pump" is commonly used to describe any device used to deliver CMP slurry under pressure to the polisher. There are enough variations, several of which do not use conventional "pumps," that a better term might be slurry dispense engine. Most polishers use a peristaltic pump to measure and deliver slurry to the polisher platen. Peristaltic pumps are well suited to this task and operate comfortably in the normal polisher consumption range of 100–1000 ml/min. For a single R&D polisher, they provide an easy-to-operate, easy-to-understand, low-budget local delivery system. However, higher capacity dispense engines are required for bulk slurry distribution systems. While there will always be exceptions, especially with homebuilt systems, Table I lists the most common types of slurry dispense engines and their normal use. Manufacturing or pilot-line dispense systems have global circulation, may have multiple polishers, and may have many peripheral devices, as shown in Fig. 1, while a local dispense system will serve only one polisher with minimal extra features.

The choice of dispense engine is currently a hot commercial issue, with variations available from several suppliers. Of the worldwide leaders in

TABLE I
Slurry Dispense Engines and Normal Use

Slurry dispense engines	Use
Pump: diaphragm, bellows	Bulk or pilot line
Vacuum pressure	Bulk or pilot line
Pump pressure	Bulk or pilot line
Peristaltic pump	Local dispense
Gravity	Local dispense

TABLE II
Slurry Dispense System Suppliers

Slurry dispense system suppliers	Dispense engine
BOC Edwards (formerly FSI International)	Vacuum pressure
MegaSystems and Chemicals	Pump
BOC-Edwards	Pump pressure

CMP slurry dispense systems in 1999 (Table II), different dispense engines have been chosen as the flagships for various technical [1] or patent reasons [2]. Each dispense engine has some stronger or weaker features compared to the alternatives, and the benefit varies with the slurry. The generally accepted comparative virtues and shortcomings of dispense engines are shown in Table III. The choice of slurry dispense engine or slurry distribution system supplier involves many other variables, however, some of which are intangible or emotional. Included are dispense engine technology, applicability to intended slurry, flexibility for R&D, reliability for manufacturing, blending technology, peripheral equipment, interface with factory monitoring system, price, customer service and support, and corporate agreements.

VII. Slurry Blending Technology

CMP slurries are shipped to user sites in ready-to-use strength or as a concentrate, which must be diluted to use strength or mixed with one or more chemicals to create the final CMP polishing slurry. Examples of widely used oxide and tungsten polishing slurries, which require various blending ratios, are listed in Table IV.

TABLE III
Advantages and Disadvantages of Dispense Engines

Dispense engine	Advantages	Disadvantages
Vacuum pressure	Very low shear on slurry Low wear on internal parts	Loss of volatile constituents
Pump	Comfortable technology	High shear on slurry Abrasive wear of pumps
Pump pressure	Low shear on slurry Low wear on internal parts	Poor pressure tank agitation Dissolved gasses effervesce Some loss of volatile constituents

TABLE IV

EXAMPLE OF SLURRY BLENDING REQUIREMENTS

Slurry	Diluting or blending chemical	Mix ratio by volume abrasive to part B
Oxide polishing slurry		
Cabot SC-112	Ready to use	
Cabot SC-1	UPW	1:1.79
Cabot SS-12	Ready to use	
Cabot SS-25	UPW	1:1.17
Cabot D7000	Ready to use	
Rodel ILD 1300	Ready to use	
Rodel Klebosol 1501	Ready to use	
Rodel Klebosol 1508	Ready to use	
Tungsten polishing slurry		
Cabot SS W-2000	30% H_2O_2	7:1 for 4% [H_2O_2]
Rodel MSW-1500	Ready to use	
Rodel MSW-2000	Chemical B	1:5.5
STI polishing slurry		
Rodel STS-1000	Chemical B	1:5

Ready-to-use slurries require no blending. The slurry can be dispensed directly from the slurry supply drum to the polisher or it can be transferred to a daytank, which is part of a global delivery loop, as shown in Fig. 1. Manufacturing-scale delivery systems typically transfer the slurry to a daytank to make continuous circulation easier, facilitate supply drum changes, ensure against interruptions in supply by sizing a daytank appropriately for the planned consumption rate, and gain flexibility in case the slurry brand is changed and requires a different blending or stirring approach.

Blending or mixing slurry components is not as simple as it seems at first glance. Furthermore the blending requirements and techniques will vary widely as one moves from single polisher R&D to continuous manufacturing. While the skilled CMP polishing researcher may mix a batch in an open container using a lab stirrer, more repeatable techniques must be used for manufacturing; nevertheless, many of the fundamental principles apply to both R&D and manufacturing.

VIII. Slurry Measuring Techniques

Commercial suppliers of bulk slurry delivery equipment have taken several different approaches to measuring slurry constituents. In addition, there are different approaches to replenishing a daytank as its slurry is consumed by the polishing process. To further complicate things, measurement of various slurry parameters may be used to monitor and adjust the mix ratio. All this just to blend the slurry for the daytank. All techniques described can and do work; each has advantages and disadvantages depending on the slurry and manufacturing philosophy of the user. However, regardless of the approach to measure, mix, or replenish slurry, the most important goal is uninterrupted delivery of uniform-quality slurry to the polisher:

Measuring techniques include

- Volumetric
- Weight scale
- Flow meter times time
- Metering pump times strokes
- Slurry mix ratio parameter endpoint
- Point of use mixing

In "volumetric" mixing, slurry components are repeatably measured by adding a fixed volume of one constituent to a fixed volume of one or more additional constituents. The desired mix ratio, by volume, is determined by the slurry manufacturer or user and then fixed for endless repetition by the slurry dispense equipment. The consistency of the final mixture is dependent on the consistency of the incoming slurry supply and the repeatability of the measuring process. Generally, the process is very repeatable as long as the sensors used to measure the amount of each constituent are cared for with preventive maintenance appropriate for the slurry and sensor type. If not cared for, slurry, which inevitably attaches even to Teflon® vessels, can affect the reliability of the sensor performance. While the advantage of this technique is high repeatability, the disadvantage is that changing to a different mix ratio requires moving the sensor and requalifying the new mix ratio. As implemented by BOC Edwards (formerly FSI International), the principal practitioner of this approach, small batches (~ 5 liters) are continuously added to a much larger daytank. By using batch averaging, small random errors in level sensing from any one batch are canceled by the succeeding batches, giving very uniform slurry in the daytank and at the polisher.

Weight scale measuring uses a medium- to large-size mixtank mounted on a weight scale to weigh in the constituents. The desired mix ratio, by weight, is determined by the slurry manufacturer or user and programmed into the controls for the filling tubing. Generally, the mix batches are $\frac{1}{4}$ to $\frac{1}{2}$ the size of the daytank, so that any inconsistency in the blend precision may have a significant effect on the consistency of the slurry in the daytank. The advantage of this approach is the flexibility of easily adjusting the mix ratio.

Disadvantages come from the measurement error, which is driven not so much by the scale error as by the connecting tubing from the supply drum to the mix tank. This approach was practiced by Systems Chemistry, Inc. (now BOC), but has been replaced by other approaches.

In "flow meter times time" measuring, individual constituents are measured into a medium-size "batch tank" by flowmeters. The desired mix ratio, by volume, is determined by the slurry manufacturer or user and programmed into the controls for the flowmeters. As with weight scale blending, the mix batches are typically $\frac{1}{4}$ to $\frac{1}{2}$ the size of the daytank, so that any inconsistency in the blend precision may have a significant effect on the consistency of the slurry in the daytank. The advantage of this approach is the flexibility of easily adjusting the mix ratio. The disadvantage is the reliability of the flow meters. Since CMP slurry is very abrasive, it can wear the moving parts in the flowmeter causing drift in the volume measured and hence drift in the assay of the slurry. In addition, agglomerated slurry can clog the meter, especially during periods of inactivity, again causing errant measurements. Preventive maintenance of the flowmeters and piping system can minimize both of these effects. The "flow meter times time" approach is practiced primarily by MEGA Systems & Chemicals.

In "metering pump times strokes" measuring, individual constituents are measured into a "batch tank" using a metering pump, which dispenses a fixed volume for each stroke of the pump. The desired mix ratio, by volume, is determined by the slurry manufacturer or user and programmed into the controls for the metering pumps. While the mix ratio is limited to incremental ratios of full pump strokes, in practice the pump strokes are small compared to the mix batch size, so that any normal mix ratio can be achieved. Furthermore, the ratio of strokes of each constituent will not vary, giving a repeatable mix ratio. The advantage of this approach is the flexibility of easily adjusting the mix ratio. The disadvantage is the reliability of the metering pumps. This approach, practiced by BOC, is new to the manufacturing arena, at the time of this writing. The advantages, disadvantages, and reliability remain to be fully tested.

"Slurry mix ratio parameter endpoint" measuring uses neither volume nor weight to mix a fresh batch of slurry. In this approach, the approximate mix batch volume is determined by experiment and experience, but the actual mix ratio is controlled by measuring some endpoint parameter, generally defined by the user. This approach is useful when some parameter, which can be reliably measured, is sensitive to changes in mix ratio. It is also useful when some reliably measured parameter is crucial to the polishing process. The usefulness of the commonly measured slurry health parameters, such as pH or density, varies widely from slurry to slurry, as shown in Fig. 2 [3] for several commercially available slurries. A more generally applicable, but

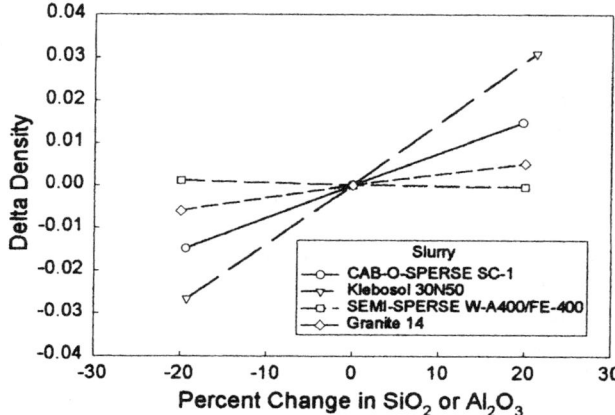

FIG. 2. Change in density (normalized) as function of percentage of abrasive.

rarely used parameter is conductivity, as shown in Fig. 3 [4]. This technique is employed by BOC Edwards as the primary approach for determining mix ratio. These same metrology parameters are used to monitor slurry health in mixed slurry in a daytank. Care should be taken to check the appropriateness of a particular parameter, since not all measures give meaningful information for a given slurry.

Point-of-use (POU) mixing has long been viewed as the solution for many of the challenges associated with blending slurry, especially for handling slurries with a short postmix potlife or to offer the possibility for changing the mixture of a slurry during a polishing cycle. However, it has never been commercialized in a bulk delivery system. The difficulties in implementation outweigh the hoped-for benefits. The usual perceived benefit is that the slurry will always be fresh and users can avoid deterioration of slurry activity as the slurry ages. However, the practical challenges can be quite sobering. Controlling mix ratio at POU is difficult. Currently, users of POU mixing with custom-built equipment employ peristaltic pumps, which both measure and dispense, to draw ingredients from a global distribution loop and deliver them to a common final tube dispensing to the polisher platen. Flow rate through the peristaltic pump will change as the peristaltic pump tubing ages and due to changes in the supply pressure from the global loop. Pressure in the global supply loop will vary as neighboring polishers turn on and off and as filters age. While these same factors affect a single stream global supply loop, they will affect two supply streams in different ways, causing mix ratio to vary. Maintaining uniform mix ratio to all polishers is impossible. Even if flow and pressure in the global distribution system are

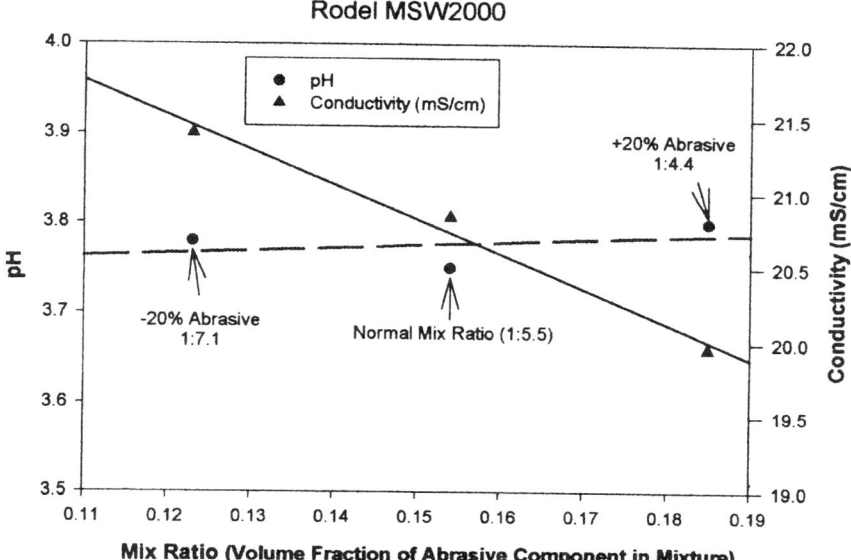

FIG. 3. Sensitivity of pH and conductivity to changes in mix ratio for Rodel MSW-2000.

perfect, variation in peristaltic pump settings and drift in performance will produce varying mix ratios. The larger the number of polishers, the larger the number of unequal polishers and the wider the variation. With a single stream of slurry supplied from a daytank, the mix ratio to all polishers will be identical. Drift in the amount of slurry delivered due to peristaltic pump tubing aging or filter aging will have a smaller effect than normal drift due to pad wear. Thus normal removal rate monitoring will compensate for these small changes. Unpredictable changes in mix ratio can produce effects that can not be compensated for simply by monitoring removal rate. Selectivity and defectivity may also be affected. Schemes may be developed in the future to make point-of-use mixing a viable alternative to bulk mixing, but at this writing they are not realistic.

IX. Daytank Replenishment

As slurry is consumed by the polishing process, the reservoir of new slurry must be replenished. The previous sections outlined various styles of measuring the slurry components, but there are additional choices that must be made regarding philosophy of daytank replenishment. As with the choice of dispense engine and the approach to blending, all of the following approaches to daytank management can and do work. However, some are more suitable than others, depending on slurry postmix potlife, consumption rate, and individual philosophy.

Daytank replenishment methods include

- Dual daytanks
- Periodic addition of large (relative to the daytank size) batches
- Continuous addition of small (relative to the daytank size) batches
- Mix order

The use of two separate daytanks or dual daytanks is easy to understand and fairly easy to implement. It has some advantages for certain slurries, but some distinct disadvantages. In this approach, two equal-size daytanks are used. One daytank supplies slurry to the fab while the second daytank is being cleaned, refilled, and the fresh batch of slurry is being qualified for use. This approach is most commonly used when a slurry has a short postmix potlife. When the on-line daytank is empty, the slurry delivery machine automatically switches to the new daytank and begins dispensing the fresh slurry. Unfortunately, the slurry in the first daytank is never completely consumed to make an uninterrupted transition from one daytank to the other; this causes a waste of expensive slurry, not to mention the capital cost of the second daytank. Second, since the slurry has a decreasing activity with time (short postmix potlife), at the time of the daytank changeover, there may be an abrupt change in the polishing characteristics of the slurry. These effects may be compounded by any differences in the mix repeatability during makeup and by the variable amount of time the backup daytank sits before being brought on line. The approach does facilitate frequent cleanout of old chemical and dried slurry and it satisfies the emotional concern about decreasing activity of "old" slurry.

Another common approach is the periodic addition of large (relative to the daytank size) batches to a daytank, sized fror the consumption rate of the fab, commonly ranging from 200 to 2000 liters (50 to 500 gal). Replenishment slurry is homogeneously blended in an off-line mixtank, typically 100–200 liters (25–50 gal). The fresh slurry may or may not be analytically qualified, depending on the bells and whistles on the slurry mix module. The analytical qualification measurement used may or may not be appropriate for the slurry being mixed. Generally, the mix batches are $\frac{1}{4}$ to $\frac{1}{2}$ the size of the daytank. Any inconsistency in the blend precision may have a significant effect on the consistency of the slurry in the daytank, since the volume of fresh slurry added is large compared to the total volume in the daytank.

Continuous addition of small (relative to the daytank size) batches has been developed to provide uniform quality slurry to the polisher. Typically, small batches (~ 5 liters) are added to a large daytank (200–2000 liters). While the volumetric measuring of small batches (~ 5 liters) has opportunity for random batch-to-batch variation similar to other measuring approaches, the addition of a small increment to a large daytank will have minimal effect on the average concentration of any constituent in the daytank. Batch averaging of many small batches will minimize the effect of random error in any one batch. Changes due to variation in the incoming slurry as supply drums are changed will be buffered by the relatively large volume in the daytank. The disadvantage of this approach is that the small batches may not be homogeneously mixed before being added to the daytank, which makes analytical qualification of each batch difficult. Complete homogenization will occur as soon as the small batch is added to the daytank.

All the techniques described can and do work; each has advantages and disadvantages depending on the slurry and manufacturing philosophy of the user. However, regardless of the approach to measure, mix, or replenish slurry, the most important goal is uninterrupted delivery of uniform-quality slurry to the polisher.

X. Mix Order

Regardless of the measuring technique, the sequence of mixing slurry constituents may be important. A number of phenomena can occur that are detrimental to the final suspension stability of the slurry. While a detailed description is beyond the scope of this chapter, these phenomena will be collectively called "chemical shock," also referred to as "pH shock." In simple terms, the electrostatic repulsive layer surrounding each suspended abrasive particle is weakened by dilution with a second constituent so that the particles no longer have sufficient charge to repel each other. Agglomeration may occur, creating large particles that may settle, clog piping systems, or cause wafer scratches.

In general, the rule of thumb is to add the diluting component to the abrasive component. There are, however, exceptions to this rule of thumb, so experience and the recommendations of the slurry manufacturer should be considered. In the case of silica-abrasive oxide-polishing slurries, this means slurry first, water second. Even when practiced in this manner, slurry diluted at the user site will typically have more agglomerates than slurry diluted to use-concentration by the slurry manufacturer. The primary reason for the difference is in the timing of slurry filtration, not the number of filtration steps or the filter pore size used. The slurry manufacturer will typically filter the slurry to remove oversized agglomerates just before packaging. Since the dilution process itself creates some of these agglomerates, slurry made by dilution at a user site will contain more agglomerates. Filtration at the user site can fully counteract this particle growth. There are also several mixing stategies that may also help minimize the creation of agglomerates.

The simplest way to minimize chemical–pH shock from dilution with water is to use pH-adjusted water. Unfortunately, water of the correct pH is rarely available as a typical fab pressurized supply chemical; it must be mixed off-line and supplied as a separate drum chemical or else a new bulk chemical supply stream must be created. Chemical shock can also be minimized by adding small amounts of pure constituents directly to a large daytank of use-concentration slurry. In this approach, the pure components never come in contact with each other. The makeup slurry is added to the upper portion of the daytank while slurry is drawn from the bottom of the daytank for distribution to the fab. A mechanical stirrer at midlevel or slightly lower will ensure complete homogenization. The increments added should be less than $\sim 10\%$ of the daytank volume to avoid fluctuations in the use-strength slurry withdrawn for delivery to the fab. This approach is ideal for slurries such as Rodel MSW-2000 tungsten polishing slurry, which is very viscous when the first portion of the oxidizer component is added to the abrasive; adding small amounts of the pure constituents to a large volume of use-concentration slurry avoids the situation altogether.

XI. Piping Systems

The design and choice of components for a global slurry delivery system have evolved from UPW and high-purity chemical delivery systems. The requirements are far more challenging since CMP slurry is not necessarily chemically or physically stable. At best the materials are suspensions, not solutions, and the behavior of one slurry may require different treatment than that of another, even when the two are otherwise substitutes, performing the same polishing step. The biggest concerns are slurry settling and slurry shear, which can form agglomerates. While slurry piping systems have evolved from UPW and high-purity piping systems, they are sufficiently different to require a far more thorough understanding of the complex chemical and physical property considerations, as well as more conventional hydraulic requirements. Designs continue to evolve as new components, developed to meet the demands of CMP, enable the construction of newer designs.

XII. Piping System Variations

The design of a piping system will be very site-specific. Factors that must be considered include

- Vertical rise and horizontal run of piping
- Slurry consumption
- Number and type of polishers
- Pressure and flow budget from slurry dispense machine
- Type of slurry — linear flow velocity (settling)
- Individual philosophy toward reliability, redundancy, and sanity of putting all polishers on a single delivery system.

Regardless of the needs dictated by these factors, the piping system itself will fall into a few general variations as shown in Fig. 4. Dead-head style piping systems are unusual and can only be used with a very well-suspended slurry. This type of system can minimize the loss of volatile components, such as ammonia from a slurry like Rodel Klebosol, since there is no circulation through a daytank or any other head space that would permit ammonia evaporation. It can also be attractive to minimize slurry turnovers to a shear- or gelling-sensitive slurry. Cabot SS-25 is well-suspended, but if

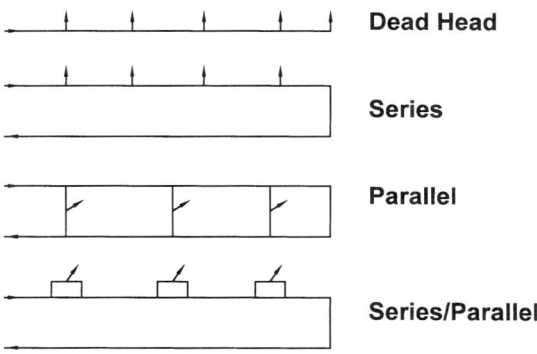

FIG. 4. Generalized piping system variations.

circulated continuously, it can gel in the piping system. The prime disadvantage to a dead-head system is that most slurries do settle eventually and thus require circulation. If a system is installed as a dead-head system for a specific slurry, conversion to use for a slurry requiring circulation may be difficult and expensive.

Series piping is the most common. It most closely resembles UPW or high-purity chemical style systems. This system can be used for well-suspended slurries and poorly suspended slurries. However, there are important differences that must be properly factored into the design to accommodate shear sensitivity and settling characteristics. This simple style can be designed with a multiway valve at the slurry takeoff. The multiway valve achieves a clean, uninterrupted flowpath for the main loop, with a small diameter takeoff to the polisher. The polisher feed tubing can be flushed from the multiway valve to the polisher to eliminate stagnant slurry or prevent slurry settling.

Parallel piping systems were very popular for accommodating fast-settling slurries, but have become less popular recently as component selection and changes in understanding of slurry behavior have improved. Balancing the flow through all sections of the piping is very delicate and consumes a great deal of the pressure–flow budget. The challenge of adjusting flow in the crossover legs to accommodate the intermittent flow of many polishers has made this approach the exception rather than the rule.

Series–parallel piping systems have been used to keep fast-settling slurries dispersed. As with parallel piping systems, balancing the flow through all sections of the piping is very delicate and consumes a great deal of the pressure–flow budget. Adjusting flow in the tool loop legs while maintaining flow in the main loop creates a rapid pressure drop through the system as each polisher loop imposes a very large pressure drop. While this style has achieved some success, newer piping components, combined with the delicacy required from complicated piping systems, has turned current favor toward the simplicity of the series design.

XIII. Materials of Construction

The most common material of construction is PFA Teflon®. This high-purity material has become the material of choice due to its universal chemical resistance, nonstick surface properties, availability and compatibility of parts, and general historical and emotional acceptance. Other materials, such as polyethylene and polypropylene, are becoming more widely used, principally for cost-saving in slurry dispense equipment, but have made little penetration into the fab piping arena. Alternative materials should be carefully considered for the consequence of switching slurry due to process changes. Materials must be robust to exposure to acids, bases, oxidizers, abrasives, and other unknown and proprietary chemicals. Metal piping such as steel should not be used for fear of metallic contamination. Materials with leachable organics, such as polyvinylchloride (PVC), should also be avoided.

XIV. Slurry Settling

In addition to being liquid sandpaper, all CMP slurries will settle in time. Oxide slurries, using silica as the abrasive, are the best suspended due to the zeta potential on the particle and lower density ($\sim 2.2 \, \text{g/cm}^3$) compared to alumina ($\sim 4.0 \, \text{g/cm}^3$) or ceria ($\sim 7.6 \, \text{g/cm}^3$) [5]. Since all slurries have a distribution of particle sizes, even very well-suspended slurries will contain some large particles, which will settle in areas without sufficient turbulence to keep them dispersed. Fortunately, the practical requirements of flow rate through a system will provide enough turbulence to keep all oxide slurries dispersed. A common, but often misused, rule of thumb is that slurry should flow faster than 1 ft/sec (30 cm/sec) to maintain dispersion. The real linear velocity required is different for each slurry and must be measured experimentally. In general, oxide slurry will maintain suspension with linear velocity as low as 1 m/min (3 ft/min). However, this lower limit is not practical, sincde the flow rate (~ 400 ml/min in 1-in.-diameter tubing) would not be enough to supply a manufacturing process. At a more practical flow rate of 8 liters/min (2 GPM), the linear velocity in 1-in. tubing would be 34 cm/sec (1 ft/sec). It seems that the 1 ft/sec figure creeps back into use, whether justified or not. Flow rate and linear flow velocity for various tubing sizes are shown in Table V. For fast settling tungsten slurries, such as Rodel MSW-2000, more turbulence is required to maintain dispersion. For these slurries, a linear velocity $\geqslant 1$ ft/sec is normally sufficient.

In summary, oxide slurry flow should be kept low, while still maintaining sufficient flow and pressure to supply the polishers. High turbulence is not required to maintain dispersion, but high flow through shear locations (point sources of high pressure gradient) can cause shear agglomeration of the slurry. Fast settling tungsten slurries, which fortunately tend to be less shear sensitive, should be kept moving faster than 1 ft/sec to maintain dispersion.

TABLE V

FLOW RATE AND LINEAR VELOCITY FOR VARIOUS TUBING SIZES

Flow rate		Flow velocity (cm/sec)					
		Tubing diameter (OD in inches)					
(liter/min)	(gal/min)	0.25	0.375	0.50	0.75	1.0	1.25
1	0.3	136	52	23	8	4	3
2	0.5	271	104	47	17	9	6
3	0.8	407	157	70	25	13	10
4	1.1	542	209	93	34	17	13
6	1.6	813	313	140	50	26	19
8	2.1	1084	418	186	67	34	26
10	2.6	1355	522	233	84	43	32
12	3.2	1626	627	279	101	51	39
15	4.0	2033	784	349	126	64	48
20	5.3	2710	1045	466	168	86	64
30	7.9	4065	1567	698	252	129	97

XV. Slurry Room Location

Traditionally, the slurry room, housing supply drums of slurry plus dispense equipment, has been located on the fab ground floor, as is common with high-purity chemicals. While there are some benefits in ease of drum handling in this location, process advantages can be achieved by locating the delivery equipment and daytank on the same floor as the polishers, or, better yet, above the polisher floor. The energy required to raise the slurry to a higher floor consumes part of the systems' pressure–flow budget, while the back-pressure devices used to provide pressure to the polishers can cause shear damage to the slurry. A daytank located above the polisher floor would help maintain more uniform pressure on all polishers in a loop and minimize the shear caused by restrictive back-pressure devices.

XVI. Pressure and Flow Consistency

One goal of a bulk delivery system is to provide constant pressure to a CMP polisher, so that the peristaltic metering pump in the polisher can provide constant slurry flow to the polisher platen. While many schemes have been devised to minimize the fluctuations in pressure, there are many sources of pressure drop, which will cause pressure to vary from polisher to polisher and at a given polisher over time. Competent hydraulic engineers can easily calculate the pressure loss along a piping design, given the slurry parameters. Unfortunately, these same designers, used to working with water or high-purity chemicals, which are stable materials, rarely account for changes due to fluctuating demand from polishers, the progressive loading of filters, or the shear sensitivity of slurry. Given all these changing conditions, users must accept some instantaneous fluctuation and longer term drift in pressure. Alternatively, as more contrived back-pressure schemes are developed, more components, which can malfunction, are introduced.

XVII. Back-pressure Devices

Historically, bulk delivery systems have used manual diaphragm valves located at the return end of the global piping system, another carryover from high-purity chemical systems. However, the shear sensitivity of many slurries has fostered the creation of other back-pressure approaches, including

- Adjustable, manual diaphragm valves
- Adjustable, manual pinch valves
- Active pressure monitors with feedback
- Fixed, tapered reducers with small-diameter tubing
- Dispense unit and daytank above polisher floor

Shear sensitivity has been a challenge rarely encountered by high-purity chemicals. The normal approach for applying back pressure for most high-purity chemicals has been a manual, adjustable diaphragm valve. Unfortunately, with the location of most slurry rooms below the polisher floor, the restriction created by a diaphragm valve causes a very high shear point as slurry under the gravity pressure head plus system pressure head squeezes through a tiny orifice. The drop from the polisher floor to the slurry room is typically 5–10 m, and greater than 20 m in some places. The convoluted pathway in the diaphragm valve aggravates the shear sensitivity of slurry. The fundamental problem for the slurry is the high pressure gradient, which causes shear; the problem for the operator is to achieve consistent pressure to the polisher.

Pinch valves were introduced in 1996 as an improvement in flow path compared to diaphragm valves. They have achieved some success, especially when used with less shear-sensitive tungsten slurries, but they do not change the physics of the problem: a large pressure drop across a very small distance.

Pressure monitors at the return end of the global loop, coupled with active feedback to an adjustable valve have been used to compensate for fluctuations in line pressure. Unfortunately, the valves used have been diaphragm or needle valves, which have all of the shear challenges. While pressure consistency may be improved for the polishes nearest the control device, the shear challenge has not been addressed. Furthermore, the pressure fluctuation in the middle section of a global loop, as polishers turn on and off, can not be corrected by a device at the end.

A solution for the shear sensitivity can be achieved by using a section of small-diameter tubing at the end of the global loop to provide a low-shear, fixed back pressure. When coupled with tapered reducers to reduce the typical 1-in.-diameter global loop tubing to a section of $\frac{3}{8}$-in. or $\frac{1}{4}$-in. tubing, the pressure gradient, and hence shear, is dramatically reduced. Unfortunately, this restriction is not easily adjustable, which has been perceived as a drawback. Adjustment can be made simply by changing the length or diameter of the reducing tubing. This would probably require a system shutdown, which is undesirable. Although adjustment is probably not necessary for slurry delivery, greater flow capability may be desirable for system maintenance.

A paradigm shift in the location of the delivery system has been employed by some fabs. The delivery system and daytank have been located on the floor above the polishers, letting gravity provide a large fraction of the supply pressure and back pressure. This approach reduces the supply pressure requirement for any dispense engine. With most of the back pressure supplied by gravity rather than any type of restrictive orifice, the shear imposed on the slurry is dramatically reduced. The pressure consistency is improved since there is a true head pressure supplied from both ends of the system. There will still be fluctuation in supply pressure to the polishers since the effect of polishers turning on and off in the middle of the tool loop can not be controlled totally from the ends of the global loop.

XVIII. Slurry Consumption Ramp

One of the ongoing challenges is the fact that most slurry delivery systems are designed and built for final steady-state use at full-scale manufacturing. Unfortunately the needs during startup may be very different from the needs at full buildout. Compromises must be made. There are no firm rules to cover this dilemma, but recognition of this fact will help deal with the unpredictable transitions of scale. Parameters that will undergo evolution include

- Total slurry consumption
- Slurry turnovers
- Slurry lifetime after mixing
- Decomposition or evaporation of components

XIX. System Redundancy

Redundancy of slurry delivery systems or system components is an emotional and user-specific decision. While all manufacturers strive to make dispense systems and all peripherals as reliable as possible, compromises must be made for cost, space, slurry lifetime, and manufacturing philosophy. In the end, it usually comes down to how much one is willing to pay for 99.99% uptime versus 99% uptime. Some of the common strategies employed include

- *Dual daytanks.* Dual daytanks are especially useful for concerns about slurry life or assay. One daytank is consumed to completion and then the tank may be cleaned before a fresh batch is mixed. The second daytank can be blended, checked for assay, and sit ready to use after the first tank has been consumed.
- *Redundant piping system.* Redundant piping systems can be installed, especially in conjunction with dual daytanks, to enable a complete cleanout of the delivery system between slurry batches. When this degree of redundancy is chosen, a totally separate slurry dispense machine is included to permit a backup for all major components.
- *Redundant dispense engines.* Redundant dispense engines have become fairly common in both pump- and non-pump-based slurry dispense systems. Since slurry is both chemically and mechanically aggressive, a backup dispense engine enables system maintenance without interruption of supply to the fab.

XX. Valve Boxes

As shown in Fig. 1, polishers are typically connected to a global delivery piping system by a valve box. The valve box enables polishers to be added or removed from the global system without shutting down the entire system to cut into the main distribution loop. In addition to isolation, flushing and cleaning of the tool loop can be implemented. Two typical valve box styles are shown in Fig. 5.

The "traditional" style (Fig. 5a) is used with a parallel piping system, as described in Fig. 4. This style enables continuous circulation of slurry through a polisher with a very small deadleg in the polisher, which can be largely flushed via a three-way valve inside the polisher and controlled by the polisher. Balancing flow through a large number of polishers in this configuration is very difficult due to the intermittent flow to the polishers. This style has generally been replaced by the "alternative" style, representing a "series" piping style. The arrangement in Fig. 5b enables the polisher flushing controls to flush the tool loop at user-defined intervals. The manual isolation valve in the valve box permits replacement or maintenance of

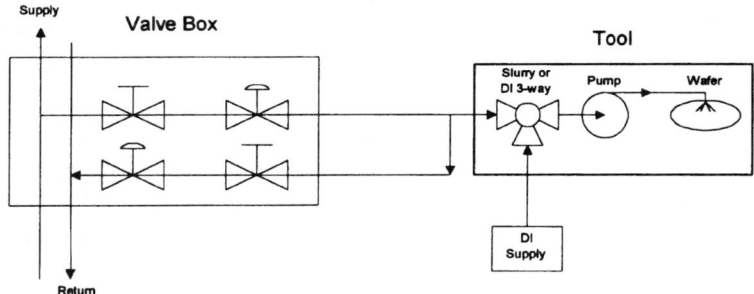

(a) Traditional Slurry Distribution Configuration

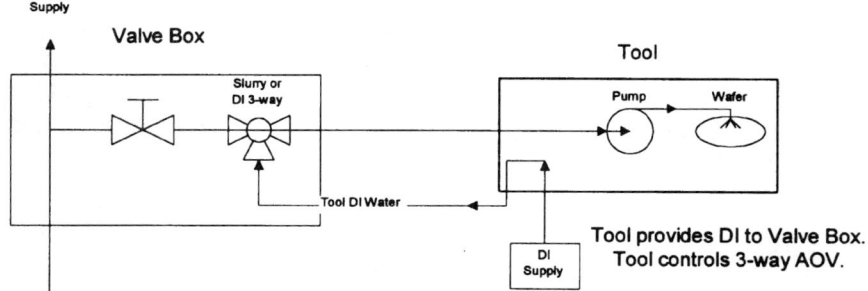

(b) Alternate Slurry Distribution Configuration

FIG. 5. Valve boxes for global slurry delivery system.

polishers or three-way valves. A variation on this arrangement would have a multiway valve in the global slurry loop, replacing the "T" and three-way valves. Water can flush the entire tool loop when slurry was not requested. A manual isolation valve should be retained to enable installation or removal of polishers.

In summary, there is no perfect piping system that combines uniform consistency of slurry, totally stable pressure to all polishers, low shear to certain slurries, turbulence for reliable dispersion of settling slurries, and no deadlegs for slurry settling. For oxide slurry, large-diameter tubing, low flow and flow velocity, and minimal shear points are desirable. For fast settling (tungsten) slurry, higher flow velocity is required.

XXI. Storage Tanks

Storage tanks, or daytanks, are used to hold a supply of slurry for circulation to a fab via a global delivery loop. Since slurry must be kept moving, rather than delivered in a deadhead delivery fashion, a slurry dispense engine will typically draw from a storage tank and circulate the slurry back to that storage tank, as depicted in Fig. 6.

As with many aspects of delivery systems, there are different philosophies surrounding daytanks, especially size. Size typically ranges from 200 liters (50 gal) to 2000 liters (500 gal). A large daytank will provide a buffer against failure to replace a supply drum after it becomes empty. A large daytank will also provide a very consistent slurry, since fluctuations in mix accuracy of freshly added slurry are buffered by the large volume. Large daytanks make sense for a full-scale manufacturing facility. For smaller facilities, small daytanks make sense to avoid unnecessarily aging the slurry or exposing it to environmental abuses. Typically, daytanks are sized to hold from 1 day to 1 week's worth of slurry. Users may also be concerned about the number of turnovers a slurry endures during its life, since each turnover exposes the slurry to various abuses including decomposition of constituents, valves, restrictive orifices, pumps, air–nitrogen exposure, vacuum, filters, and stagnation. A common misconception is that the size of the daytank affects the number of turnovers in a slurry's life. In reality, the number of turnovers is a function only of the rate of slurry consumption and the flow rate through the system. A larger daytank will last longer, but will turn over more slowly, which exactly cancels the effect of daytank size as shown in the equation:

Turnovers = Flow Rate (liter/min)

$\div$ Consumption Rate (liter/day) $\times$ 1440 (min/day)

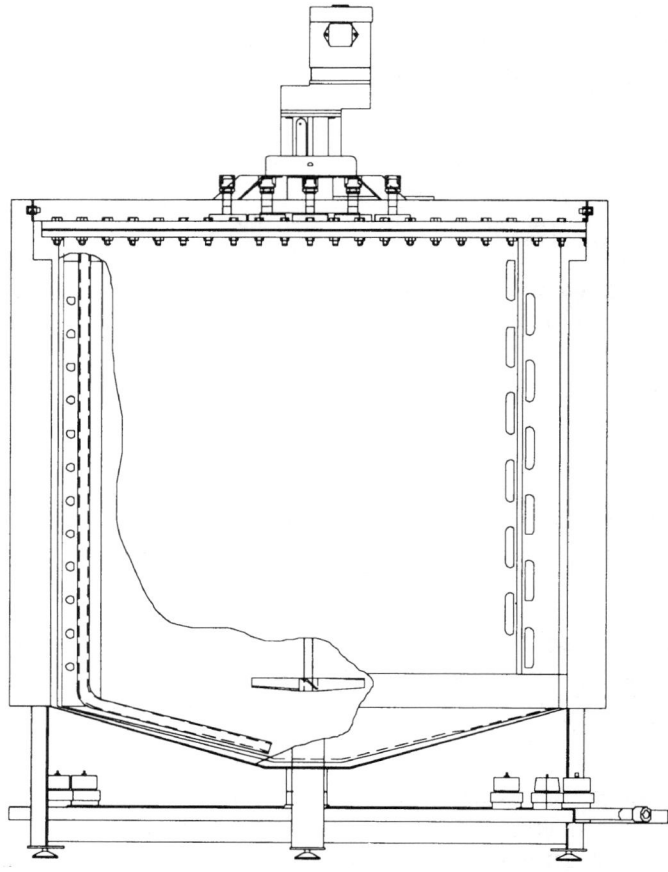

Fig. 6. Slurry storage tank or daytank (Courtesy of BOC Edwards).

For example, if system flow rate = 8 liter/min and consumption rate = 100 liters/day

Turnovers = 8 liters/min ÷ 100 liters/day × 1440 min/day

= 115 turnovers in slurry life

The size of the daytank will affect the life of the slurry, measured in days, but not in turnovers. Since some slurry health measures are a function of time and others a function of turnovers, one must decide which is more important.

Daytank size is also a challenge during manufacturing rampup. Many users prefer to use a low daytank volume during system startup and change to a full daytank at full-scale manufacturing. Slurry return tubes and mechanical stirrer paddles must be adjusted so that slurry does not splash in the daytank. Splashing can permit slurry to accumulate and dry on the tank walls; these dried particles can fall into the slurry, causing wafer scratches. While there are solutions to all the problems, including humidification, filtration, changing dip tube, and stirrer position, users must recognize the implications of something as innocent sounding as changing the working level of a daytank.

Daytanks are generally constructed of PFA Teflon®. As with piping systems in general, this high-purity material has become the material of choice due to its universal chemical resistance, nonstick surface properties, and general historical and emotional acceptance. Other materials, such as polyethylene and polypropylene, are becoming more widely used, principally for cost savings. Most slurries are shipped from the manufacturer in polyethylene containers, so there is precedent for this material. Alternative materials should be carefully considered for the consequence of switching slurry due to process changes. Materials must be robust to exposure to acids, bases, oxidizers, abrasives, and other unknown and proprietary chemicals. Metal piping such as steel should not be used for fear of metallic contamination. Materials with leachable organics, such as polyvinylchloride, should also be avoided.

Storage daytanks come in many shapes, but the most common shape is round with a cone bottom and bottom draw. Square tanks fit nicely into available space, but hinder stirring. Flat-bottomed daytanks have also been commonly used with a dip tube inserted from the top, extended to near the bottom. The inserted dip tube prevents drawing material that may have settled to the bottom. If agitation is insufficient, or system cleanout is not often enough, agglomerated slurry can accumulate and suddenly be drawn into the dip tube, causing a catastrophic increase in defect-causing particles. Bottom draw tanks eliminate the possibility for this step change in slurry quality.

The headspace in a daytank should be kept free of oxygen and full of water vapor. This means a humidified nitrogen blanket. The intent is to prevent oxygen absorption and also to prevent formation of dried slurry. Humidified nitrogen can be supplied from many sources, but the simplest is a bubbler (see Fig. 1) or atomizer in the incoming nitrogen supply. Humidity greater than 90% can be easily achieved. Although most users install a humidified nitrogen blanket directly into the daytank, this creates a nitrogen purge as well as a blanket. A better connection would be through the vent line connecting the daytank to atmosphere, which is intended to prevent the daytank from becoming either pressurized or evacuated. If connected through this vent tube, the daytank will draw in or expel humidified nitrogen only as the daytank level rises or falls. There will be no continuous purge through the daytank. Continuous purge can have two devastating effects: (1) volatile components such as ammonia can be stripped from a slurry and (2) dried slurry can accumulate near the vent exit as the purging nitrogen, which is less than 100% humidified, carries minute slurry quantities of slurry, deposits it near the daytank vent exit, and dries it; the dried particles can then fall into the liquid slurry, causing process problems.

XXII. Agitation

Agitation is normally required for slurry components and use-strength slurry, in both supply drums and daytanks. Supply drum stirring is recommended for all slurry, since some settling can occur for even the best-suspended slurries and the time spent since manufacture in shipping and storage is very unpredictable. Some tungsten-polishing slurries, such as Rodel MSW-2000A, settle fast and require vigorous stirring for ~ 30 min. Other, better suspended slurries, such as Cabot SS-12 or Rodel Klebosol 1501/1508, exhibit only modest settling and require only gentle intermittent stirring. Furthermore, some slurries, such as Cabot SS-25, are subject to gelling with overagitation and should only be stirred intermittently. While stirring is sometimes carried out with a circulation system using a pump or other type dispense engine, by far the most common technique is a drum stirrer (Fig. 7).

The stirrer can be controlled by the slurry dispense unit or operate independently. Various propeller styles are available, typically coated marine-style propellers or metal folding propellers. Most supply drums have two 2-in. openings, which limit the number and size of stirrers and dip tubes that can be inserted. Stirring a full drum to ensure homogeneity is quite straightforward. However, as the slurry level in the drum approaches the level of the stirrer propeller, splashing can occur, which can produce dried slurry on the drum walls. Many schemes have been devised to minimize the effect of nonstirring versus splashing, including varying propeller speed as the drum empties or transferring an entire drum to a daytank before settling can occur. Generally, some sort of compromise must be made since it is simply not possible to provide uniform stirring during the entire life of the drum. Furthermore the stirring requirement is highly slurry-specific. In all cases, obtain slurry-specific recommendations from both slurry manufacturer and distribution equipment manufacturer.

3 FACILITIZATION

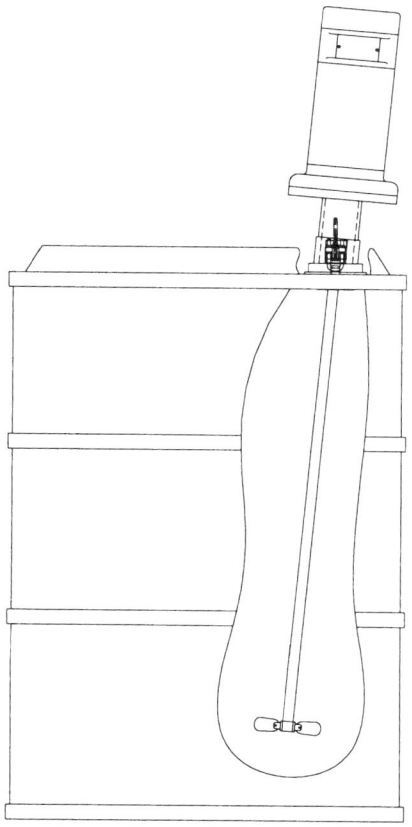

FIG. 7. Supply drum stirrer (Courtesy of BOC Edwards).

Daytank stirring involves many of the same challenges as supply drum stirring, but the stirrers are generally more permanent and do not need to be removed as each drum is consumed. As shown in Fig. 6, stirrers can be mounted on the top of a daytank. Since the amount of agitation energy required depends on the slurry and the size of the daytank, techniques other than propeller stirrers may be used. For well-suspended slurries or small daytanks, the return flow from the circulation of the global delivery loop may provide sufficient agitation. Regardless of the agitation technique, care must be taken to avoid splashing the slurry. Splashing can cause slurry to collect on tank walls where it can dry, despite humidification, and create wafer-scratching particles.

XXIII. Metrology

Monitoring slurry health—its stability and contamination level—through in-line or off-line metrology is important to decrease the possibility that slurry is a major contributor to CMP process variability. The most common parameters for monitoring slurry health—pH, density, and percentage of solids—may not always be appropriate [3, 6]. For any metrology technique, equipment reliability and durability must be considered. All analytical equipment requires periodic calibration or replacement. For example, a 6-month life expectancy for an in-line pH probe is typical; yet rarely are the probes replaced. Recognition of the useful accuracy is almost unrelated to the number of digits displayed in the output.

The common and not-yet common metrics of slurry health include

- pH
- Density (or specific gravity)
- Percentage of solids
- Conductivity
- Viscosity
- Component assay
- Ion-selective analysis
- Oxidation-reduction potential (ORP)
- Particle size distribution (PSD)
  (full distribution PSD or large particles only)
- Zeta potential

Figure 2 shows dramatic differences among slurries with regard to the sensitivity of density for monitoring mix ratio, ranging from useful to worthless. Figure 8 [7] shows the sensitivity of pH and density to changes in mix ratio of a Cabot SS W-2000, a commercially available tungsten polishing slurry, which contains hydrogen peroxide.

While density does change over a wide range of hydrogen peroxide concentration, it does not change enough to measure small changes in hydrogen peroxide concentration. pH barely changes over the entire concentration range and provides virtually no information about hydrogen peroxide assay.

In cases like this, assay determination of hydrogen peroxide concentration would be the most sensitive metric. Titrametric determination will give a reliable measure of the specific ingredient that is most crucial to the polishing process. Assay by titration of low-concentration ingredients, such as hydrogen peroxide or ammonium hydroxide, both of which degrade over

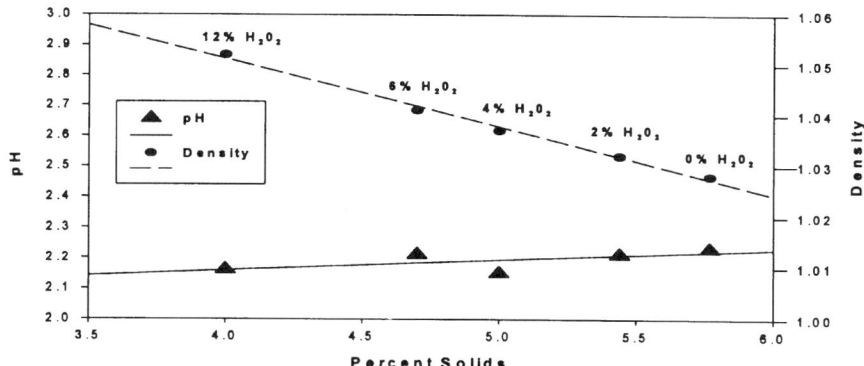

FIG. 8. Sensitivity of pH and density to [H_2O_2] for Cabot SS W-2000.

time through decomposition or evaporation, is an excellent choice. Manual titration is simple and requires inexpensive equipment available in any analytical lab. Automatic titration combined with automatic replenishment units are available from most slurry delivery equipment suppliers.

Particle size distribution analysis has become more popular as CMP users become more sophisticated. While there is certainly a distribution of particle size in any slurry, and some manufacturers claim benefits from a mixture of different sized particles, detection and removal of agglomerates that can cause wafer scratches is more important [8, 9]. Figure 9 shows a typical PSD for a silica slurry. While the average particle size will affect polishing performance, wafer scratching will be primarily caused by the presence of a few large particles [8, 9]. PSD analyzers that focus on the main particle fraction can not reliably detect the few large particles whose concentration is more than 10 orders of magnitude smaller, while PSD analyzers that focus specifically on the large particles are flooded by bulk distribution. Furthermore, a small number of large particles can cause many wafer scratches without affecting the overall particle size. Consequently, users must carefully choose analytical instruments that can detect the desired effect.

Figure 10 shows an example of the large particle population distribution for an abused silica slurry. Steady growth in slurry agglomerates can be detected by looking at the small number of large wafer-scratching particles. This information can be combined with other metrics to assess overall slurry health.

Metrology for monitoring slurry health has evolved from pH and density to include particle size and slurry-specific parameters such as component assay and conductivity. Other parameters mentioned earlier are occa-

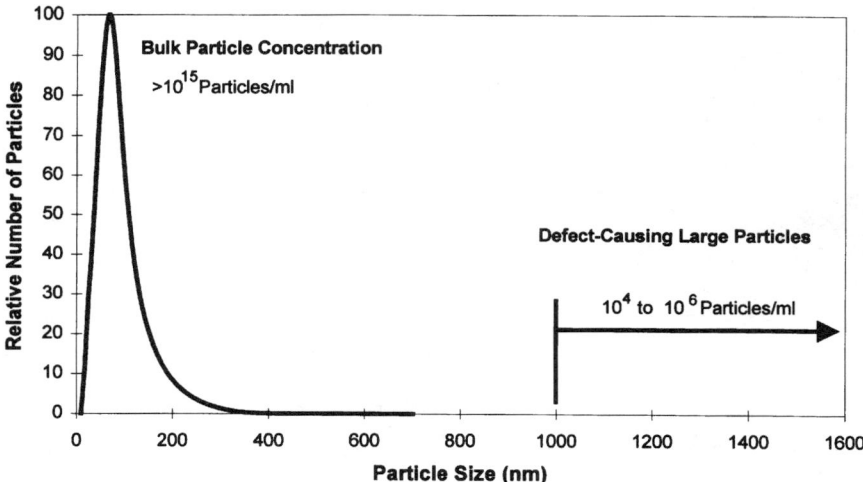

FIG. 9. Typical bulk particle size distribution in silica slurry (Courtesy of Millipore Corporation).

sionally used, while the entire CMP community searches for additional analytical techniques or slurry properties to help measure, predict, and maintain more consistent slurry.

XXIV. Filtration

On-site filtration of CMP slurry is one of the remedial approaches used to reverse the effects of changes to slurry particles between manufacture and delivery to the polisher platen. Figure 11 shows the locations that can be considered for slurry filtration. The candidate locations include

- Supply drum filtration before transfer to daytank or blending
- Filtration of diluted or blended slurry before transfer to daytank
- Global circulation loop
- Point-of-use filtration at the CMP polisher
- Polishing of daytank slurry

While each location can be used, the vast majority of filter installations are in the global circulation loop or at the point of use (polisher).

The development of filters for CMP by the filter manufacturers is relatively recent. As recently as 1995, filter manufacturers questioned whether

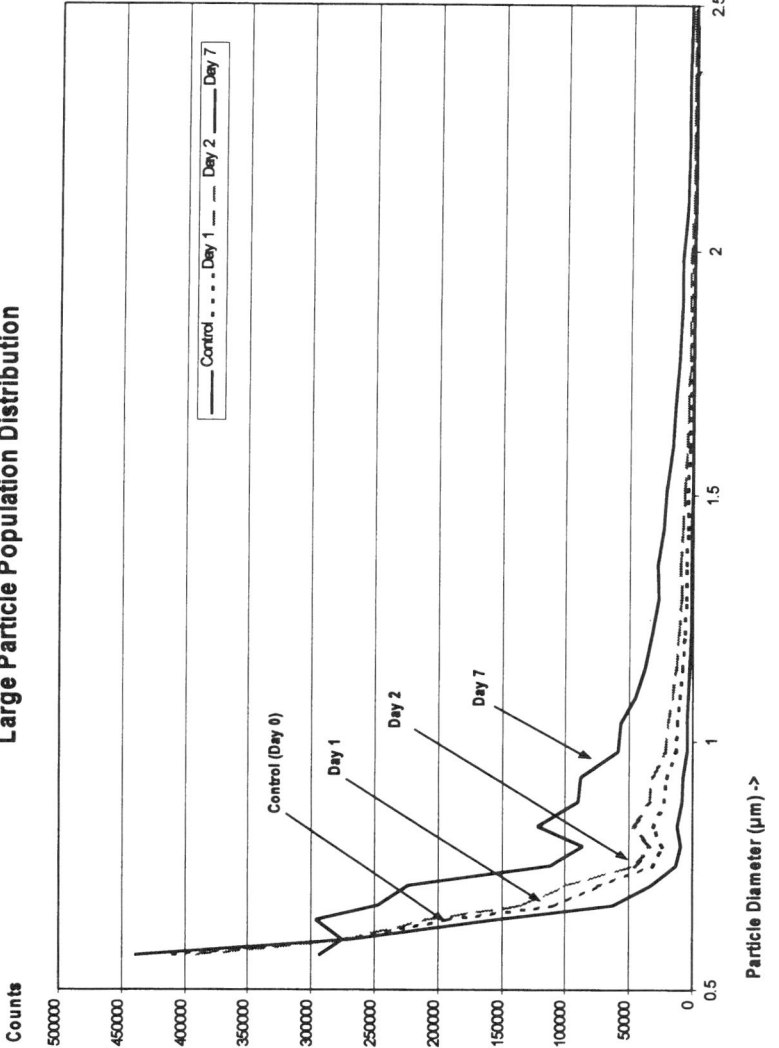

Fig. 10. Large particle distribution for abused silica slurry (Courtesy of Particle Sizing Systems).

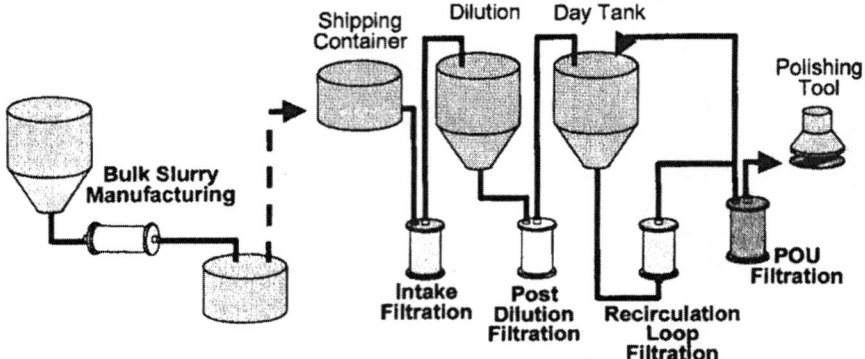

FIG. 11. Locations for slurry filtration in a slurry delivery system (Courtesy of Millipore Corporation).

CMP could be successfully filtered at all without dramatically affecting the desirable particles in the slurry. Consequently, there is great confusion over what filter pore size ratings mean or whether pore size ratings should be used at all. All the usual filter manufacturers supply CMP filters, but the filter ratings are not directly comparable. If in doubt, begin with a more open filter for any application. After all, some filtration is better than none, but interrupted delivery from a clogged too-tight filter is not a good thing. Experience will help balance performance versus filter cost versus labor to maintain the filters.

Supply drum filtration sounds like a great idea, especially for removing foreign contaminants and really big particles such as dried slurry. However, concentrated slurry is much more difficult to filter than use-strength slurry. Intermittent flow through the filter causes pressure surges and each surge pushes some of the previously removed material through the filter and into the daytank. Fast settling alumina or ceria slurries can settle out during periods of no flow and coat the filter medium or clog the filter housing. Finally, most slurry dispense modules are set up with all facilities inside the machine. This means that the pump or vacuum vessel drawing the slurry from the supply drum will be pulling the slurry through the filter rather than pushing it through the filter. If supply drum filtering is used, use a very open filter ($>100\ \mu m$) to remove only the very large foreign particles and consider putting the supply drums on an elevated platform to let gravity alleviate some of the aforementioned problems.

Filtration of diluted or blended slurry before transfer to daytank avoids some of the pitfalls just described, but still suffers from intermittent flow between batches. Since the slurry receives only one-pass filtration with this approach, it is rarely practiced. A medium pore-size filter (5–10 μm) is recommended.

Global circulation loop filtration is very popular. The filters are constantly challenged with steady flow. All slurry gets the same filtration, so that all polishers receive uniform quality slurry. Since CMP slurry filtration is a game of percentages, typically using graded density filters rather than absolute filters, multiple passes through a loose filter (20–40 μm) will remove particles much smaller than the rating of the filter. As described in the section on daytank turnovers, a typical number of turnovers in the life of a slurry is around 100. This large number of passes through a large-pore-size filter will be as effective as a single pass through a much tighter filter. Filter boxes, as shown in Fig. 12, which contain pressure gages and automatic filter swapping plus changeout without interrupting flow to the fab are available from many equipment suppliers. The primary drawback to circulation loop filtration is that the pressure to the polishers may slowly decline as the filter ages; this can change the flow rate to the polisher since the volume of slurry drawn by the peristaltic pump varies with the pressure supplied to the polisher. A simple scheme has been employed by some users to minimize this pressure drift while improving filtration and prolonging filter life. Two filters can be installed in the circulation loop: a tighter filter (10–30 μm) on the return line to the daytank to provide primary filtration and aid in providing loop backpressure combined with a relatively open "rock catcher" filter (60–100 μm or more) at the supply end of the circulation loop. If the two filters age at approximately the same rate, or the supply end filter slightly slower than the return end filter, two-stage filtration can be achieved and pressure drift will be minimal.

Point-of-use filtration permits use of the tightest filters (1–5 μm) and will give one-time filtration just before the slurry flows onto the polisher platen. This arrangement is also very popular. While giving excellent filtration protection just before slurry is delivered to the polisher, it suffers from all the problems of intermittent flow and slurry settling described earlier. Since space inside or on the polisher is limited, POU filters rarely have pressure gauges to monitor filter aging or any other safety features. Without vigilant attention to changing before filters exceed their useful lifetime, filters can clog totally and starve a polisher of slurry. A few trashed wafers can make the cost of a properly designed filter box look inexpensive.

Daytank polishing with a separate circulation engine can avoid most of the problems described earlier, using the same type filters used for circulation loop filtration (20–40 μm). Unfortunately, it can not protect against a large particle being delivered to a polisher on its first pass through the

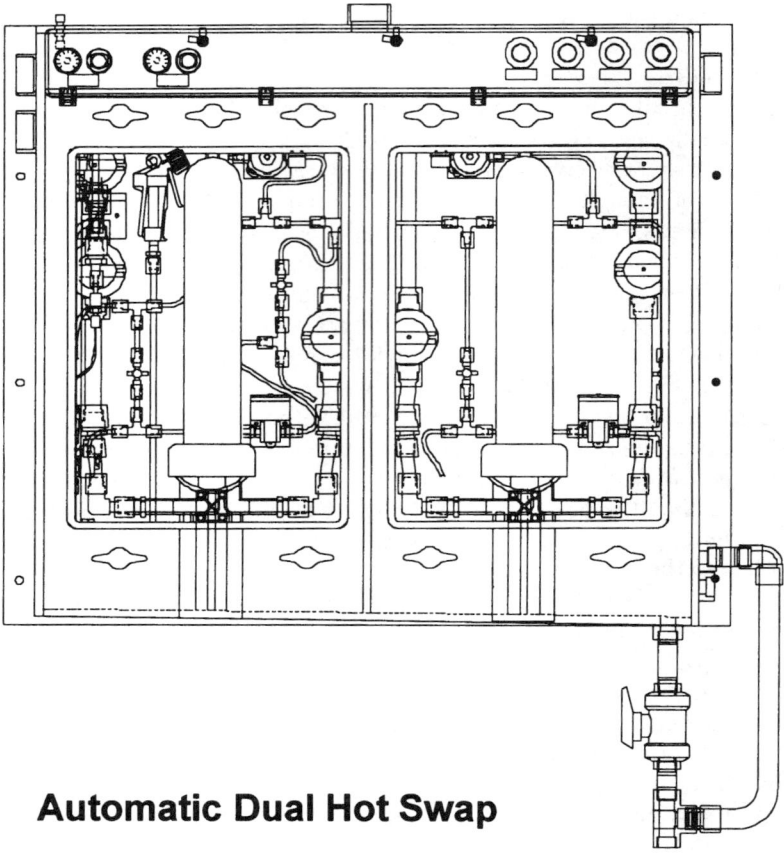

FIG. 12. Filter box with automatic changeover (Courtesy of BOC Edwards).

distribution system. And since filtration is a game of percentages, this will inevitably happen.

In summary, filtration is a good thing and implementation is growing. There are benefits, drawbacks, and compromises to all filtration styles, so it is imperative to work with filter manufacturer, slurry manufacturer, and delivery equipment manufacturer. Improper selection or installation of filters can have surprising unintended side effects. A cautious, eyes-open approach using more open filters in the beginning will be more rewarding than a quantum leap into an arena that is very slurry- and site-specific.

XXV. Slurry System Maintenance

Slurry delivery systems require maintenance and cleaning more frequently than high-purity chemical delivery systems. Despite improvements by slurry and equipment manufacturers, the best CMP process results are achieved by those willing to perform regular preventive maintenance. The exact type and frequency of maintenance depends on the slurry used, the style of dispense equipment, the piping design, and other system-specific features. The following guidelines will help identify types of maintenance areas common to most slurry delivery systems. For guidelines for a particular brand or style of equipment, the manufacturer should be consulted.

Removal of coatings or deposits from any surface is highly dependent on the specific slurry. Users commonly flush with DI water, but this may not be the best choice. Dl water can cause chemical shock to a slurry and create a precipitate rather than flush out the slurry. Use of the slurry mother liquor, such as potassium hydroxide solution or ammonium hydroxide for oxide slurries, may be better for a first rinse and also to dissolve hard-to-remove deposits. Most acidic tungsten- and copper-polishing slurries can be better rinsed with an acid like nitric acid. In all cases, the concentration is not critical. Typical concentration would be approximately 1–3% by weight or 0.2–1.0 M. Subsequent rinses can be with water, but this is not necessary if the chemical used is already a slurry component. If a change to a new slurry is part of the reason for the cleanout process, then more attention should be given to multiple rinses, including water and the the mother liquor of the new slurry. A warm chemical can be used to remove tough deposits, but be careful not to thermal cycle the plastic components, as the screwed-together fittings can work loose and create leaks.

Slurry coatings or deposits on tubing can interfere with the performance of level sensors. Virtually all systems use nonintrusive-level sensors in some location: supply drum, daytank, line full–empty, vessel full–empty. Coatings can come from slurry abrasive adhering to tubing walls or penetration of dissolved slurry constituents into pores in plastic material. While water flushing, especially with warm water, can frequently remove coatings, a chemical flush may be better or even required. Ferric nitrate, a constituent of some tungsten-polishing slurries, infiltrates pores in the plastic material and must be chemically removed with nitric acid or oxalic acid. Table VI lists some common chemical cleaning agents.

Solid accumulation in low-flow sections of a piping system is another common problem. Alumina and ceria abrasive particles will settle due to their density—exactly the same reason they require supply drum or day-tank stirring. Silica abrasive particles, which are normally well suspended, may agglomerate because of some environmental abuse, and also settle in low- or no-use locations. A very common location is in a section of a valve box intended for a future polisher that has seen no flow in a very long time, if ever. Another common location is the low-flow region of a partially closed manual diaphragm valve. Thick, compacted deposits require aggressive chemical or mechanical attack to dislodge and remove them. Periodic system flushing can help prevent the accumulation from developing into a hard deposit. While the frequency will vary due to many factors, a complete system cleanout every 6 months is a general recommendation.

One of the manufacturing challenges with a full-system cleanout, is the need to interrupt manufacturing. Some fabs have installed totally redundant delivery systems, including dispense module and piping. Some adopt a scheduled downtime for system maintenance. Most simply refuse to recognize the need until disaster strikes in the form of unscheduled interruption of slurry delivery or dramatic increase in wafer defects.

XXVI. Waste Disposal

CMP slurry is not particularly hazardous, especially compared to many other materials in a semiconductor fab. In some regions, it is not even (yet) considered a hazardous material. Nevertheless, as the sheer volume of use grows, waste treatment is becoming an issue for various regulatory, economic, or good citizen reasons including

- Water consumption minimization
- Suspended solids
- Dissolved metals, especially copper
- Hazardous waste minimization

TABLE VI

CHEMICAL CLEANING AGENTS FOR DELIVERY SYSTEMS

Slurry	Chemical cleaning	
	Chemical	Concentration
Oxide-polishing slurry	Potassium hydroxide	1–3 (% by weight)
	Ammonium hydroxide	1–3 (% by weight)
Tungsten- and copper-polishing slurry	Nitric acid	1–3 (% by weight)
	Hydrochloric acid	1–3 (% by weight)
	Oxalic acid	1–3 (% by weight)

3   FACILITIZATION

Solids typically generated in CMP waste streams are

- Silica — used for oxide, tungsten, and polysilicon polishing
- Alumina — used for tungsten polishing
- Ceria — used for polysilicon polishing
- Tungsten — trace quantities
- Copper — trace quantities

In addition to solids, there are dissolved materials that are often overlooked, but fortunately fit nicely into conventional waste treatment schemes because they are ions commonly present from other fab process. They include:

- Potassium hydroxide
- Ammonium hydroxide
- Ferric nitrate
- Potassium iodate
- Trace amounts of many soluble proprietary organic and inorganic acids and salts

Water consumption by the CMP process is enormous. Estimates vary widely, but range up to a total of 30–40% of total fab consumption [10]. Whatever the value is, water consumption by CMP will increase for the foreseeable future as CMP becomes more widely implemented and as new wafer fabrication schemes make greater use of this process.

Most R & D and prototype manufacturing facilities have taken the simple approach of commingling CMP waste with the main acid waste stream from the fab. Since the largest volume of CMP has been oxide polishing using slurries with a pH of 10.5–11.0, this approach has generally worked. However, the suspended solids are often small enough to survive the normal settling or filtration steps and escape into the municipal sewer system. To date, most anecdotal reports of regulatory issues involve suspended solids. Whether or not CMP slurry is well suspended in its use strength, after dilution and commingling with other waste streams, few abrasive particles will remain colloidally suspended. Flocculation, agglomeration, or loss of surface charge (Zeta potential) will cause particle settling. Settling may occur in fab piping systems, and a lot will settle in conventional fab waste treatment systems, but more likely settling will occur in municipal sewer systems, where there is insufficient fluid velocity to completely flush out the solid particles. Even if the particles can flush through the sewer system piping, the increased load on the POTW is an enormous strain.

A number of companies have made entries into the CMP waste treatment market, using a variety of technical approaches. So far there has been no consensus on the preferred waste treatment scheme and no clear market leader. A limited number of commercial installations are in service, in addition to home-built systems by some large CMP users (see Table VII).

Slurry reuse or recycling has often been proposed as a way to minimize costs and waste. Slurry reuse is routinely practiced in related industries such as hard disc and silicon raw stock polishing. However, two big issues repeatedly arise as objections. First, most users find that quality control of incoming virgin material, combined with the challenge of mixing, blending, delivering, and monitoring, is daunting enough. The challenge of blending used slurry, containing unknown new impurities, with fresh slurry, is more of a challenge than users are up to at this time. Small amounts of metal added to oxide-polishing slurry can introduce imperfections to the insulating oxide layer; agglomerates accumulated can introduce scratches. Second, there is the issue of accountability. Many consumable and equipment suppliers influence the health of fresh slurry. Many more would influence the health of recycled slurry. There is nothing to prevent the waste material from being reused in less demanding applications outside the semiconductor industry, whether as filler for asphalt or, with a little added perfume, as kitty litter.

While the path to waste treatment remains uncertain, the goals remain:

- Reduce the total water consumption
- Reuse the water reclaimed by any treatment technology for refueling the UPW system or for less demanding applications
- Reclaim the spent slurry for use in less demanding industries [10].

The exact solution will vary from user to user depending on the slurry or

TABLE VII

COMPANIES MARKETING CMP WASTE TREATMENT CAPABILITY [10]

Company	Technology
Asahi	EnTek filtration
Ebara	Vacuum accumulation and disposal
IPEC	Slurry filtering and reuse
Kinetico/EPOC	Microfiltration, optional polymer addition
Kurita	Chemical flocculation, filtration
Lucid Treatment Systems	POU water reuse, ultrafiltration
Microbar/EnChem	Polymer addition, microfiltration
Millipore	Ionic filtration
Pall	Ultrafiltration
Toshiba Ceramics	Ceramic filtration

slurries being used, the state of the art of the local treatment works, and the local regulations regarding water and effluent.

REFERENCES

1. J. P. Bare, B. Johl, and T. A. Lemke, "Comparison of Vacuum-Pressure vs. Pump Dispense Engines for CMP Slurry Distribution," *Proceedings of SEMICON/West Contamination in Liquid Chemical Distribution Systems Workshop* (July, 1998).
2. E. W. Bernosky, J. T. Geatz, E. T. Ferri, G. A. Roberson, U.S. Patent No. 5,370,269.
3. J. P. Bare and T. A. Lemke, "Monitoring slurry stability to reduce process variability," MICRO, pp. 53–63 (September 1997).
4. J. P. Bare and B. Johl, "Aging and Handling Evaluation of Tungsten CMP Slurry in Bulk Delivery System," *Proceedings of Fourth Int'l Chemical-Mechanical Polish Conference* (CMP-MIC) (February 1999).
5. CRC Handbook of Chemistry and Physics, D. R. Lide, editor, 77th Edition, CRC Press, Boca Raton, FL, pp. 4-38, 4-50, and 4-83.
6. J. P. Bare and T. A. Lemke, "Parameters for Monitoring CMP Slurry Stability and Contamination," *Proceedings of SEMICON/West 1997 Workshop on Contamination in Liquid Distribution Systems* (July 1997).
7. J. P. Bare and B. Johl, "Evaluation of Manufacturing Handling Characteristics of Hydrogen Peroxide-Based Tungsten CMP Slurry," *Proceedings of IEEE International Electronics Manufacturing Technology Symposium* (October 1998).
8. Nagahara et al., "The Effect of Slurry Particle Size on Defect Levels for a BPSG CMP Process," *Proceedings of 1996 VMIC Conference* (June 1996), p. 443.
9. Capitanio et al., "Defect Reduction During Chemical Mechanical Planarization by Incorporation of Slurry Filtration," *Proceedings of SEMICON/West Contamination in Liquid Chemical Distribution Systems Workshop* (July 1998).
10. G. Corlett, "Can CMP Waste Treatment Ever Be Environmentally Friendly?" *J. Adv. Appl. Contamin. Control* (December 1998).

CHAPTER 4

Modeling and Simulation

Duane S. Boning and Okumu Ouma*

MASSACHUSETTS INSTITUTE OF TECHNOLOGY
CAMBRIDGE, MASSACHUSETTS

I. INTRODUCTION	89
II. WAFER-SCALE MODELS	90
1. *Macroscopic–Bulk Polish Models*	90
2. *Sources of Wafer-Scale Nonuniformity*	92
3. *Empirical Approaches to Wafer-Level Modeling*	97
4. *Status of Wafer-Level Modeling*	97
III. PATTERNED WAFER CMP MODELING	98
1. *Feature-Scale Models*	100
IV. DIE-LEVEL MODELING OF ILD CMP	104
1. *Topography Modeling: Layout Density Dependence*	105
2. *Modeling of Thickness Evolution*	106
3. *Planarization Length and Response Function*	108
4. *Characterization—Determination of Planarization Length*	113
5. *STI CMP Modeling*	118
6. *Models for Step Height Reduction*	120
7. *Applications of Density Models*	124
V. MODELS FOR METAL POLISHING	125
1. *Tungsten CMP Modeling*	126
2. *Tungsten CMP—Contact Wear Model*	128
3. *Copper CMP Modeling*	131
VI. SUMMARY AND STATUS	132
REFERENCES	133

I. Introduction

Modeling and simulation approaches for chemical mechanical polishing (CMP) are essential to support a variety of needs ranging from exploration and understanding of physical mechanisms, to process optimization, process integration, and manufacturing control of CMP. In this chapter, we focus

---

*Currently with Bell Laboratories, Lucent Technologies Inc., Murray Hill, New Jersey.

on modeling approaches that have a practical application for simulation of the process at three key length scales. These include empirical and mechanistic models of wafer-scale polishing performance, modeling, and characterization methods useful to predict die-scale effects, and models to describe feature-scale effects. These models are often dependent on the material system undergoing polishing, and we discuss models for oxide, dual dielectric polish (e.g., as arising in shallow trench isolation), and metal polishing. Formulation of alternative models and simulation tools for CMP continues to be an active area of research and new methods are rapidly emerging and evolving. As a result, we do not attempt to be comprehensive or survey the complete existing work; rather, we focus on the fundamental issues being addressed through simulation efforts.

In Section II, we focus first on wafer-scale models, including macroscopic or bulk polish models (e.g., via Preston's equation), as well as mechanistic and empirical approaches to model wafer-scale dependencies and sources of nonuniformity. In Section III, we turn to patterned wafer CMP modeling and discuss the pattern-dependent issues that have been examined; we also discuss early work on feature-scale modeling. In Section IV, we focus on die-scale modeling efforts and issues in the context of dielectric planarization. In Section V, we examine issues in modeling pattern-dependent issues in metal polishing. Summary comments on the status and application of CMP modeling are offered in Section VI.

II. Wafer-Scale Models

The modeling of polishing effects in CMP begins with two key issues: what are the process-related dependencies in the rate of removal of exposed surface material during polishing, and on what does the wafer-scale uniformity of that polish depend? In this section, we begin with the modeling of polish or removal rate, and then consider models for the effects that impact the commonly observed nonuniformity in polish across the wafer.

1. Macroscopic–Bulk Polish Models

Chemical mechanical polishing appears to consist of two cooperating physical mechanisms [9]. First, chemical interaction of the slurry with material at the surface of the wafer weakens the surface to be polished. Second, the weakened surface is mechanically removed by a combination of slurry particles, polish pad asperities, and hydrodynamic effects. The extent of each component is not well known although several hypotheses exist for both metal and dielectric CMP. In oxide CMP, for example, it is believed that the following reaction takes place [9, 20, 47]:

$$H_2O + Si\text{—}O\text{—}Si \leftrightarrow Si\text{—}OH + OH\text{—}Si \tag{1}$$

The forward reaction is favored by the alkaline slurry solutions which result in breakage of the Si—O bonds. In metal CMP, oxidizing slurries are often used, resulting in faster removal rates. Since the contributions of the chemical and mechanical components are not well known, modeling efforts have focused on empirical approaches guided by physical intuition of process mechanisms.

In a typical modeling approach, the material removal rate is modeled as a function of easily controlled process parameters. The most basic model is one that predicts the bulk rate of material removal in a macroscopic fashion. An empirical observation by Preston is widely used, in which the rate of material thickness reduction is proportial to the product of (a) the relative velocity between the wafer and the polish pad and (b) the pressure on the surface of the wafer:

$$\frac{dz}{dt} = K \cdot \frac{N}{A} \cdot \frac{ds}{dt} \tag{2}$$

where z is wafer thickness, t is time, N/A is the pressure due to normal force N on the area A, and s is the distance some point on the wafer travels in contact with the pad (or other abrading material). Preston's equation also conventionally appears as

$$R = K_p \cdot P \cdot V \tag{3}$$

where K_p is the "Preston coefficient," R is the material removal rate, P is pressure, and V is relative velocity. It should be emphasized that this relationship is empirical, and a great deal of effort has been focused on either understanding how such a result is generated by more fundamental physical action of mechanical–chemical contact wear on the one hand [22], or seeking to identify aspects of CMP as used in microelectronics that might result in non-Prestonian behavior (e.g., in the dependency on pressure and velocity). For example, it has been observed that a general linear relationship between removal rate and PV product is usually best fit with a nonzero value of the removal rate for zero PV product, as shown in Fig. 1. Experimental studies at very low pressures and velocities appear to indicate

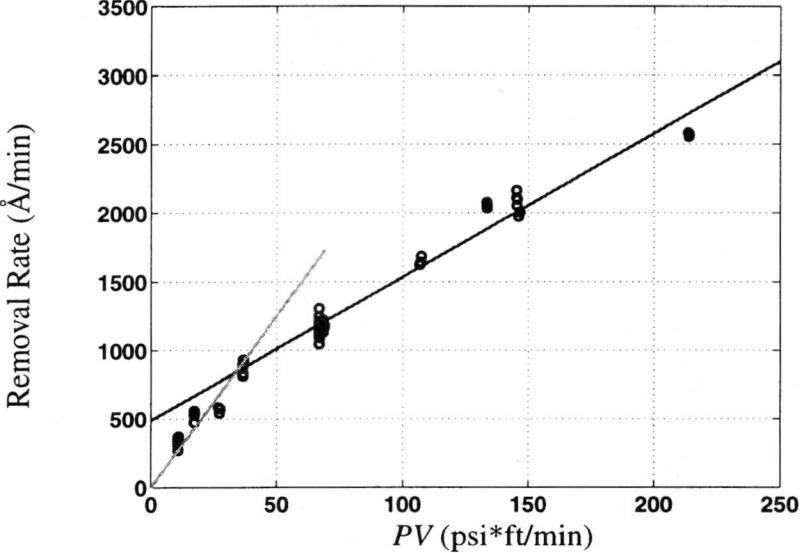

Fig. 1. Non-Prestonian effect in oxide CMP.

a slightly nonlinear or different behavior in these regimes. At the range of velocities and pressures in common use in CMP, such "non-Prestonian" behavior is relatively unimportant, although such behavior may be crucial in elucidating the physical mechanisms behind CMP.

Based on mechanistic models of particle-based wear, other nonlinear (nonunity power) dependencies on pressure and velocity have also been proposed [44, 61]. Clear experimental evidence for such dependencies remains to be found, particularly to isolate or remove the influence of other factors (e.g., slurry flow) that might also indirectly be influenced by pressure and velocity (and thus change K_p rather than the primary PV product dependency).

2. Sources of Wafer-Scale Nonuniformity

a. Relative Velocity Mismatch Across the Wafer

The Preston relationship provides a "pointwise" dependency of removal rate based on the relative velocity and pressure within some region of the wafer. A straightforward application of Preston's model is to study the wafer-level nonuniformity due to dependencies in the relative velocity arising from machine configurations. Indeed, Preston's original work contains an analysis of the relative velocity arising from rotary or table based polishing apparatus [36]; similar work reexamining the kinematics have also appeared [42, 66]. Figure 2 shows a typical rotary machine configuration commonly used for CMP. In the figure, O is the center of the polish table and P is the axis of rotation of the wafer carrier. The circle of radius R centered at P is a ring of points on the wafer; the goal is to determine the relative velocity at point Q on the wafer and the effective relative velocity at any point on the ring. The carrier usually oscillates along OP to ensure even pad degradation and improved uniformity. The average offset (average distance of OP), e, is used in the derivation.

Let the angular rotation of the wafer and table be ω_2 and ω (in radians per minute) respectively. Preston has obtained the relative velocity between pad and point Q on the wafer as

$$v = v_s(1 + R_1^2 \omega_s^2 + 2R_1 \omega_s \cos\theta)^{1/2} \qquad (4)$$

where $v_s = e\omega$ is the relative velocity between any wafer position and the pad if the wafer and pad are synchronized (i.e., $\omega_2 = \omega$), $R_1 = R/e$, and $\omega_s = 1 - \omega_2/\omega$. Eq. (4) may be reduced to:

$$v = v_s(1 + x^2 + 2x\cos\theta)^{1/2} \qquad (5)$$

where $x = R_1 \omega_s$. Given v at any point on the ring, we can determine the effective (time-averaged) relative velocity at any point at a distance R from

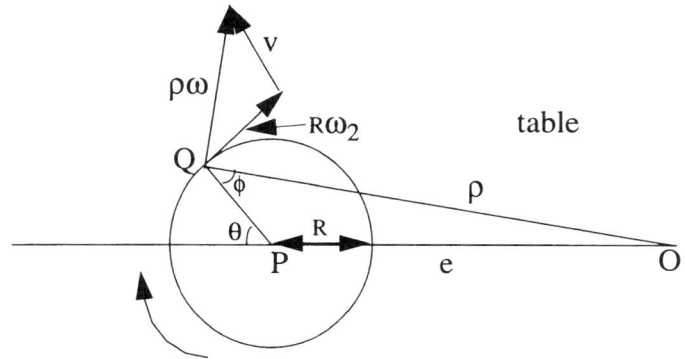

FIG. 2. Polishing machine setup detailing the velocity vectors of interest.

the center as

$$v_e(R) = \frac{v_s}{\pi} \int_0^\pi (1 + x^2 + 2x \cos \theta)^{1/2} \, d\theta \qquad (6)$$

This formulation assumes radial symmetry in the polishing process. With the substitution $\theta = 2\varphi$ and $\cos \theta = 1 - 2 \sin^2 \varphi$, the effective radial velocity becomes

$$v_e(R) = \frac{v_s(1 + x)}{\pi} \int_0^{\pi/2} \left[1 - \frac{4x}{(1 + x)^2} \sin^2 \varphi \right]^{1/2} d\varphi \qquad (7)$$

which involves a complete elliptic integral of the second kind and is easily evaluated numerically. Figure 3 shows the typical variation of relative velocity from the center of the wafer for different carrier and table speeds. An interesting result is that if the carrier and table rotation speeds are synchronized, then the relative speed is constant across the entire wafer surface. On the other hand, when the table and carrier speeds are mismatched, the center of the wafer always experiences a reduced velocity compared to the outer edges of the wafer, as shown in Fig. 3. This has significant implications for wafer-level uniformity and control of the polishing process. For typical machine settings, center-to-edge variation in relative velocity is comparatively small. For example, if the table angular velocity is

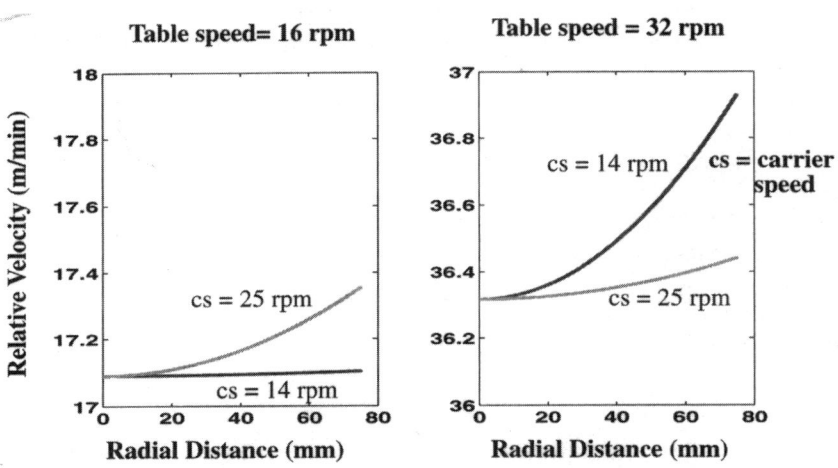

FIG. 3. Relative velocity dependence on carrier and table speeds.

twice the carrier angular velocity and the average carrier offset e is 170 mm, the percentage velocity change from center of the wafer to the edge is 1.2% for a 6-in. wafer and 2.2% for an 8-in. wafer. Center-to-edge variation in removal rate is usually much larger than this due to the edge and other effects to be outlined.

Active development of alternative wafer-polishing tool configurations (e.g., linear polishing, rotary configurations) has gone hand in hand with examination and modeling of the dynamics (i.e., relative velocities) resulting from various tool configurations and movement paths.

b. *Wafer-Level Pressure Variation*

Preston's equation indicates a pressure dependency and if the pressure distribution across the surface of the wafer is not uniform, one expects a wafer-level removal rate dependency. Runnels *et al.*, for example, report a model incorporating pressure dependencies to account for wafer scale nonuniformity [42]. The distribution of applied force across the surface of the wafer is highly dependent on the wafer carrier design, and significant innovation in head design to achieve either uniform or controllable pressure distributions is an important area of development.

A related issue has to do with the initial wafer-level uniformity (wafer thickness, wafer warp and bow, thicknesses of thin films across the wafer surface, uniformity of stress in such thin films across the wafer) and the subsequent impact on wafer-level polish performance. Some examination has been made of the impact of wafer warp and bow on the polish performance [68], where it was found that the initial warpage can have significant impact (with the implication that reclaimed wafers may not be appropriate monitors of wafer-level polish performance). Other work has considered inherent variation due to Von Mises stress concentrations at the edge of the wafer (conceptually, a downward pressure on the wafer causes lateral stress buildup near the edge of the wafer) [64].

The consideration and modeling of local pressure differences near the edge of the wafer has also been reported. These approaches examine discontinuities in the pad compression due to the edge of the wafer (either statically or dynamically including leading and trailing edge issues), to understand such effects as slow or fast polish in the several millimeters long "edge exclusion" at the wafer edge [1]. These explorations have contributed to the design of "active retaining ring" heads, where the wafer carrier has a separately pressurized or controlled retaining ring to precompress the pad and enable uniform pressure distributions further out on the edge of the wafer.

c. *Wafer-Level Slurry Flow*

A number of additional wafer-level variations in the removal rate may result from the variation in the Preston coefficient K_p across the surface of the wafer. Because K_p depends on the slurry, pad, material, and process parameters in a poorly understood fashion, it is not clear where all of these dependencies ultimately reside.

Runnels and Eyman [41] report a tribological analysis of CMP in which a fluid-flow-induced stress distribution across the entire wafer surface is examined. Fundamentally, the model seeks to determine if hydroplaning of the wafer occurs by consideration of the fluid film between wafer and pad, in this case on a wafer scale. The thickness of the (slurry) fluid film is a key parameter, and depends on wafer curvature, slurry viscosity, and rotation speed. The traditional Preston equation $R = KPV$, where R is removal rate, P is pressure, and V is relative velocity, is modified to $R = k'\sigma\tau$, where σ and τ are the magnitudes of normal and shear stress, respectively. Fluid mechanic calculations are undertaken to determine contributions to these stresses based on how the slurry flows macroscopically, and how pressure is distributed across the entire wafer. Navier–Stokes equations for incompressible Newtonian flow (constant viscosity) are solved on a three-dimensional mesh:

$$\vec{u} \cdot \nabla \vec{u} = -\frac{1}{\rho}\nabla p + \frac{\mu}{\rho}\nabla^2 \vec{u}$$
$$\nabla \cdot \vec{u} = 0 \tag{8}$$

where ρ is density, μ is the dynamic viscosity, p is the pressure, and $\vec{u}$ is the vector velocity in the flow.

While the resulting model is not quantitatively predictive, important observations can be made based on parametric simulation studies. It is proposed that changes in viscosity due to wafer temperature may be as large as 30%, and that such viscosity dependencies can have significant impact on fluid film thickness and transitively on removal rate. The importance of other process parameters, such as wafer curvature, is also indicated by the model.

Other fluid dynamic models of slurry flow have also been developed by other workers [57]. Coppeta, Rodgers, Radzak, and coworkers examined slurry flow, both from a simulation point of view, and from an experimental angle [6, 10, 11]. A special test apparatus is used consisting of flourescent injections of die that is entrained beneath a glass "wafer" enabling observation of slurry flow patterns and residence time. Such studies are instrumental in understanding the macroscopic flow of the slurry, the potential for degradation of the chemical activity of slurry as it ages, and the design of alternative slurry introduction patterns or pad structures (perforations and grooves).

d. Pad-Related Wafer Nonuniformity

With the current generation of polyurethane based polishing pads, the wear of the pad concurrently with the wear of the wafer surface is observed. Because each radial "band" of the pad comes into contact with a different surface area of the wafer, nonuniform wear of the pad is generally observed wherein the region of the wafer "track" near the center of the wafer carrier wears faster than those regions of the pad near the edges of the wafer. The dependence of the removal rate on pad wear is poorly understood. Indeed, there appear to be various time-dependent effects such as pad-breakin and pad recovery (e.g., after overnight soaking) in addition to the polish-time or removed material relationship. Some modeling has been done seeking to relate pad conditioning effects to the wafer level uniformity [26].

3. EMPIRICAL APPROACHES TO WAFER-LEVEL MODELING

Empirical or experimental design based approaches to wafer-scale CMP modeling are often used very effectively for practical process design, optimization, and control. For a particular tool or family of tools sharing configuration, consumable, or process parameters, the impact of some number of design factors on wafer-level uniformity, removal rate, and other important design considerations (e.g., surface damage) can be studied through experimental runs together with response surface modeling. This approach is illustrated in Fig. 4, where the response of the removal rate and wafer-level uniformity is shown as a function of carrier speed and downforce across a limited design space [4]. While such experimental designs are relatively straightforward, care must be taken in statistical modeling of uniformity as a function of process conditions [50].

4. STATUS OF WAFER-LEVEL MODELING

The modeling of wafer-level polishing effects has primarily been done in the context of "blanket" wafers with a uniform (unpatterned) single material thin film, and that modeling is usually considered independently of the

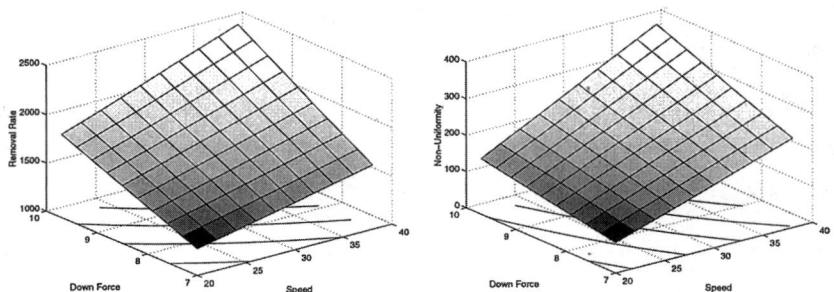

FIG. 4. Response surface modeling of removal rate and nonuniformity [4]. Copyright 1996 IEEE.

particular material. Most of the models are thus appropriate to understanding or studying the dependence in material removal rate across the wafer for oxide or other dielectric polishing as well as for metals. More recent work has begun to consider details of chemical interactions, fluid flow both on the wafer-scale and on the microscopic scale (e.g., in pad grooves and between the pad and wafer) [25], and thermal effects [39] particularly in copper polishing; these are promising areas for future understanding and simulation.

The desire to improve wafer-level uniformity continues to drive the study of wafer-scale CMP dependencies. Effective and practical means now exist for empirical simulation of wafer-level polish rate across the entire wafer; however, substantial enhancement remains necessary to develop predicitive models with sufficient detail and physical bases for exploratory tool design and optimization.

III. Patterned Wafer CMP Modeling

While wafer-scale modeling of unpatterned wafers has some direct application (e.g., to model polishing or thinning of raw silicon wafers), the more typical use of CMP in integrated circuit fabrication is on patterned wafers. CMP is often used to "planarize" topography created during typical microfabrication through lithographic patterning, etching, and deposition of thin films on the surface of the wafer. The goal of planarization is to preferentially remove "raised" material so as to reduce height differences across short to long length scales. CMP is also often used to "polish" away unwanted material at the surface of the wafer. When used to form shallow trench isolation or inlaid metal features, the goal is to completely remove a thin film except in patterned trench or via structures while simultaneously avoiding the creation of topography such as dished features in the trench areas.

In an idealized picture of CMP, planarization would result in a perfectly flat oxide or other thin film regardless of the starting topography or underlying patterns. Similarly, damascene or inlaid processes would result in a perfectly flat surface across both trench material (e.g., copper, tungsten, or oxide) and field material (e.g., oxide in metal polishing, or nitride in STI). Much of the modeling effort, however, is focused on the actual, nonideal outcome from CMP, as illustrated schematically in Fig. 5. In the case of oxide planarization, one finds that local steps (created by underlying metal lines, for example) can be effectly removed; however, the difference in pattern density across the chip can result in the creation of global nonplanarity or significant height differences across the chip [35]. In some cases, the details of the remaining local step height may also be important. In shallow trench isolation (STI) and metal damascene polish processes, both dishing into the feature and the erosion of surrounding support material can occur, as well as nonideal rounding of features.

Much confusion exists in the literature when discussing the planarization and polishing of pattern features primarily because the length scales in question are often ambiguous. Figure 6 illustrates some of the length scales that are important in interlevel dielectric planarization, for example. On the scale of the chip (typically over several millimeters), different regions (cells or functional blocks) on the chip will have different designed densities of patterns (raised material). Within such blocks local steps are created by the deposition of oxide over individual lines or structures. In the traditional oxide CMP process, local steps on the patterned feature scale (a "feature" being an individual patterned metal line, for example) are usually very

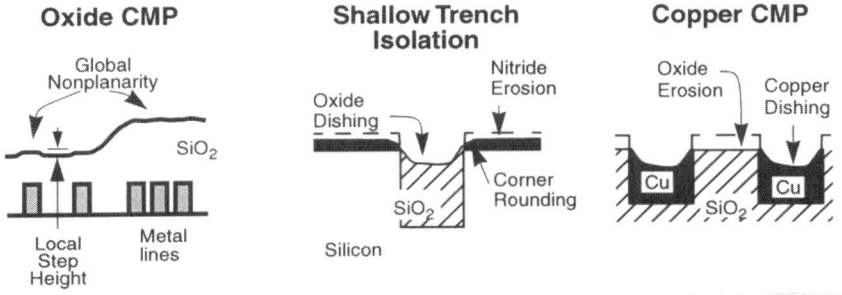

FIG. 5. Pattern dependent issues in oxide planarization and polishing of shallow trench isolation and metal damascene structures.

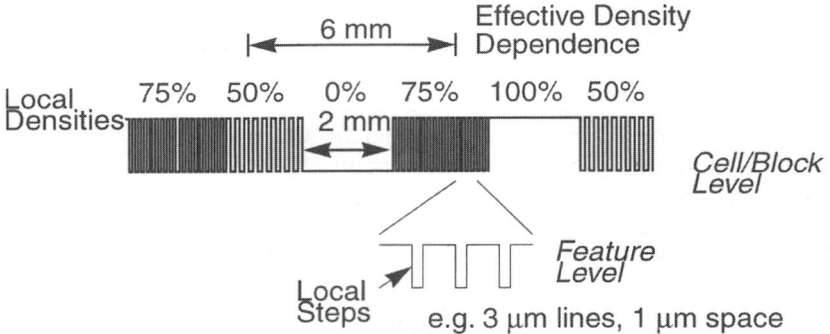

FIG. 6. Length scales in pattern-dependent CMP simulation.

effectively removed. Important modeling questions at the feature scale include how the oxide thickness evolves over time in the course of the polishing both over "up" areas (i.e., the oxide over the underlying metal line) and within "down" areas between patterned features. The term "step height reduction" is also commonly used to denote the evolution in time of the height difference between such local up and down areas. At the conclusion of typical oxide polishing, the local steps have been removed, but unfortunately global nonplanarity is often introduced. By global nonplanarity we mean the difference in oxide thickness at the "block" or cell level; one typically finds variation in remaining oxide thickness across length scales of several millimeters. The modeling of this "chip level" effect is also very important since the variation impacts process control windows (setting of minimum and maximum oxide thickness limits across the chip) that influence, for example, lithographic depth of field and subsequent etch steps (e.g., via etch latitudes).

In this section, we first discuss some of the early empirical work on "feature" polishing. Models and methods appropriate for die-level modeling and prediction are presented in Section IV.

1. FEATURE-SCALE MODELS

Burke developed an empirical model which gives the polish rate of the "down" and "up" areas [7]. The down area polish rate D is given by

$$D = \left[1 - (1 - D^0)\frac{S}{S^0}\right]U \qquad (9)$$

where U is the blanket polish rate, D^0 is the normalized down polish rate, S is the step height, and S^0 is the step height value with D^0 as the initial step height. The change in step height with time is given by

$$-\frac{dS}{dt} = \left(\frac{1 - D^0}{S^0}\right) SU \tag{10}$$

from which the step height reduction can be obtained given the initial conditions.

The model is empirical and can be used to explain aspects of measured data. Characterization or extraction of parameters would include D^0, which is obtained as the average rate over a raised area divided by the blanket rate. Equation (9) is then used to trace the actual thickness evolution in conjunction with step height reduction in Eq. (10). These equations define the feature polish characteristics for relatively large features. For a given array of patterned features, the "up" polish rates follows the following relationship:

$$\text{Rate} = \frac{U}{P_F} \tag{11}$$

where P_F is the pattern factor. One issue not addressed by the model is how the pattern factor is determined (such as over what area to consider) since a die pattern is typically irregular.

Warnock proposed a phenomenological model that mathematically captures the polishing process but that did not directly seek to incorporate the physical phenomena in CMP [63]. The model has three parameters that can be used to fit measured data. The surface is divided into n discrete points each with x, y, and z coordinates. For each point i in the set of n points, the polish rate P_i is defined as

$$P_i \sim \frac{K_i A_i}{S_i} \tag{12}$$

where K_i is the kinetic factor or horizontal component in the polish rate, A_i is an accelerating factor associated with points that protrude above their neighbors, and S_i is a shading factor describing how the polish rate is decreased by the effect of neighboring points protruding above point i. These definitions are schematically illustrated in Fig. 7. There is a reciprocity relationship between the set of S_i and A_i. When a given point S_i is sheltered or shaded (i.e., $S_i > 1$) by a neighboring point j, the polish rate

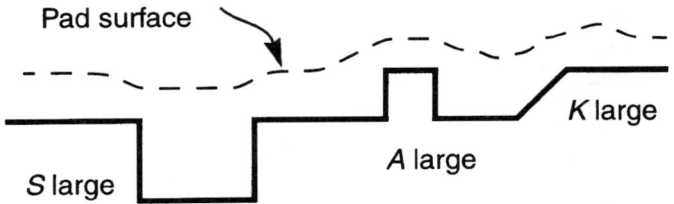

Fig. 7. Definition of "shading," "acceleration," and "kinetic" factors used in Warnock's pattern evolution model [63]. Reproduced by permission of the Electrochemical Society, Inc.

decrease at point i will be compensated by a corresponding increase in rate at point j. This condition is satisfied by the following relationship over the set of all n points:

$$\sum_{i=1}^{n} \frac{A_i}{S_i} = n. \tag{13}$$

The analysis is accomplished by first obtaining an expression for S and then using the reciprocity condition to obtain A. Then K is obtained independently, and S_i is assumed to be of the form:

$$S_i = \exp\left(\frac{\overline{\Delta z_1}}{z_0}\right) \tag{14}$$

where z_0 is a model parameter setting the length scale in the vertical direction, and z_1 is determined by how much the surrounding topography protrudes above point i. It is given by

$$\overline{\Delta z_1} = \frac{1}{2\pi} \int_0^{2\pi} z(r_m, m) W(r_m) \, d\theta \tag{15}$$

where $W(r)$ is a weighting function that captures the effect of the surrounding topography and r_m maximizes W. The functional form of W is given by

$$W(r) = \frac{1}{\cosh(r/r_0)}. \tag{16}$$

where r_0 is the deformation length scale, which is a model parameter. The particular choice of W is claimed not to be very important but it must be well behaved at the origin. In addition to the two model parameters z_0 and r_0, there is another model parameter associated with the kinetic factor, K_i. The kinetic factor is given by

$$K_i = 1 + K_0 \tan(\alpha_i) \tag{17}$$

where K_0 is the third model parameter and α_i is the local angle that the polish surface makes with the horizontal.

A third model for feature-scale polish was proposed by Runnels [40], and focuses on stresses created by flowing slurry on feature surfaces under continuum mechanics. The model incorporates fracture mechanics and chemistry through empirical means. The geometry of a typical structure under study is shown in Fig. 8.

The model is composed of two parts. In the first part, steady state two-dimensional Navier–Stokes equations for incompressible flow are used to relate local velocity u and pressure p:

$$\rho u_i \frac{\partial u_i}{\partial x_i} = -\frac{\partial p}{\partial x_j} + \mu \frac{\partial^2 u_j}{\partial x_i \partial x_i}$$
$$\frac{\partial u_i}{\partial x_i} = 0 \tag{18}$$

where ρ is the slurry density and μ is the slurry viscosity. For typical pad velocity ($u = 0.42\ m/s$), slurry density ($\rho = 997\ kg/m^3$) and slurry viscosity ($\mu = 0.8908 \times 10^{-3}\ kg/ms$), a Reynolds number for a characteristic length L of 1 μm is $Re = \rho u(L/\mu) \approx 0.5$, which indicates symmetrical fluid flow and antisymmetric pressure.

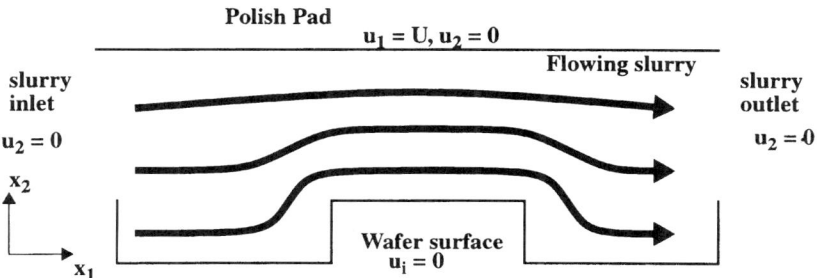

FIG. 8. Feature-scale geometry illustrating fluid flow near the wafer surface in Runnels model [40]. Reproduced by permission of the Electrochemical Society, Inc.

The second component of the model then relates fluid flow to local stresses at the surface, giving a stress tensor

$$\sigma_{ij} = -p\delta_{ij} + \mu \left(\frac{\partial u_i}{\partial x_j} + \frac{\partial u_j}{\partial x_j} \right) \qquad (19)$$

from which normal and shear stresses can be determined. The model indicates that surface stresses may enhance the otherwise slow natural etching action of the (alkaline) slurry by breaking material off of surface features. Assuming the particles that are removed are small enough that a continuum model remains valid, Runnels then models the motion of the eroding surface boundary using a surface normal erosion rate based on an empirical relationship

$$V_n = f[\sigma_n(t), \sigma_t(t)] \qquad (20)$$

where $\sigma_n(t)$ and $\sigma_t(t)$ are the normal and shear stresses, respectively. The resulting model results in feature profiles that roughly match empirical periodic structures under planarization processes. In terms of physically verified understanding of the process, the contributions of such hydrodynamic effects compared to chemical and pad–slurry interactions with the surface remain in doubt, and further research into the underlying physics is needed.

Models such as those proposed by Burke, Warnock, and Runnels can be effective in simulating the evolution of step height and film thicknesses around particular features. One limitation with feature based models is that they are difficult to apply over the die scale. In many cases, the features are so small that an attempt to trace their polish evolution is computationally expensive and may be difficult to apply to the entire die. In the next section, we focus on models which seek to address pattern dependencies observed over large regions of the chip or across the entire chip.

IV. Die-Level Modeling of ILD CMP

Planarization and polish behavior within the die depends strongly on the topography and the material being polished. In this section, we first consider the development of models for single material polishing, most prevalently applied to the planarization of deposited oxide over other topographies (metal lines, for example), and then we examine the approaches and models reported for dual-material polishing (e.g., nitride and oxide) arising in other CMP applications. Models reported for metal polishing are discussed in Section V.

1. Topography Modeling: Layout Density Dependence

Layout pattern density is the dominant pattern factor in oxide CMP [47, 54]. This follows from Preston's equation, which postulates that the polish rate is a linear function of the local pressure, which is related to pattern density by the effective wafer–pad contact area. Figure 9 illustrates the significance of pattern density in die thickness evolution. The mask consists of blocks of regular patterns which are arranged in rows. Each row consists of five design blocks of 4 × 4 mm with each block consisting of vertical lines and spaces. The blocks in rows L1–L3 have a fixed pitch of 100 μm. The designed block density—defined here as the ratio of line width to pitch—is achieved by varying the line width and space in the block. Row L1 has blocks with the following densities: 10, 20, 40, 60, and 80%. Row L2 has densities of 30, 50, 70, 90 and 100%. Row L3 consists of step density blocks with densities of 10, 90, 30, 70, and 50%, whereas row L4 consists of blocks with a constant density of 50% but the pitch ranges from 10 to 200 μm. The final thickness profiles of the oxide over the patterned metal features for three different polish times are shown on the right. The optically measured final thicknesses are circled.

The different polish characteristics of each of the traces demonstrate the significance of layout pattern dependencies in oxide CMP. For example, the L3 profile traces the large step density transitions with dramatic variations in the resulting oxide thickness. In contrast, the blocks along L4 polish at the same rate, underscoring the fact that layout pitch is not a first-order determinant of the polish rate. This is the case for a very large range of pitch

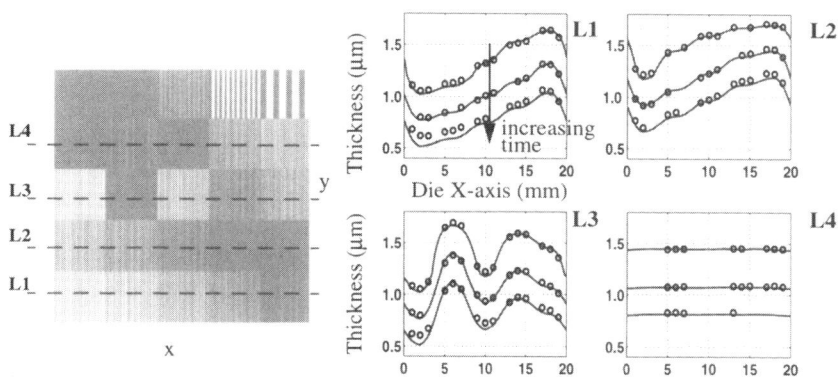

FIG. 9. Polish thickness evolution profiles illustrating the significance of layout pattern density.

values examined (10–200 µm). Studies which confirm small pitch dependency for oxide polishing have been reported for as large as 500-µm pitch structures [28]. The conclusion to be drawn from this and most studies is that for oxide CMP, practical modeling of the polish evolution of the raised oxide over patterned regions need not focus on the feature polish characteristics. Rather, the thickness evolution can be effectively modeled as a function of the pattern density distribution across the die.

Several models have been proposed to account for pattern effects in CMP [7, 38]. A detailed review of some of the early models including those based on feature erosion has been provided by Nanz and Camilletti [27]. Other models based on contact mechanics around features has also been proposed, such as the work of Runnels *et al.* [43] or Chekina *et al.* [8]. In this section, our goal is to present an example of the class of models that support thickness profile prediction across the whole die. Models in this category have been proposed by Hayashide *et al.* [17] and Stine *et al.* [54] with extensions in Ouma [30], Takahashi *et al.* [58], and Tung [62]. There is a common principle behind these models in that they start from Preston's equation and determine a way to evaluate the effective pattern density. The differences arise in the methods used to account for neighboring topography when considering the removal at any given spatial location.

2. Modeling of Thickness Evolution

Stine *et al.* proposed a simple analytic model for patterned feature removal incorporating an "effective density" determination step [54]. Figure 10 defines terms used in the model, which reformulates Preston's equation [Eq. (2)] as a function of blanket rate K and effective pattern density $\rho(x, y, z)$:

$$\frac{dz}{dt} = -K_p PV = -\frac{K}{\rho(x, y, z)} \tag{21}$$

The equation is then solved for the oxide thickness z under the assumption that no "down area" polishing occurs until the local step, z_1, has been removed, after which the local rate is equal to the blanket polish rate. This is captured by expressing the effective density as

$$\rho(x, y, z) = \begin{cases} \rho_0(x, y) & z > z_0 - z_1 \\ 1 & z < z_0 - z_1 \end{cases} \tag{22}$$

In the determination of the as-patterned effective density $\rho_0(x, y)$, the effect

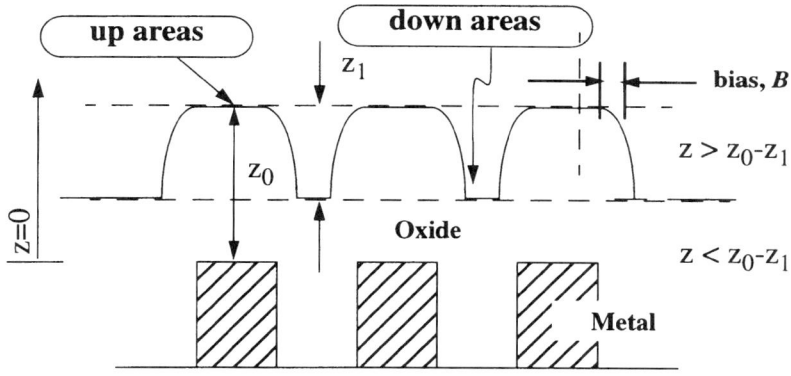

Fig. 10. Definition of terms used in the basic model.

of lateral deposition is accounted for by adding a bias term B to the metal lines, which constitute the mask layout pattern. This ensures that the effective density is that of the final film profile and not the initial mask layout. The bias is positive for conformal films and negative for high-density plasma CVD oxides. The assumption that local pattern density is independent of film thickness before local planarity approximates the actual deposition profiles with a vertical profile. In reality for nonvertical deposited topography, the effective density of the exposed surface depends on height; it is possible to "time-step" the profile evolution to account for such a time-varying density (as in Tung [62]), but such detail is typically unnecessary for final oxide thickness prediction. Experimental results have indicated that an approximate constant effective density works extremely well so long as a proper choice of the bias is used to account for total increase or decrease of oxide volume due to the deposition profile. The assumption makes it possible to express the final film thickness for any time t in a closed-form as

$$z = \begin{cases} z_0 - \left(\dfrac{Kt}{\rho_0(x,y)}\right) & t < (\rho_0 z_1)/K \\ z_0 - z_1 - Kt + \rho_0(x,y)z_1 & t > (\rho_0 z_1)/K \end{cases} \quad (23)$$

Before local planarity is achieved (i.e., while local step height still exists), the final film thickness is inversely proportional to the effective local density. The film is assumed to polish linearly at the blanket rate afterward. The key model predictive power arises from the fact that differences in initial effective pattern density across the die result in postpolish global thickness nonuniformities, although local planarity (complete step height removal) will still be achieved in all regions of the die given sufficient polish time. In reality, raised areas are not strictly preferentially polished, and model extensions or modifications to address accurate prediction of down area polish and step height reduction are needed as discussed in Subsection 6 of Section IV.

Although the model above is relatively straightforward, its accuracy depends on the correct determination of the effective local pattern density $\rho_0(x, y)$. A key element of the model is that neighboring topography is accounted for by employing an appropriate weighting function that takes the pad and process interaction with nearby raised regions into account. The effective density is thus a nonlocal parameter spanning a relatively long length scale characterized by the "planarization length" of the process. The planarization length is the length scale over which neighboring topographies affect the polishing at the point of interest (that is to say, the planarization length defines the area over which the effective density for any point of interest should be calculated). Each layout will produce a range of effective densities across the die, resulting in a range of global oxide thicknesses due to different polishing rates across the die. Longer planarization lengths will "average" topography more effectively and thus decrease the total range of perceived density within the die. A longer planarization length thus results in smaller global oxide thickness variation across the die.

3. Planarization Length and Response Function

The effective pattern density at any spatial location on the mask is evaluated within a specific area of the mask to capture the localized region over which the polish pressure is distributed. An important relationship is believed to exist between the pad mechanical properties (and other properties of the slurry and process conditions) and the length scale over which raised topography interact. The key assumption is that the size of this area (defined in extent by the "planarization length") over which effective density is calculated is intrinsic to the pad and process; once it is known for a given pad and process, then the effective density may be determined across the entire mask layout for arbitrary layouts. While the planarization length is indeed a function of many pad and process parameters, the mechanical deformation of the pad is conceptually the dominant effect: intuitively, the stiffer the pad, the larger the planarization area since a stiffer pad resists deformation and thus distributes pressure over a larger area. The use of the same planarization area to determine the density at all spatial locations on the wafer simplifies the computation of effective density significantly as there is no longer the need to focus on the specific deformation caused by specific features on the mask: rather, one can directly consider the net effect of density variations on polish performance.

In the simplest case, a square area can be used to determine the effective density across the mask, as shown in Fig. 11. Density to be assigned to the coordinates at the center of the window is equal to the ratio of raised to total area of the square window. The length of each side of the square is then defined as the planarization length; this square region approximates the deformation characteristics of the pad and process. The size of the square (or the planarization length) is determined experimentally by varying the square size until the effective density calculation results in predicted thickness values that best fit experimentally measured polish data when used in the thickness evolution model.

The density weighting function can be considered as the "planarization impulse response" of the pad and process, and the planarization length may be appropriately defined as a number which fully characterizes the applicable impulse response function [2]. For the square window, the length of the side of the square fully defines the planarization response. Alternative weighting functions may be more appropriate, as pictured in Fig. 12, that may have both spatial symmetry, and that weight nearby local features differently than farther away structures. One challenge has been to determine the appropriate density weighting function that captures the elastic properties of the pad and to use an appropriate length parameter to define it.

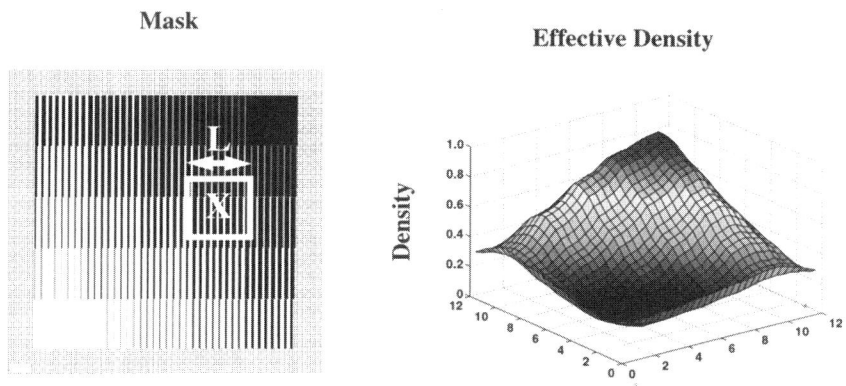

FIG. 11. Layout mask and effective density based on a 3-mm-square window.

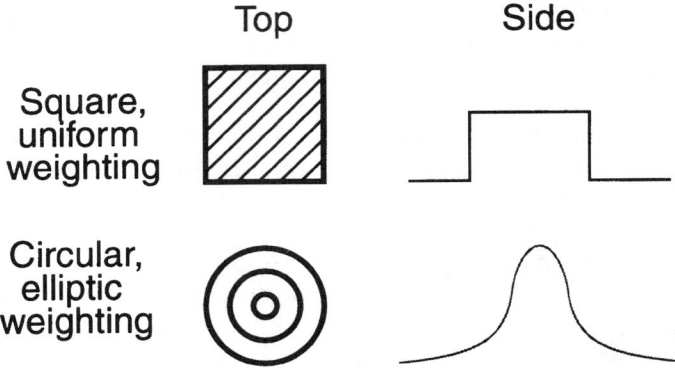

Fig. 12. Window shapes and weighting functions used in calculation of effective density from a given layout.

a. Pad Deformation During CMP — Density Weighting Function

Determining the pad deformation characteristics in CMP is a difficult problem for two primary reasons. The first problem arises from the material properties of the pad: the polyurethane from which the polishing pads are made is a nonlinear elastic material. The deformation characteristics therefore depend on the applied loads, relative speeds, and the material thicknesses. The second problem is that the films are usually composite, making it difficult to obtain analytic expressions for the deformation profiles even for static deformation loads. Due to these problems, several approximations have to be made when determining the deformation profile and the appropriate density weighting function. The nonlinearity of the materials is accommodated by instituting a characterization phase in which actual polish data is used to determine the exact characteristic length of the weighting function once the general shape of such a function is determined.

Common dielectric CMP polish pads consist of a stack of two materials with the stiffer material at the top. The deformation of the pad by the wafer features may be inferred by examining the deformation profile based on a localized force over a region of the pad. The localized constant force (or pressure) may be applied by a mechanical indenter or through any other scheme that facilitates the application of a centralized force. As shown by Boggy and Shield [45, 46], even this simple configuration does not lend itself to a suitable analytic solution for both the deformation (displacement) and pressure distribution. Numerical solutions involving integral transforms must be used to determine the final profile. Since our goal is to capture the essence of the pad deformation, focus can be shifted to the single-layer case for which an analytic solution for the deformation exists. The deformation for the one material case is very similar to the two material case thus justifying this simplification [45, 46].

For a one-material case, analytic solutions exist for both the deformation profile of the elastic material as well as the pressure and stress distributions for the indenter (approximating wafer features). Consider a single-layer pad that is thick relative to the vertical deformation and has a deformation force applied over a circular region of radius a. The deformation is given by a set of two equations that represent deformations within and outside the circular radius over which the force is applied. The deformation at any radius r less than a is given by [59]:

$$w(r) = \frac{4(1-v^2)qa}{\pi E} \int_0^{\pi/2} \sqrt{1 - \frac{r^2}{a^2} \sin^2\theta} \, d\theta \qquad (24)$$

where q is the load, v is the Poisson ratio, and E is Young's modulus of the pad material. The deformation profile for $r > a$ is correspondingly given by

$$w(r) = \frac{4(1-v^2)qr}{\pi E} \left[\int_0^{\pi/2} \sqrt{1 - \frac{a^2}{r^2} \sin^2\theta} \, d\theta - \left(1 - \frac{a^2}{r^2}\right) \right.$$
$$\left. \times \int_0^{\pi/2} \frac{d\theta}{\sqrt{1 - (a^2/r^2)\sin^2\theta}} \right] \qquad (25)$$

These two equations show that the pad deformation is proportional to the applied load and inversely proportional to the pad material stiffness (represented by the Young's modulus). The equations involve complete elliptic integrals that can be readily evaluated numerically. The relative deformation represented by Eqs. (24) and (25) is plotted in Fig. 13, where the inner region corresponds to Eq. (24) and the outer corresponds to Eq. (25).

An important observation regarding the equations representing the deformation is that the shape of the deformation (with only vertical scaling factors) is independent of the material constants and only dependent on the area of force application. The deformation shape is therefore similar for different material properties. The material properties and applied force control only the actual amount of the deformation. This is ideal for our initial goal that was to determine the typical deformation profile of the polish pad. All that remains is the determination of the appropriate length scale to use to represent the deformation.

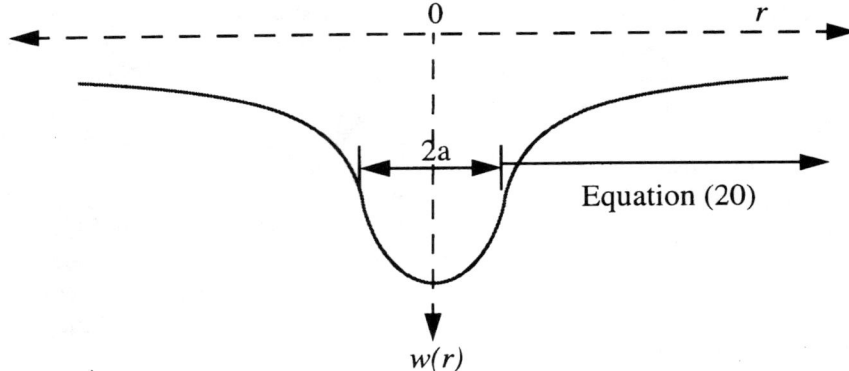

FIG. 13. Relative material deformation of an elastic material if the load is distributed over a circular region or radius a.

A way to characterize the deformation may be obtained by examining the maximum deformation as well as the deformation amount at the edge of the force application area ($r = a$). The maximum deformation occurs at the center ($r = 0$) and is given by

$$w_{max} = \frac{2(1 - v^2)qa}{E} \tag{26}$$

The deformation at the edge of force application area is

$$(w)_{r=a} = \frac{4(1 - v^2)qa}{\pi E} \tag{27}$$

The ratio between these relative deformations is $2/\pi$ and can be used to define the deformation profile or length scale. Due to the presence of a softer back pad, more deformation is expected for the stacked pad but the shape, which is the main concern, will be approximately similar [45, 46]. The deformation is relatively small compared to the region of application of the force. Using approximate material properties for the IC1000 pad (Young's modulus of 2.9×10^7 Pa [41] and approximate Poisson ratio of 1/3) with force applied in a circular region of radius 2 mm, and a local pressure of 7 psi, the maximum deflection is about 6 μm. This deformation is referenced to the origin as illustrated in Fig. 13. It is also important to note that the transition shape is very gradual and this sets the polish limit for the down areas.

b. Optimal Weighting Function

The weighting function, described as elliptic because it is obtained by solving elliptic integrals, is defined by Eqs. (24) and (25). From these equations, the general shape is determined. The appropriate length scale must be determined through characterization procedures. Any given family of such functions can be characterized by the width (which corresponds to the area over which the deformation force is applied in the original deformation equation). The function width can be selected to match the deformation profile of interest. Formally, the width is defined where the deformation is $2/\pi$ of the peak deformation as shown in Fig. 14. The width which best captures the deformation profile of polished data is formally defined as the planarization length for the elliptic weighting function. To account for the different material stiffness, stiffer pads (which deform less) are assigned longer characteristic planarization lengths to account for the spread of the polish pressure over a larger region. In density calculations, the relative (not the absolute) weight is used hence the most important factor is the spread which now intuitively relates to the extent of the deformation due to the surrounding raised features.

4. Characterization — Determination of Planarization Length

a. Characterization Masks

A key component in CMP process characterization is the choice of test layout mask. For planarization length extraction, the test mask should

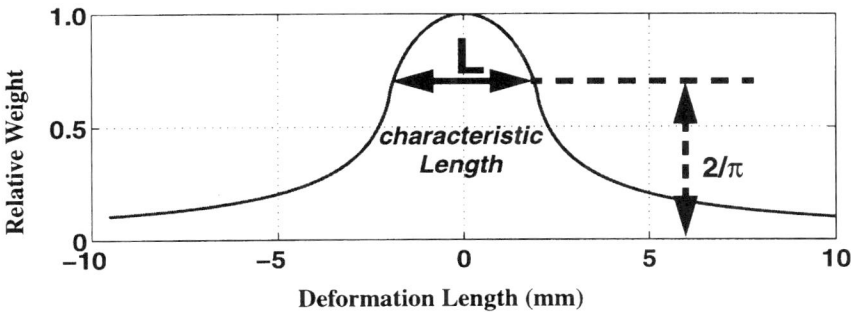

Fig. 14. One-dimensional cross section of an elliptic weighting filter. The characteristic length is defined as the section length when the relative weight has dropped to $2/\pi$. The filter shape corresponds to the deformation profile of an elastic material under distributed load in a circle of radius $L/2$.

consist of sections with abrupt pattern density changes that result in large post-CMP thickness variations. The step density pattern should be arranged such that the model fit across a long spatial stretch is easily visualized. The test mask may also include structures that are not necessary for planarization length extraction but are otherwise useful for assessing and confirming other aspects of the planarization process (such as the polish characteristics of down areas which require the use of constant density but varying pitch structures). One such mask is shown in Fig. 9.

b. *Calculation of Effective Pattern Density*

The following three steps are followed in determining the effective density across a chip: (i) layout biasing, (ii) local discretized density evaluation, and finally, (iii) effective density evaluation using a weighting filter of appropriate size. In the biasing phase, it is recognized that the drawn mask layout (from the layout tool) can be significantly modified by the deposited film. This effect is more serious for small features. Depending on the deposition film, the density at any local area of the die may be larger or smaller than the drawn density. The drawn features must therefore be enlarged or shrunk accordingly before effective density calculation.

The deposition cross section can be substantially different for different types of deposited films or processes. For example, densified oxone SACVD (subatmospheric pressure CVD), like TEOS, is conformal to the layout features [31]. Small feature arrays such as 1 μm pitch structures are filled such that their top is at the same level with large solid areas. The local density is thus approximately 100% for a large range of layout densities. For HDP CVD, on the other hand, complete gap fill is achieved for small spaces but the top of the film is now level with the field areas except for a small triangular peak. A region on the wafer consisting primarily of fine features can thus be treated as nearly 0% local density in comparison to regions with large features, where a substantial raised or thick oxide results. The adjustment value for an effective bias compared to the layout can be obtained from scanning electron micrography (SEM) cross sections; if the deposition mechanism is well known, the bias may be obtained through deposition simulation. The bias phase is very important especially for realistic layout masks with small features. If the step is omitted and the drawn layout is used, the effective density obtained could be completely invalid. This is illustrated in Fig. 15, which shows the effective density across a section of a mask (assuming the same weighting function length) for the three cases [31].

The next stage is to discretize the local density into small cells to facilitate profile simulation. The mask is divided into small square cells in a regular

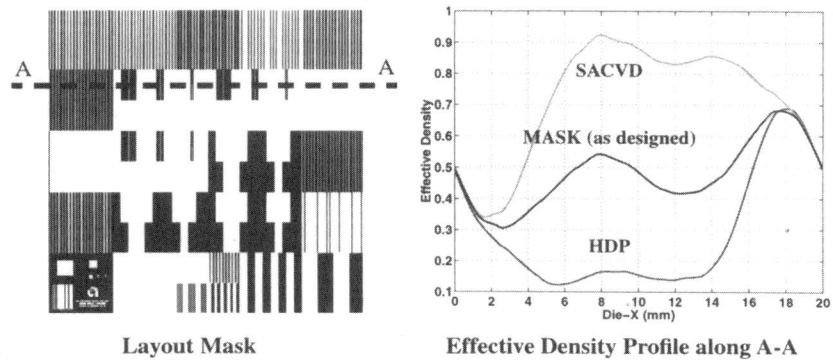

Layout Mask   **Effective Density Profile along A-A**

FIG. 15. Illustration of need for bias in effective density calculation [31].

grid. The density in each cell is defined as the ratio of raised to total area of the cell. Cell sizes of 40 × 40 μm have been shown to be sufficient [29]. Smaller cells increased computation time without increasing the model accuracy. The cell sizes should not be very large because in the next phase when the effective density is calculated, the weighted area of each cell is summed within a certain radius. Fig. 16 shows a 3-D plot of the local cell densities across a mask with conformal TEOS deposition.

The final density calculation step is the stage at which the 2-D weighting density window function is first employed. The 2-D filter of the correct

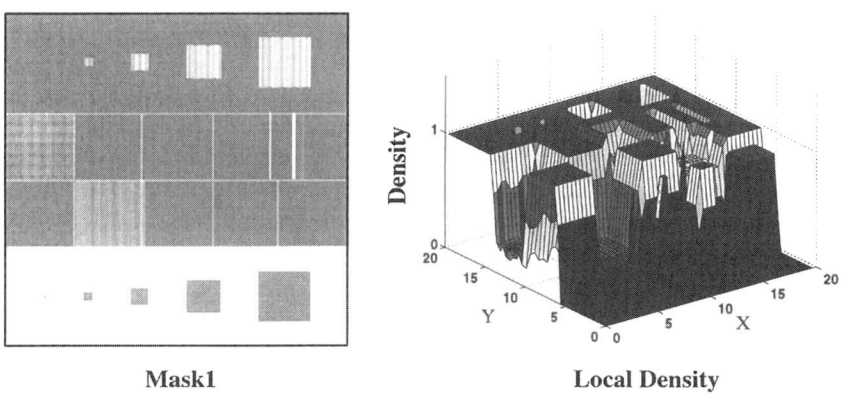

Mask1   **Local Density**

FIG. 16. Local pattern density across a die evaluated in 40-μm cells.

length scale (i.e., correct planarization length) is used to determine the effective density for each cell by weighting the influence of nearby cell densities. If the discretized mask density is given by $d(n_1, n_2)$, the discretized weighting function by $f(n_1, n_2)$, then the effective discretized density is given by the convolution sum:

$$\rho(n_1, n_2) = \sum_{k_1 = -\infty}^{\infty} \sum_{k_2 = -\infty}^{\infty} f(n_1, n_2) \, d(n_1 - k_1, n_2 - k_2) \qquad (28)$$

The actual limit of the summation is the extent of the weighting filter. Zero padding is used to ensure that the discretized matrices have sizes which are a power of two so that the computation can be done in the frequency domain using fast fourier transform (FFT) techniques. The effective discretized density, $\rho(n_1, n_2)$, is then given by

$$\rho(n_1, n_2) = \text{IFFT}\{\text{FFT}[d(n_1, n_2)] \cdot \text{FFT}[f(n_1, n_2)]\} \qquad (29)$$

The use of the FFT reduces the number of computations significantly. If the number of cells is $N \times N$, and the discretized filter is of size $M \times M$, Eq. (28) requires $N^2 M^2$ computations, while Eq. (29) requires $NM \log_2(NM)$ computations. The computation time is decreased by orders of magnitude for typical large dies where the padded cell densities will be 512×512 for a 20- $\times$ 20-mm mask.

The computation in Eq. (28) must recognize that the die is periodically repeated across the wafer so that density calculation at the edge of the die accounts for the effects of features from the next die. The dies at the edge of the wafer should be treated differently if the edge is not patterned since the periodicity no longer exists. Because of the existence of a step density transition between dies bordering the nonpatterned region near the wafer edge, outermost dies may have very different polish characteristics from those away from the wafer edge. A good practice is to pattern the whole wafer and to discard the outermost dies (which occupy the area normally known as the edge exclusion region) which will suffer from wafer edge effect.

Figure 17 shows the effective density using the elliptic filter with a characteristic length of 2.9 mm. The optimal length must be determined for each consumable set and process conditions since the planarization length is dependent not only on the polish pad type but also on the polish process conditions, notably the down force.

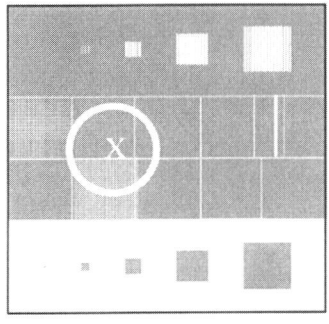
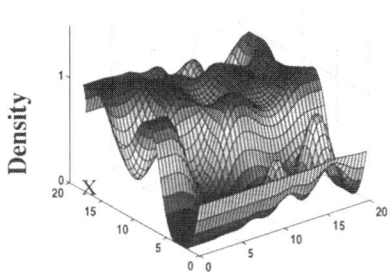

Discretized Density Map **Effective Density**

Fig. 17. Effective pattern density obtained with an elliptic weighting filter.

c. Model Calibration — Determination of Planarization Length

The calibration phase focuses on the determination of the planarization length itself. This is a crucial characterization phase since once the planarization length is determined, the effective density, and thus the thickness evolution, can be determined for any layout of interest polished under similar process conditions. The determination of planarization length is an iterative process. First, an initial approximate length is chosen. This is used to determine the effective density as detailed in the previous subsection. The calculated effective density is then used in the model to compute predicted oxide thicknesses, which are then compared to measured thickness data. A sum of square error minimization scheme is used to determine when an acceptably small error is achieved by gradient descent on the choice of planarization length.

An important part of the calibration is the data set. The planarization length is fully defined when enough material has been removed and the pad conforms to the locally planarized features across the die. For best results, the data should exhibit large variation over a long spatial range, preferably with a step change in thickness arising from a step change in density. Layout test masks with step densities are therefore ideal for planarization length extraction. Thickness measurements should be taken along an horizontal scan, as shown in Fig. 9. A minimum of three equally spaced measurements is needed in each of the 4 × 4-mm density blocks. More measurements are preferred but little is gained after taking more than five measurements per block. Note that only "up" areas measurements are needed as the goal is to characterize global planarity. Planarization length extraction is based on data from one die, so only the center die needs to be probed on a wafer. More dies can be probed near the center to confirm the robustness of the methodology. In any case, the correct blanket rate should be used in the model. This is easily determined by recognizing that all data point are shifted by the same amount if the wrong polish rate is used.

5. STI CMP Modeling

Shallow trench isolation (STI) has emerged as the isolation technique of choice as the drive for dense memory and logic applications has intensified. The effectiveness of CMP at achieving good planarity over a large area has enabled the use of an inlaid process, such as STI, to satisfy the high-device-density requirements of modern integrated circuits. However, due to pattern dependency which still remains over a global range, a two-mask process is typically used in STI CMP. This increases the process complexity and cost; a single mask process is therefore of widespread interest. The modeling efforts are therefore focused on the single-mask CMP STI process.

Pattern dependency concerns arise at two levels in STI CMP [67]: during the oxide overburden polish phase, and when the nitride layer is exposed. In the first stage, the process is similar to interlevel dielectric (ILD) CMP and the characterization and modeling methodologies presented in the previous section are applicable. Once the nitride is exposed, two different materials exist at the same level, and pattern dependency manifestation is more complex.

Pattern dependency in STI CMP may be complicated by the choice of consumable, particularly slurry. Since polishing should stop as soon as nitride is exposed, slurries with high oxide to nitride selectivity are typically used. Unfortunately, the selectivity is usually based on blanket wafer polish results, but may also depend on pattern density [31]. Such problems make STI CMP modeling a serious challenge, and one must worry about consumable selection in much greater detail than in pure oxide or ILD CMP. The complication is that for some slurries, the efficient pattern dependent models for the pure oxide phase may need modifications. In addition to the slurry related problems, the impacts of other process conditions on planarization length and wafer-level uniformity must still be accounted for. Modeling of the pattern effect in STI therefore requires a complete understanding of a wide range of process and consumable effects.

The bulk nitride polish model is used when the nitride is exposed. At this phase, the goal is to predict the polish rate of the bulk composite material. Just before reaching the nitride, the surface is usually completely planar such that the local polish rate is the blanket oxide polish rate. When the nitride is initially exposed, the local density remains 100%. This scenario continues until the oxide begins to recess (dish). However, the recess remains small, such that the local density is still approximately 100%. The removal rate of the composite material is dominated by the lower polish rate material so that the effective polish rate at which the nitride surface recedes is given by

$$K_{\text{eff}} = K_{\text{nitride}} \tag{30}$$

For low nitride density, nitride domination is weak and for effective densities below 30%, a weighted average blanket rate should be used instead of Eq. (30). This is given by

$$K_{\text{eff}} = (1 - \rho)K_{\text{oxide}} + \rho K_{\text{nitride}} \tag{31}$$

where ρ is the effective density using the planarization length obtained for the process, but with the cell densities corresponding to the nitride layout patterns, not the biased layout. This is very important since once nitride is exposed, there is no longer the need to use a bias. Since the effective density range should be minimized across the layout, Eq. (30) will apply in most cases. For nitride to oxide selectivity of 4:1 (blanket oxide polishes four times faster than blanket oxide) we empirically determine that above 40% nitride density, Eq. (30) is sufficient. The model fit based on the data shown in Fig. 15 is shown in Fig. 18.

Erosion of nitride across an array can also occur due to excess overpolish. If only a small degree of nitride erosion exists (as in typical practice), the erosion can be neglected and the model approach described in the previous paragraph for the oxide trench thickness in the presence of the nitride regions is adequate. For large amounts of nitride erosion, on the other hand, an alternative approach can be used. In this case, one can assume that the oxide has substantially dished due to the excess overpolish. Thus, one can treat the array of nitride active areas as "up" features, and use a nitride effective density dependent model with the nitride blanket rate rather than the oxide rate to model the continued polish of the nitride region. While these approaches achieve good approximate results for modeling STI polish, the dishing and erosion occurs over a small vertical scale that is comparable to the error in the basic oxide CMP model described here. In order to achieve more accurate results, additional model development is required. In particular, step height dependencies should be accounted for as described in the next section.

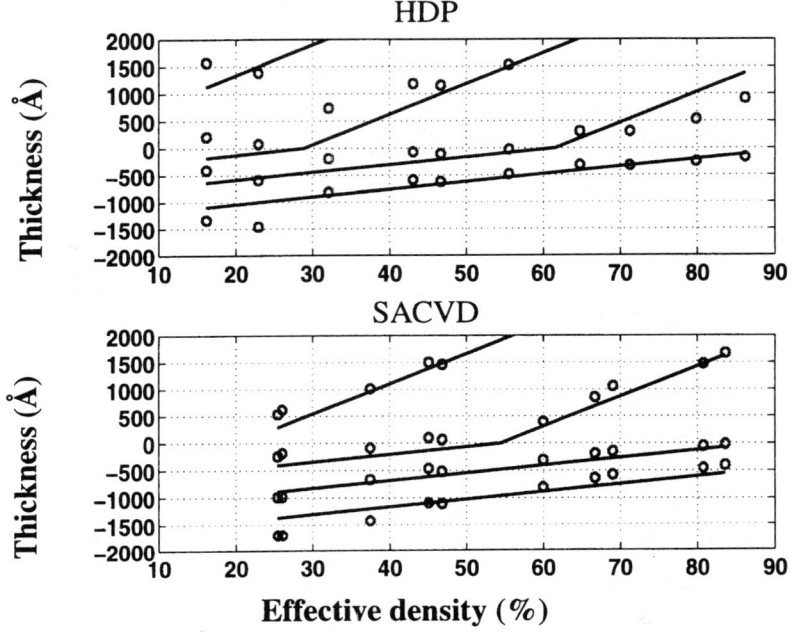

FIG. 18. Model fit for nitride phase of shallow trench isolation polish [31].

6. Models for Step Height Reduction

The die-level models described earlier focus on the need to predict the remaining oxide thickness over metal (or other) features at the completion of processing. The prediction of oxide polishing rates in "down" regions compared to those in "up" regions is also of interest, and is often studied in a combined fashion where the parameter of interest is the remaining local step height. This is pictured in Fig. 19, where it is important to note that while "up" area polishing dominates, significant amounts of down area polish also occur and must be taken into account if one is to correctly predict the remaining step height, or target the total deposition thickness required.

A model by Tseng *et al.* [60] applied an analysis of local pressure differentials in up and down regions, assuming that the pad is always in contact with both regions. This results in a differential rate of step height reduction that is proportional to the remaining step height; the solution gives an exponentially decaying step height in time. The pressure differential

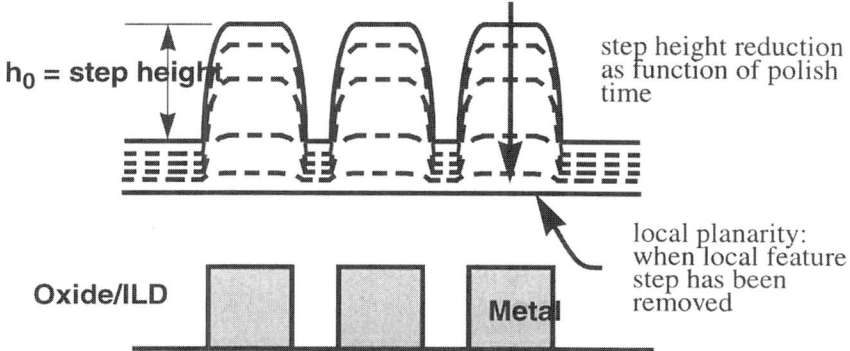

FIG. 19. Schematic of local steps and step height reduction as planarization proceeds.

results from a pad compression or bending analysis, enabling one to (roughly) relate the rate of step height reduction to pad modulus.

Grillaert et al. [14, 15] considered a model where the pad is not necessarily in contact with the down areas. Instead, two different regimes of pad contact apply, as illustrated in Fig. 20. Assuming that for large step heights the pad will not be in contact, a "linear" polishing regime exists where all of the applied load is borne by the raised areas, leading to a simple density dependent polish rate that results in a linear (in time) reduction in step height. At some "contact time" or "contact height," the pad begins to touch the down area and the apportionment of forces related to step height again occurs leading to an exponential decay in step height from that time forward. These two scenarios are pictured in Fig. 21.

In the linear regime, an incompressible pad model gives a step height h as a function of time

$$h = h_0 - \frac{r}{a} t \tag{32}$$

where r is the blanket rate, and a is the percentage of raised area (or density). Experimental results in this linear regime are shown in Fig. 21.

In the exponential regime, a compressible pad model results in a step height dependence as

$$h = h_0 e^{-t/\tau} \quad \text{with} \quad \tau = \frac{LP}{Er}, \tag{33}$$

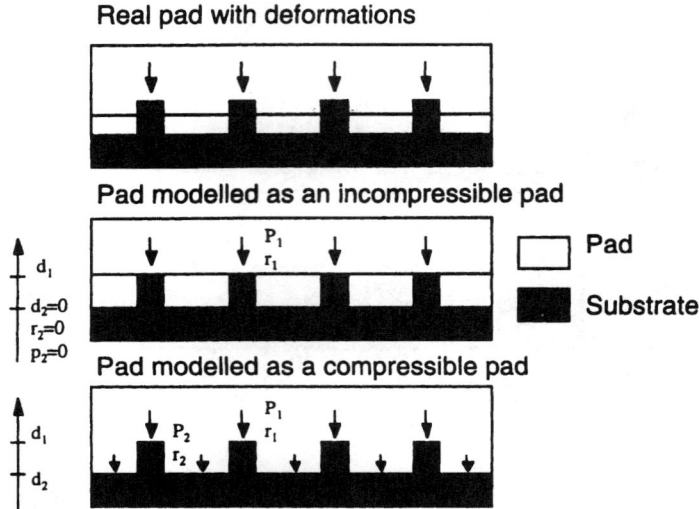

Fig. 20. Two-regime model for pad–wafer contact. In the "linear" regime the pad is modeled as an incompressible pad which only contacts raised areas. In the "exponential" regime, the pad is assumed to be compressible and able to also contact down areas [14].

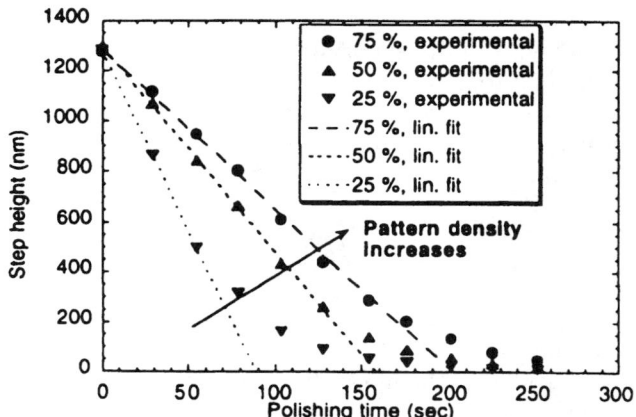

Fig. 21. Step height reduction in the linear regime [14].

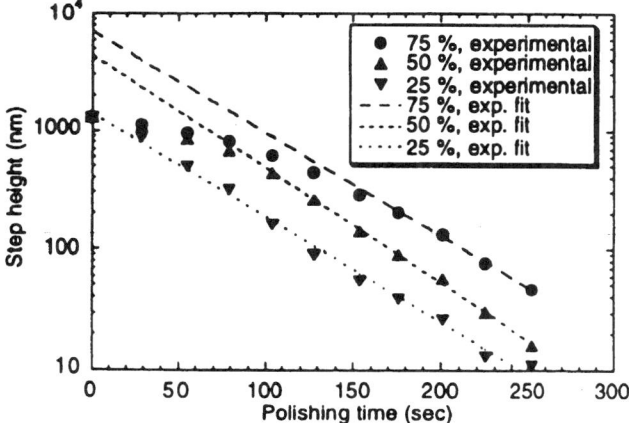

FIG. 22. Step height reduction in the exponential regime [14].

where L is the pad thickness, E is the Young's modulus for the pad, and τ is the exponential decay time constant related to the properties of the pad. The match with experimental results in this regime where one is seeking to remove the last 1000–2000 Å of step height is shown in Fig. 22. Further work also examined the dependencies of contact time on process and layout parameters, suggesting that the contact time will also be a function of pattern density [15].

The model by Grillaert et al. addresses step height dependencies and includes a density dependence. Smith et al. [48] integrated the effective pattern density model described earlier with the time and step-height dependent model of Grillaert et al. to accurately predict both up and down area polishing. The resulting analytic expression for the up area amount removed (AR) is

$$AR_u = \begin{cases} t_p K/\rho & t_p \leqslant t_c \\ t_c K/\rho + K(t_p - t_c) + (1-\rho)\dfrac{h_1}{\tau}(1 - e^{-(t_p - t_c)/\tau}) & t_p > t_c \end{cases} \quad (34)$$

and correspondingly for the down area amount removed:

$$AR_d = \begin{cases} 0 & t_p \leqslant t_c \\ K(t_p - t_c) - \rho\dfrac{h_1}{\tau}(1 - e^{-(t_p - t_c)/\tau}) & t_p > t_c \end{cases} \quad (35)$$

where K is the blanket removal rate, h_0 is the initial step height, ρ is the

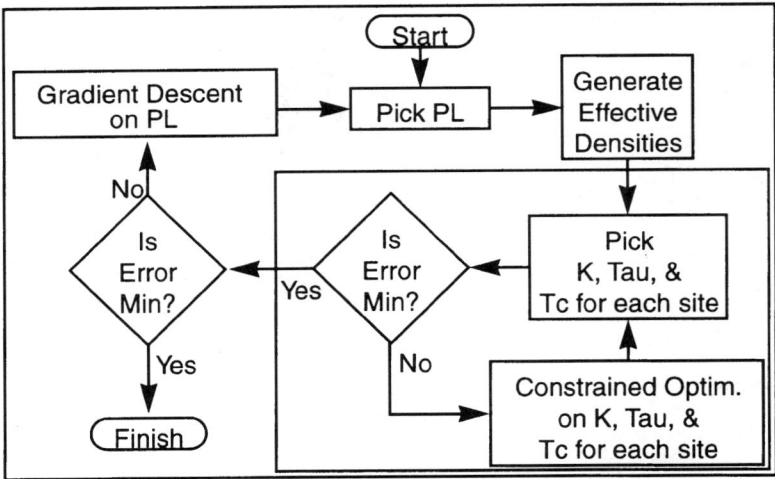

FIG. 23. Integrated density and step-height-dependent model parameter extraction approach. The outer loop finds the planarization length that best captures the density dependence, while the inner loop find the step-height model parameters that best explain up and down area polish data [48].

effective pattern density, τ is the exponential time constant, t_p is the polish time, t_c is the time of contact with the down area, and $h_1 = h_0 - (K/\rho) t_c$ is the transition or contact step height. A constrained nonlinear optimization procedure illustrated in Fig. 23 is used to determine the planarization length (used to calculate effective density for a given measurement point) and the preceding parameters. Comparison of the resulting model (assuming the contact heights to be a function of density $h_1 = a_1 + a_2 \cdot e^{-\rho/a_3}$) and experimental results are shown in Fig. 24, where RMS (root mean square) errors under 100 Å have been achieved.

7. Applications of Density Models

The integrated modeling methodology is useful for several applications. These include the ability to determine the optimal amount of material to deposit before CMP, the provision of an effective characterization scheme through the use of planarization length as a process performance monitor [29, 55], and the correct prediction of post-CMP ILD thickness variation, which is useful for assessing the impact of such variation on circuit performance [24, 56].

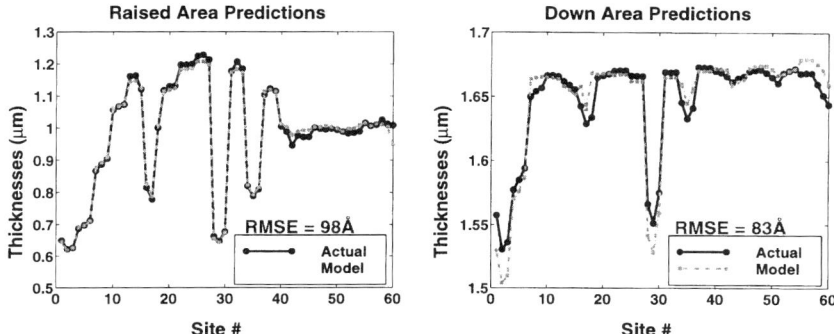

FIG. 24. Comparison of model and experimental results using an integrated density and step height dependent model [48].

A key benefit of accurate CMP models that needs emphasis is the capability to optimize layout design before polishing. Post-CMP ILD thickness variation is a serious concern from both functionality and reliability concerns. An effective method of minimizing this effect is the use of dummy fill patterns that lead to a more equitable pattern density distribution across the chip. Evaluation of such schemes before actual product implementation has become a major use of CMP modeling [53]. Dummy fill is also being investigated for front-end processes where shallow trench isolation CMP suffers from substantial pattern dependencies.

V. Models for Metal Polishing

Pattern dependencies in metal polishing have also been identified as a substantial concern, both from manufacturability and performance points of view. In this section, we briefly review the key issues that have attracted modeling attention, including dishing within the metal feature (a via, a line, or a larger metal region) and erosion of the surrounding oxide or dielectric support. Modeling and simulation challenges include the need to address the simultaneous polishing of two or more materials, interactions of chemical and mechanical polish mechanisms, and the importance of polish effects at very small dimensions (e.g., within an individual via).

In the typical metal polish application recessed features are patterned in a supporting oxide or other dielectric material; the goal is to have metal reside only in these recessed regions at the conclusion of the polishing process. After patterning the recessed regions, thin metal adhesion and barrier films are deposited (e.g., titanium, titanium nitride, tantalum, tantalum nitride), followed by deposition of the fill metal (e.g., tungsten or copper). The fill overburden (the metal and barrier material over field regions) must be removed by way of CMP. The necessity to polish off the metal and barrier films completely in field regions (to avoid shorts) typically requires some degree of "overpolish," during which underdesirable dishing and erosion may occur, as illustrated schematically in Fig. 25.

It should be pointed out that in the case of metal polishing, a very strong relationship exists between wafer level polishing uniformity and pattern dependent effects within the die. If a perfectly uniform clearing of the metal film occurs (that is, the overburden is removed simultaneously across the entire wafer), then the overpolish time can be minimized. If, on the other hand, some regions of the wafer clear more slowly than others, then one must continue polishing until the metal is removed everywhere; this can result in the substantial overpolishing of those regions on the wafer that were the first to clear [32].

1. TUNGSTEN CMP MODELING

A number of works have focused on the pattern dependencies in tungsten (both tungsten via and local interconnect) polishing [12, 19, 21]. To first order, empirical results seem to suggest that dishing is dominated by the width or size of the metal feature being polished: wide lines tend to suffer

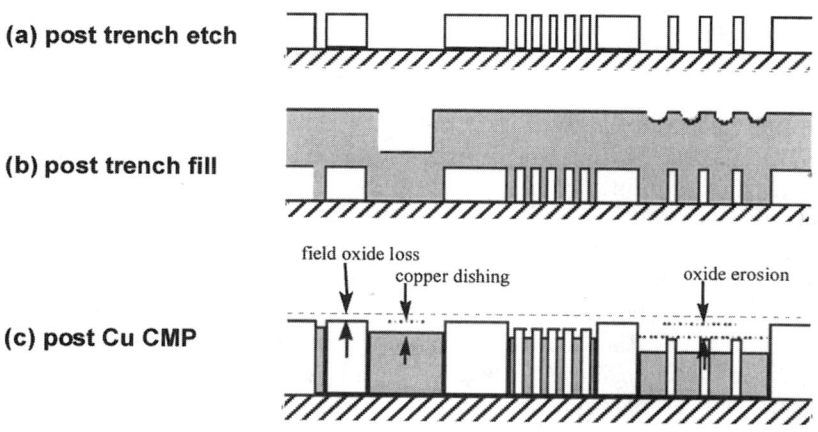

FIG. 25. Illustration of dishing and erosion concerns in copper CMP [32].

more dishing. As in the case of dielectric CMP processes, there also appears to be a relationship between the "density" of metal features and the resulting erosion characteristics. This has some intuitive justification: if the metal polishes much more quickly than the oxide (in typical metal slurries the selectivity is 3:1 or much greater), then erosion will be dominated by an approximately "single-material" oxide polish characteristic.

Elbel et al. [12] have studied the polish of tungsten lines in a damascene-style process. The focus of the study and model is to relate the degree of dishing and erosion to the layout patterns, including both density and linewidth dependencies. A simple model is used to relate the amount of erosion (ε) to the density of metal and the overpolishing time:

$$\varepsilon = \alpha \frac{\Phi}{1 - \Phi}(t - t_0) \tag{36}$$

where Φ is the tungsten pattern density, α is the erosion rate at 50% density, and $t - t_0$ is the overpolish time. This relationship is based on the empirical observation shown in Fig. 26. Here the time t_0 is the hypothetical time at which clearing occurs.

In addition to erosion, an interesting model for dishing within metal features is proposed, based on the concept of a "maximum dishing," which

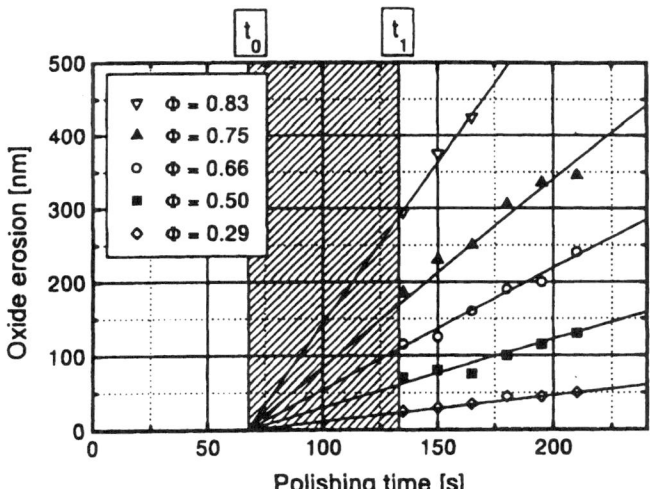

FIG. 26. Degree of oxide erosion as a function of polishing time in tungsten CMP [12]. Reproduced by permission of the Electrochemical Society, Inc.

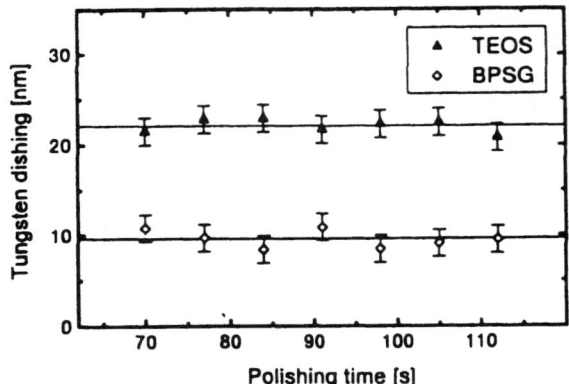

FIG. 27. Maximum dishing within an array of contacts (of size 0.5 μm) observed in tungsten CMP overpolish [12]. Reproduced by permission of the Electrochemical Society, Inc.

occurs when the rate of polish within a dished feature matches the rate of erosion of the surrounding oxide material. Dishing then reaches a maximum d_{max} irrespective of overpolish for given feature size as illustrated in Fig. 27. In the non-equilibrium case (more typical of polish conditions), the observed dishing is expressed as usual

$$d = d_{max}\left(1 - \frac{\kappa_{ox}}{\kappa_w} \cdot \frac{1}{1-\Phi}\right) \qquad (37)$$

where κ_{ox} and κ_w are the Preston coefficients for the rate of polish for the oxide and the tungsten materials, respectively, defining the selectivity of the polish. The dependency on density is illustrated in Fig. 28, and indicates that the d_{max} value depends strongly on the feature size.

The physical mechanisms (e.g., chemical interactions, slurry interactions, and pad mechanics) governing dishing and erosion remain unclear, and accurate modeling awaits further experimental studies both to empirically characterize and physically identify important effects.

2. TUNGSTEN CMP — CONTACT WEAR MODEL

Chekina et al. [8] have applied contact wear methods to the modeling of surface evolution in both oxide and dual material (tungsten–oxide) CMP to predict erosion and dishing or recess. The formulation uses calculation of

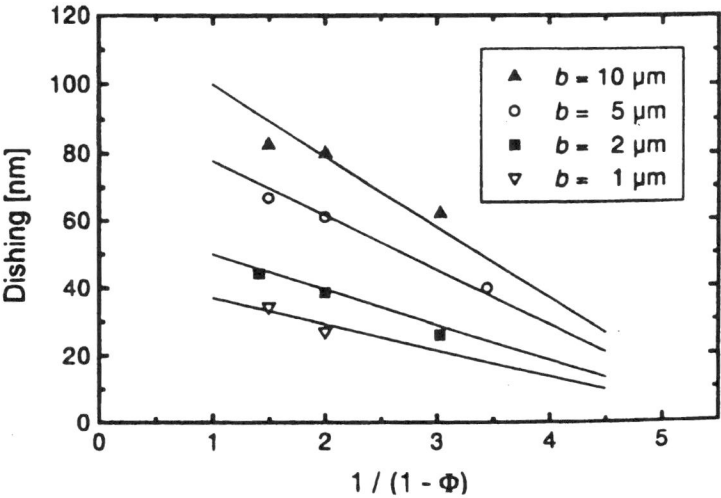

FIG. 28. Dishing observed in tungsten polishing as a function of metal density [12]. Reproduced by permission of the Electrochemical Society, Inc.

contact pressure and area of pad–wafer contact as a function of the loading conditions and surface geometry. Considering the wafer surface as a rigid body with a surface shape described by $f(x, y)$, one can examine the displacement of an elastic pad $w(x, y)$ and the contact pressure $p(x, y)$ at each point:

$$w(x, y) = \frac{1 - v^2}{\pi E} \int_\omega p(\xi, \eta) \frac{1}{\sqrt{(x - \xi)^2 + (y - \eta)^2}} d\xi \, d\eta \tag{38}$$

Because the local pressure distribution depends on the surface shape, and the evolution of the surface shape depends on the local pressure distribution, a discretization in space of the shape profile combined with a time-step approach can be used to predict the progress of the polish process.

An example application of the contact wear model to simulation of the polish of an embedded array (assuming no material in the trenches within the array) is shown in Fig. 29. One can also observe the evolution of the pressure distribution, where clear sharp pressure concentrations at the edges of the features can be seen. Such localized pressures work to rapidly round the corners of features undergoing polish.

The contact wear approach is also applicable to modeling dual material systems such as tungsten polish. The calculated shape evolution is shown in

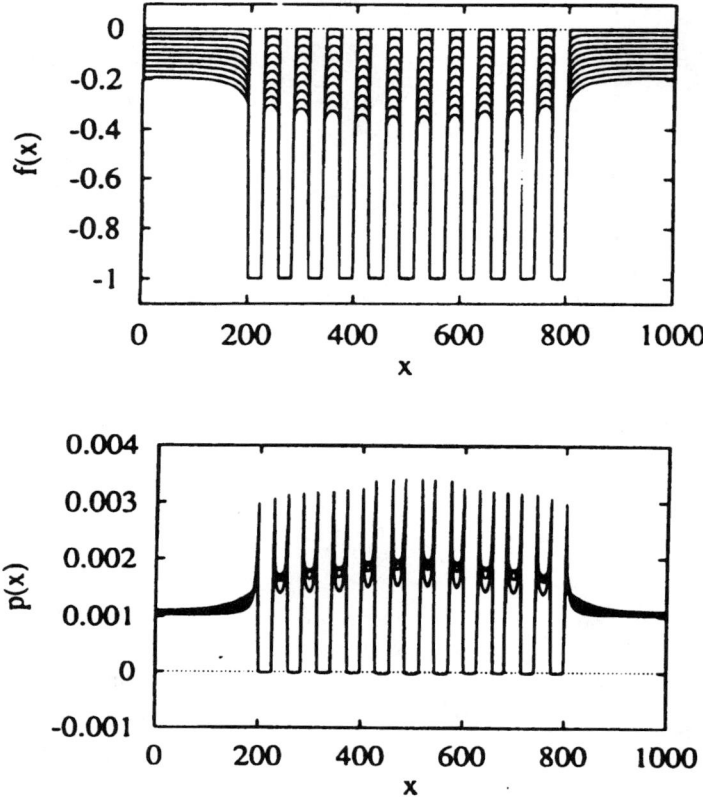

FIG. 29. Simulation of wear (and pressure) evolution in an array of trenches in a single-material system [8]. Reproduced by permission of the Electrochemical Society, Inc.

Fig. 30 for polishing of an initially flat surface composed of alternating regions of a fast polishing material and slow polishing materials (i.e., having different Preston coefficients). Again, the rounding of features as well as the evolution of both dishing and erosion can be observed. Indeed, a steady-state result where the depth of the dish below the eroded surface is constant can also be seen [8].

The contact wear calculations appear to hold great promise for detailed feature-level simulation. The die-level simulation approaches discussed earlier can be viewed as approximations to the contact-wear approach that focus on the polish or planarization results over large length scales rather than over feature-length scales.

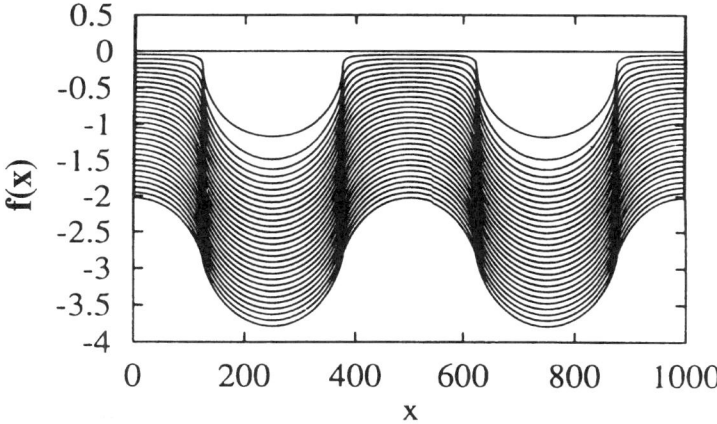

FIG. 30. Simulation of profile evolution in a dual-material (selective polish) system [8]. Reproduced by permission of the Electrochemical Society, Inc.

3. Copper CMP Modeling

Early investigation of copper polish pattern dependencies as illustrated in Fig. 25 was performed by Steigerwald *et al.* [52]. Key pattern dependencies examined included the effect of metal pattern density, as well as pitch or width of the lines and oxide spaces between copper lines. Steigerwald observed that copper dishing appeared to be most strongly a function of feature line width, becoming a substantial mechanism for copper line thickness loss for lines of the order of 20 μm or larger. Erosion, on the other hand, appeared to be primarily a function of pattern density (being relatively consistent for the 5 μm and larger features examined), where the erosion depth became substantial at metal densities of 40% and larger. Substantial similarities exist between the effects seen in tungsten polish and those in copper, and similar modeling approaches may prove applicable for study of feature-level dependencies.

To date, only limited work has been reported on models for dishing and erosion in copper structures; most work has been experimental in nature and report underlying pattern dependencies [13, 33, 51]. For example, Pan *et al.* [32] show experimental data based on electrical and profilometry measurements across a variety of test structures with different density, line width, and line spacing; an example plot is shown in Fig. 31 where both dishing and erosion is clearly present.

While no complete pattern-dependent model has been presented to date for copper CMP, one step forward is a model framework for pattern-

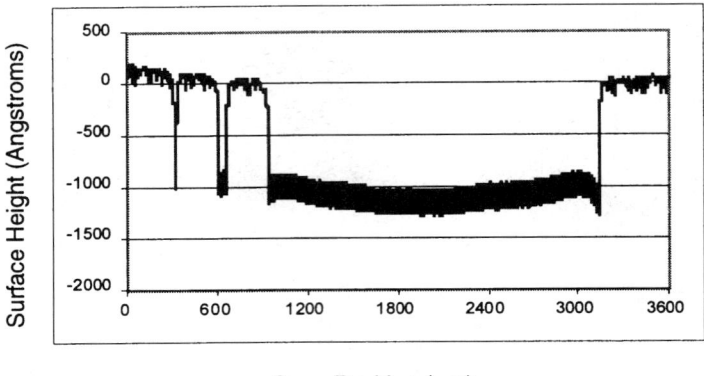

FIG. 31. Surface height after polish for a 5-μm pitch, 50% density test structure with overpolish of 30% [32].

dependent copper CMP simulation proposed in [5]. This work generalizes the integrated density and step-height model of Smith *et al.*, in combination with selectivity and dishing limits as in Elbel.

Reexamination of bulk and wafer-level modeling in the context of copper CMP is also underway. Due to the strong interaction between chemical and mechanical processes in copper polishing, consideration of the removal mechanisms as well as proper Preston-equation like modeling is being pursued [37, 65]. More work is needed to produce effective and efficient wafer-level, feature-level, and die-level models for copper CMP, particularly as the industry moves to copper interconnect systems.

VI. Summary and Status

The potential uses of accurate and efficient CMP models are numerous. To date, however, CMP models have been effectively used only in a subset of the possibilities.

It is now possible to model the wafer-level performance for most CMP processes. These models cover only some of the important tool or process design issues, such as relative velocities and pressure dependencies; additional work is needed to predict the results for other parameters such as slurry composition or particle size, temperature dependencies, pad properties, and other effects. Die-level modeling has been used effectively to identify potential problem areas (e.g., "high" and "low" regions on the die) prior to oxide polishing for a given pattern. Tools and methods for applying oxide and dielectric CMP models are starting to emerge that address the need for good dummy fill strategies or automatic fill generation. A further coupling to circuit design involves the integration of die-level film thickness predictions with circuit extraction and analysis tools. Further model development and validation is needed for predictive models of shallow trench isolation polishing. In the case of both copper and tungsten CMP, substantial model and simulation tool development, as well as characterization and model extraction methods, are needed before a predictive capability is available.

As understanding of the CMP process improves, one can expect a great deal of work in all aspects of CMP modeling and simulation. These improvements are likely to extend over many length scales spanning wafer-level polish and uniformity concerns to die-level prediction to microscopic feature, chemical, and mechanical interactions.

References

1. A. R. Baker, "The Origin of Edge Effect in Chemical Mechanical Planarization," *Proceedings of Electro. Chem. Soc. Meeting*, vol. 96, no. 22, p. 228, 1996.
2. D. Boning, J. Chung, D. Ouma, and R. Divecha, "Spatial Variation in Semiconductor Processes: Modeling for Control," in *Proc., Process Control, Diagnostics, and Modeling in Semiconductor Manufacturing II, Electrochem. Soc. Meeting*, Montreal, May 1997.
3. D. Boning, D. Ouma, and J. Chung, "Extraction of Planarization Length and Response Function in Chemical-Mechanical Polishing," *Materials Research Society 1998 Spring Meeting*, San Francisco, CA, May 1998.
4. D. Boning, A. Hurwitz, J. Moyne, W. Moyne, S. Shellman, T. Smith, J. Taylor, and R. Telfeyan, "Run by Run Control of Chemical Mechanical Polishing," *IEEE Trans. on Components, Packaging, and Manufacturing Technology — Part C*, vol. 19, no. 1, pp. 307–314, Oct. 1996.
5. D. S. Boning, B. Lee, C. Oji, D. Ouma, T. Park, T. Smith, and T. Tugbawa, "Pattern Dependent Modeling in Chemical Mechanical Polishing," *Materials Research Society 1999 Spring Meeting*, San Francisco, CA, April 1999.
6. D. Bramano and L. Racz "Numerical Flow-Visualization of Slurry in a Chemical Mechanical Planarization Process," *Proc. of CMP-MIC*, pp. 185–192, Santa Clara, CA, Feb. 1998.
7. P. A. Burke, "Semi-Empirical Modeling of SiO_2 Chemical-Mechanical Polishing Planarization," *Proc. VMIC Conf.*, pp. 379–384, Santa Clara, CA, June 1991.
8. O. G. Chekina, L. M. Keer, and H. Liang, "Wear-Contact Problems and Modeling of Chemical Mechanical Polishing," *J. Electrochem. Soc.*, vol. 145, no. 6, pp. 2100–2106, June 1998.
9. L. M. Cook, "Chemical Processes in Glass Polishing," *J. Non-Cryst. Solids*, vol. 120, pp. 152–171, 1990.

10. J. Coppeta, J. Rogers, L. Racz, C. Duska, D. Bramono, J. Lu, A. Philipossian, and F. Kaufman, "Pad Effects on Slurry Transport and Fluid Film Thickness Beneath a Wafer," *Proc. of CMP-MIC*, Santa Clara, CA, Feb. 1998.
11. J. Coppeta, R. Rogers, L. C. Racz, A. Philipossian, and F. B. Kaufman, "The Influence of CMP Process Parameters on Slurry Transport," *Proc. of CMP-MIC*, Santa Clara, CA, Feb. 1999.
12. N. Elbel, B. Neureither, B. Ebersberger, and P. Lahnor, "Tungsten Chemical Mechanical Polishing," *J. Electrochem. Soc.*, vol. 145, no. 5, pp. 1659–1164, May 1998.
13. M. Fayolle and F. Romagna, "Copper CMP Evaluation: Planarization Issues," *Microelectr. Eng.*, 37–38, pp. 135–141, 1997.
14. J. Grillaert, M. Meuris, N. Heylen, K. Devriendt, E. Vrancken, and M. Heyns, "Modelling Step Height Reduction and Local Removal Rates Based on Pad-Substrate Interactions," *Proc. of CMP-MIC*, pp. 79–86, Feb. 1998.
15. J. Grillaert, M. Meuris, E. Vrancken, K. Devriendt, W. Fyen, and M. Heyns, "Modelling the Influence of Pad Bending on the Planarization Performance During CMP," *Materials Research Society 1999 Spring Meeting*, San Francisco, CA, April 1999.
16. J. Grillaert, M. Meuris, N. Heylen, K. Devriendt, E. Vrancken, and M. Heyns, "Modelling Step Height Reduction and Local Removal Rates Based on Pad-Substrate Interactions," *Materials Research Society 1998 Spring Meeting*, San Francisco, CA, May 1998.
17. Y. Hayashide, M. Matsuura, M. Hirayama, T. Sasaki, S. Harada, and H. Kotani, "A Novel Optimization Method of Chemical Mechanical Polishing (CMP)," in *Proc. VMIC Conf.*, Santa Clara, CA, pp. 464–470, June 1995.
18. S. Hymes, K. Smekalin, T. Brown, H. Yeung, M. Joffe, M. Banet, T. Park, T. Tugbawa, D. Boning, J. Nguyen, T. West, and W. Sands, "Determination of the Planarization Distance for Copper CMP Process," *Materials Research Society 1999 Spring Meeting*, San Francisco., CA, April 1999.
19. F. B. Kaufman, D. B. Thomson, R. E. Broadie, M. A. Jaso, W. L. Guthriem, D. J. Pearson, and M. B. Small, "Chemical-Mechanical Polishing for Fabricating Patterned W Metal Features as chip interconnect," *J. Electrochem. Soc.*, vol. 138, no. 11 pp. 3460–3465, Nov 1991.
20. N. B. Kirk, J. V. Wood, "Glass Polishing," *Br. Ceramic Trans.*, vol. 93, no. 1, pp. 25–30, 1994.
21. H. van Kranenburg and P. H. Woerlee, "Influence of Overpolish Time on the Performance of W Damascene Technology," *J. Electrochem.. Soc.*, vol. 145, no. 4, pp. 1285–1291, April 1998.
22. C.-W. Liu, B.-T. Dai, W.-T. Tseng, and C.-F. Yeh, "Modeling of the Wear Mechanism during Chemical-Mechanical Polishing," *J. Electrochem.. Soc.*, vol. 143, no. 2, pp. 716–721, Feb. 1996.
23. A. Maury, D. Ouma, D. Boning, J. Chung, "A Modification to Preston's Equation and Impact on Pattern Density Effect Modeling," *Conf. for Advanced Metallization and Interconnect Systems for ULSI Applications*, San Diego, CA, Oct. 1997.
24. V. Mehrotra, S. Nassif, D. Boning, and J. Chung, "Modeling the Effects of Manufacturing Variation on High-Speed Microprocessor Interconnect Performance," *1998 International Electron Devices Meeting*, San Francisco, CA, Dec. 1998.
25. J. F. Morris, "Computed Fluid Flow and Stresses in Model CMP Films," *Materials Research Society 1999 Spring Meeting*, San Francisco, CA, April 1999.
26. B. Mullany, G. Byrne, and M. Power, "The Effect of Polishing Pad Conditioning on the Planarisation Capability of the Chemical-Mechanical Polishing Process," *Proc. of CMP-MIC*, pp. 147–150, Feb. 1999.

27. G. Nanz and L. Camilletti, "Modeling of Chemical-Mechanical Polishing: A Review," *IEEE Trans. on Semi. Manuf.*, vol. 8, no. 4, pp. 382–389, Nov. 1995.
28. D. Ouma, B. Stine, R. Divecha, D. Boning, J. Chung, G. Shinn, I. Ali, and J. Clark, "Wafer-Scale Modeling of Pattern Effect in Oxide Chemical Mechanical Polishing," *Proc. SPIE Microelectronic Man. Conf.*, Austin, TX, Oct. 1997.
29. D. Ouma, D. Boning, J. Chung, G. Shinn, L. Olsen, and J. Clark, "An Integrated Characterization and Modeling Methodology for CMP Dielectric Planarization," *International Interconnect Technology Conference*, San Francisco, CA, June 1998.
30. D. Ouma, "Modeling of Chemical Mechanical Polishing for Dielectric Planarization," Ph.D. Thesis, Elect. Eng. and Comp. Sci. Dept., MIT, Nov. 1998.
31. J. T. Pan, D. Ouma, P. Li, D. Boning, F. Redecker, J. Chung, and J. Whitby, "Planarization and Integration of Shallow Trench Isolation," *VLSI Multilevel Interconnect Conference*, Santa Clara, CA, June 1998.
32. J. T. Pan, P. Li, K. Wijekoon, S. Tsai, F. Redeker, T. Park, T. Tugbawa, and D. Boning, "Copper CMP and Process Control," *Proc. of CMP-MIC*, pp. 423–429, Feb. 1999.
33. T. Park, T. Tugbawa, J. Yoon, D. Boning, J. Chung, R. Muralidhar, S. Hymes, Y. Gotkis, 3S. Alamgir, R. Walesa, L. Shumway, G. Wu, F. Zhang, R. Kistler, and J. Hawkins, "Pattern and Process Dependencies in Copper Damascene Chemical Mechanical Polishing Processes," *VLSI Multilevel Interconnect Conference*, Santa Clara, CA, June 1998.
34. T. Park, T. Tugbawa, D. Boning, J. Chung, S. Hymes, R. Muralidhar, B. Wilks, K. Smekalin, and G. Bersuker, "Electrical Characterization of Copper Chemical Mechanical Polishing," *Proc. CMP-MIC*, Santa Clara, CA, Feb. 1999.
35. W. J. Patrick, W. L. Guthrie, C. L. Standley, and P. M. Schiable, "Application of Chemical Mechanical Polishing to the Fabrication of VLSI Circuit Interconnections," *J. Electrochem. Soc.*, vol. 138, no. 6, pp. 1778–1784, June 1991.
36. F. W. Preston, "The Theory and Design of Plate Polishing Machines," *J. Soc. Glass Technol.*, vol. 11, pp. 214–256, 1927.
37. S. Ramarajana and S.V. Babu, "Modified Preston Equation For Metal Polishing: Revisited," *Materials Research Society 1999 Spring Meeting*, San Francisco, CA, April 1999.
38. P. Renteln et al., *VLSI Multilevel Interconnect Conference*, pp. 57–63, Santa Clara, CA, June 1990.
39. P. Renteln and Ninh, "An Exploration of the Copper CMP Removal Rate Mechanism," *Materials Research Society 1999 Spring Meeting*, San Francisco, CA, April 1999.
40. S. R. Runnels, "Feature-Scale Fluid-Based Erosion Modeling for Chemical-Mechanical Polishing," *J. Electrochem.. Soc.*, vol. 141, no. 7, pp. 1900–1904, July 1994.
41. S. R. Runnels and L. M. Eyman, "Tribology Analysis of Chemical-Mechanical Polishing," *J. Electrochem.. Soc.*, vol. 141, no. 6, pp. 1698–1701, June 1994.
42. S. R. Runnels, M. Kim, J. Schleuter, C. Karlsrud, and M. Desai, "A Modeling Tool for Chemical Mechanical Polishing Design and Evaluation," *IEEE Trans. on Semi. Manuf.* vol. 11, no. 3, p. 501, Aug. 1998.
43. S. R. Runnels, I. Kim, and F. Miceli, "Implementing Large Area 3D Pattern Erosion Simulation," *Proc. CMP-MIC*, pp. 128–135, Santa Clara, CA, Feb. 1999.
44. F. G. Shi, B. Zhao, S.-Q. Wang, "A New Theory for CMP with Soft Pads," *Proceedings of International Interconnect Technology Conference*, San Francisco, CA, pp. 73–75, June 1998.
45. T. W. Shield and D. M. Bogy, "Some Axisymmetric Problems for Layered Elastic Media: Part I — Multiple Region Contact Solutions for Simply-Connected Indenters," *Transactions of the ASME*, vol. 56, pp. 798–806, Dec 1989.
46. T. W. Shield and D. B. Bogy, "Multiple Region Contact Solutions for a Flat Indenter on a Layered Half Space: Plain-Strain Case," *J. Appl. Mech.*, vol. 56, pp. 251–261, June 1989.

47. S. Sivaram, H. Bath, R. Leggett, A. Maury, K. Monning, and R. Tolles, "Planarizing Interlevel Dielectrics by Chemical-Mechanical Polishing," *Solid State Tech.*, pp. 87–91, May 1992.
48. T. H. Smith, S. J. Fang, D. S. Boning, G. B. Shinn, and J. A. Stefani, "A CMP Model Combining Density and Time Dependencies," *Chemical Mechanical Polish for ULSI Multilevel Interconnection Conference (CMP-MIC)*, pp. 97–104, Santa Clara, Feb. 1999.
49. T. Smith, S. Fang, J. Stefani, G. Shinn, D. Boning, and S. Butler, "Device Independent Process Control of Chemical-Mechanical Polishing," *Process Control, Diagnostics, and Modeling in Semiconductor Device Manufacturing III*, 195th Electrochemical Society Meeting, Seattle, WA, May 1999.
50. T. Smith, C. Oji, D. Boning, and J. Chung, "Bias and Variance in Multiple Response Surface Modeling," *Third International Workshop on Statistical Metrology*, Honolulu, HI, June 1998.
51. Z. Stavreva, D. Zeidler, M. Plotner, and K. Drescher, "Influence of process parameters on chemical-mechanical polishing of copper," *Microelectr. Eng.*, 37–38, pp. 143–149, 1997.
52. J. M. Steigerwald, R. Zirpoli, S. P. Muraka, and R. J. Gutmann, "Pattern Geometry Effects in the Chemical-Mechanical Polishing of Inlaid Copper," *J. Electrochem. Soc.*, vol. 142, no. 10, pp. 2841–2848, Oct. 1994.
53. B. Stine, D. Boning, J. Chung, L. Camilletti, F. Kruppa, E. Equi, W. Loh, S. Prasad, M. Muthukrishnan, D. Towery, M. Berman, and A. Kapoor, "The Physical and Electrical Effects of Metal Fill Pattern Practices for Oxide Chemical Mechanical Polishing Processes," *IEEE Trans. Electr. Dev.*, Feb 1998.
54. B. Stine, D. Ouma, R. Divecha, D. Boning, J. Chung, D. Hetherington, C. R. Harwood, O. S. Nakagawa, and S.-Y. Oh, "Rapid Characterization and Modeling of Pattern Dependent Variation in Chemical Mechanical Polishing," *IEEE Trans. on Semi. Manuf.*, vol. 11, no. 1 pp. 129–140, Feb 1998.
55. B. Stine, D. Ouma, R. Divecha, D. Boning, J. Chung, D. Hetherington, I. Ali, G. Shinn, J. Clark, O. S. Nakagawa, and S.-Y. Oh, "A Closed-Form Analytic Model for ILD Thickness Variation in CMP Processes," *Proc. CMP-MIC Conf.*, Santa Clara, CA, Feb 1997.
56. B. Stine, V. Mehrotra, D. Boning, J. Chung, and D. Ciplickas, "A Simulation Methodology for Assessing the Impact of Spatial/Pattern Dependent Interconnect Parameter Variation on Circuit Performance," *IEDM Tech, Digest*, pp. 133–136, 1997.
57. S. Sundararajan, D. G. Thakurta, D. W. Schwendeman, S. P. Murarka, and W. N. Gill, "Two-Dimensional Wafer-Scale Chemical Mechanical Planarization Models Based on Lubrication Theory and Mass Transport," *J. Electrochem.. Soc.*, vol. 146, no. 2, pp. 761–766, 1999.
58. H. Takahashi, K. Tokunaga, T. Kasuga, and T. Suzuki, "Modeling of Chemical Mechanical Polishing Process for Three-Dimensional Simulation," *1997 Symp. on VLSI Tech.*, pp. 25–26, Kyoto, Japan, 1997.
59. S. P. Timoshenko, J. N. Goodier, *Theory of Elasticity*, 3rd Int. ed., McGraw-Hill Book Company, New York, Chap. 12, 1970.
60. E. Tseng, C. Yi, and H. C. Chen, "A Mechanical Model for DRAM Dielectric Chemical-Mechanical Polishing Process," *CMP-MIC*, pp. 258–265, Santa Clara, CA, Feb. 1997.
61. W-T. Tseng, Y-L. Wang, "Re-examination of pressure and speed dependencies of removal rate during chemical mechanical polishing processes," *J. Electrochem. Soc.*, vol. 144, no. 2, p. L14, Feb 1997.
62. T.-L. Tung, "A Method for Die-Scale Simulation for CMP Planarization," *Proc. of SISPAD Conf.*, Cambridge, MA, Sept. 1997.
63. J. Warnock, "A Two-Dimensional Process Model for Chemimechanical Polish Planarization," *J. Electrochem. Soc.*, vol. 138, no. 8, pp. 2398–2402, Aug 1991.

64. D. Wang, J. Lee, K. Holland, T. Bibby, S. Beaudoin, and T. Cale, "Von Mises Stress in Chemical-Mechanical Polishing Processes," *J. Electrochem. Soc.*, vol. 144, no. 3, pp. 1121–1127, March 1997.
65. G. Wu and L. Cook, "Mechanism of Copper Damascene CMP," *Proc. of CMP-MIC*, pp. 150–158, Feb. 1999.
66. B. U. Yoon, Y. R. Park, I. K. Jeong, C. L. Song, and M. Y. Lee, "The Effects of Platen and Carrier Rotational Speeds on Within Wafer Non-Uniformity of CMP Removal Rate," *Proc. of CMP-MIC*, Santa Clara, CA, pp. 193–197, Feb. 1998.
67. C. Yu, P. C. Fazan, V. K. Mathews, and T. T. Doan, "Dishing Effects in a Chemical Mechanical Polishing Planarization Process for Advanced Trench Isolation," *Appl. Phys. Lett.*, vol. 61, no. 11, pp. 1344–1346, Sept. 1992.
68. Y. Zhang, P. Parikh, B. Stephenson, M. Bonsaver, J. Ling, and M. Li, "Effect of PETEOS Film Stress on Chemical Mechanical Planarization," *Proc. VMIC Conf.*, pp. 424–426, Santa Clara, CA, June 1996.

CHAPTER 5

Consumables I: Slurry

Shin Hwa Li, Bruce Tredinnick, and Mel Hoffman

EKC TECHNOLOGY
HAYWARD, CALIFORNIA

I. INTRODUCTION	139
II. ABRASIVES	140
1. *For Oxide Slurry*	140
2. *For Metal Slurry*	142
3. *Agglomeration*	143
4. *Milling*	145
III. SLURRY SOLUTION	146
1. *For Oxide Slurry*	146
2. *For Metal Slurry*	149
IV. COMPARISONS AMONG SLURRIES	150
1. *Oxide Slurries*	150
2. *Tungsten Slurries*	151
REFERENCES	153

I. Introduction

As chemical mechanical planarization (CMP) has become increasingly popular in integrated circuit manufacturing, more and more applications have been discovered. For example, initially, oxide CMP was used for the interlayer dielectric (ILD) process only; today, CMP can be used for premetal dielectrics (PMD) and shallow trench isolation (STI) as well. Metal CMP was once used for tungsten plug processing only, but it is now used for all damascene processes (tungsten, aluminum, and copper). Since different applications have different specifications, the questions that arise are whether an oxide process used for ILD is suitable, for example, for an STI application, or whether a plug process can satisfy a damascene application. Before we can answer these questions intelligently, it is necessary that the basics of slurry be understood, because CMP is a highly consumable-dominated process. The choice of slurry directly affects the materials chemically and the structures mechanically in each CMP application in integrated circuit manufacture.

Slurry consists of two major components, abrasives and solution. Depending on the material of the abrasives, the chemistry of the slurry, and the synergy between them, each kind of slurry behaves differently.

II. Abrasives

The most commonly used abrasives in CMP slurries are silica (SiO_2) and alumina (Al_2O_3). The physical appearance of these abrasives, before mixing with water, are white powders, as shown in Fig. 1. These materials must be ultrapure (>99.99%) and have nearly uniform particle shape and size to ensure a consistent and reasonable polish rate.

The distribution of the particle size must be as tight as possible, especially for hard materials such as alumina. A typical particle size distribution curve for alumina used in tungsten slurries is shown in Fig. 2a (data taken by the laser scattering method). However, from time to time the particle size distribution in a slurry may be out of control, so that a long tail will appear at the large particle end (Fig. 2b). These large particles may become a source of CMP scratches.

1. FOR OXIDE SLURRY

Commercially, only silica is used in oxide CMP slurries. The size of an abrasive particle is in the range of 500–2000 Å. There are two kinds of silica, fumed silica and "colloidal" silica. The fumed silica is formed by oxidizing

FIG. 1. Appearance of slurry powder before mixing with water.

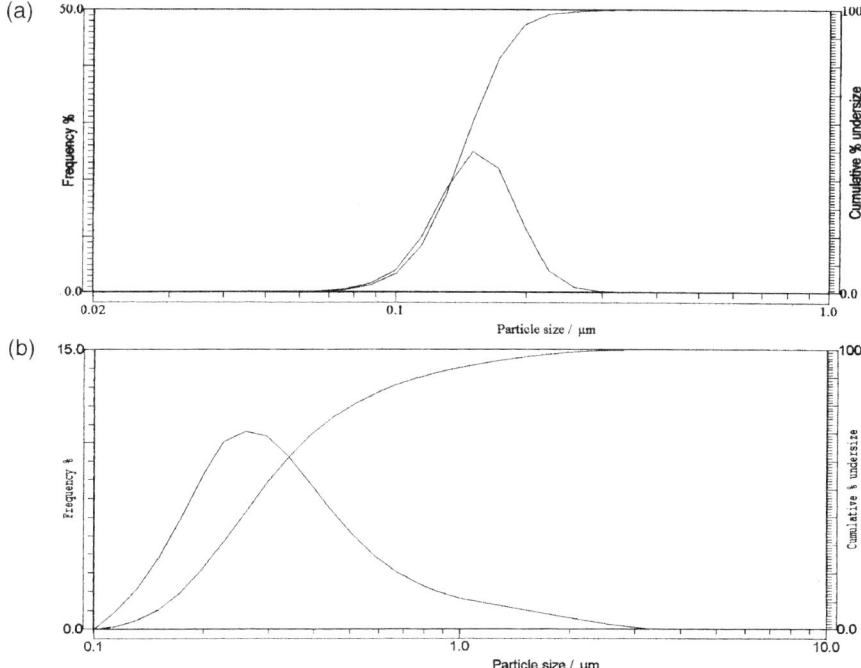

FIG. 2. Typical particle size distribution of silica in oxide CMP slurry: (a) normal and (b) abnormal with a long tail.

chlorosilane ($SiCl_4$) in a flame reaction at 1800°C [1]:

$$SiCl_4 + 2H_2 + O_2 \rightarrow SiO_2 + 4HCl$$

The silica produced from this reaction will have the particle size of approximately 300 Å (during the flame process). Then the particles cool and fuse into aggregates in length, which is irreversible. In addition, because the silica crystallizes in gas phase, the particle shape tends to be irregular, as shown in Fig. 3a.

In contrast to fumed silica, colloidal silica is synthesized in the liquid phase. The starting material is sodium silicate (Na_2SiO_3), or sometimes the sodium meta-silicate ($NaHSiO_3$), which are liquid glasses with approximately 70% SiO_2. By mixing liquid glass and water, colloidal silica crystals will be formed and suspended simultaneously. The material is then stabilized by passing it through an acid (H^+) charged ion exchange resin. As a result,

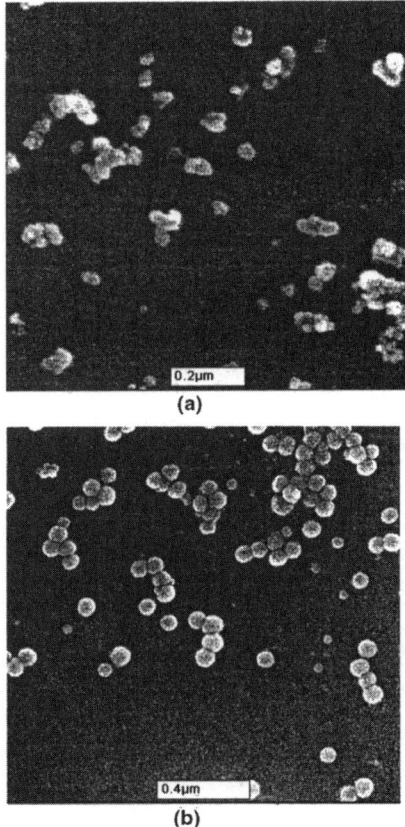

Fig. 3. Scanning electron microscopic (SEM) pictures of (a) fumed silica and (b) colloidal silica. (Courtesy of Millipore Corp., Bedford, MA.)

the shape of the colloidal silica particle is spherical. The shape of colloidal silica is shown in Fig. 3b and can be compared to the fumed silica of Fig. 3a. It is generally believed that the slurry using fumed silica is cheaper, but the risk of surface scratches or large particle growth is higher.

2. For Metal Slurry

Unlike oxide slurries, which use only one kind of abrasive (silica), metal slurries use various types and mixtures of abrasives. The particle size also

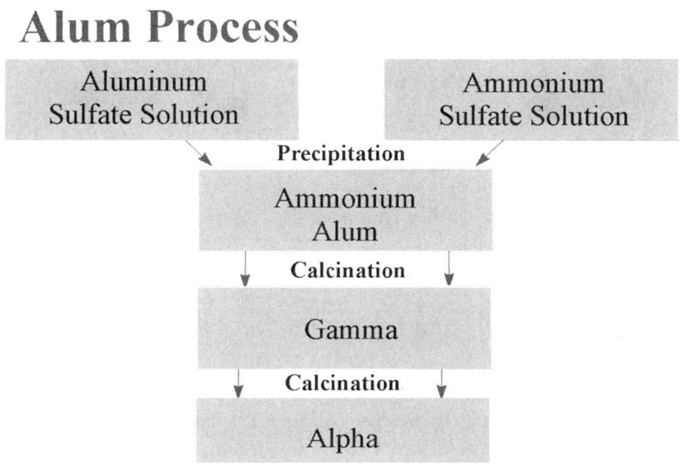

Alum Process

```
   Aluminum                    Ammonium
Sulfate Solution           Sulfate Solution
         ↘   Precipitation   ↙
              Ammonium
                Alum
              ↓ Calcination ↓
                Gamma
              ↓ Calcination ↓
                Alpha
```

FIG. 4. Synthesis of the alumina abrasives using the alum process. (Courtesy of Baikowski International Corp., Charlotte, NC.)

varies. This is because the oxide CMP process has been developed and used longer than the metal process, so that the oxide slurry formulation has been largely fixed. Another factor is that the solution (oxidizer) in metal slurry plays a more dominating role. As a result, to obtain the desired polishing performance, the selection of the oxidizer outweighs the selection of abrasives in importance. After a metal oxidizer has been selected, the type and the size of the abrasives must fit in.

Alumina is the most often used abrasive in tungsten slurries. The most common method for synthesis of alumina is called the alum process. As described in Fig. 4, the ammonium alum ($NH_4Al(SO_4)_2 12H_2O$) is precipitated by the mix of two solutions, namely, aluminum sulfate ($Al_2(SO_4)_3$) and ammonium sulfate (($NH_4)_2SO_4$). After precipitation, the precipitated alum is collected and calcinated. The ammonium alum will be converted into pure alumina. Depending on the completion of the phase change kinetically, the alumina can be either gamma or alpha phase. Alpha phase alumina is harder than the gamma phase [2, 3].

3. AGGLOMERATION

Despite the fact that these abrasives can be produced chemically in ultrapure form using the preceding synthesis methods, physically these

Fig. 5. Three different forms of silica particles [4]. (Copyright © 1979. Reprinted by permission of John Wiley & Sons.)

abrasive particles are not crystallized in the form of single particles. They can exist as discrete particles, aggregates, and agglomerates [4, 5], as shown in Fig. 5. A discrete particle is a single solid sphere or other geometric shape. An aggregate consists of multiple particles that have strong chemical or physical attachment to each other. An agglomerate is particles and/or aggregates that come together into close-packed clumps that are not sufficiently ionically charged to provide permanent suspension.

These large groups of particles are not desirable in CMP slurry. They will cause scratches and show an additional peak on the particle size distribution curve (see Fig. 6). To fix these problems, milling and/or slurry filters can be used. Milling is used at the point of slurry manufacture, and filtration is used at the point of use (filtration can also be used to fix the long tail problems mentioned in Fig. 2b), as discussed in the following.

Proper electrolyte balance can also prevent agglomeration. The tendency toward agglomeration is related to the pH value of the slurry [6], because

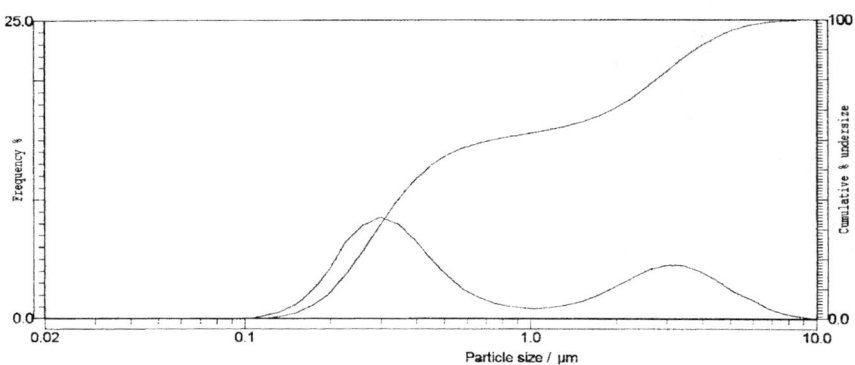

Fig. 6. Particle size distribution curve with an additional peak due to large agglomerates of particles.

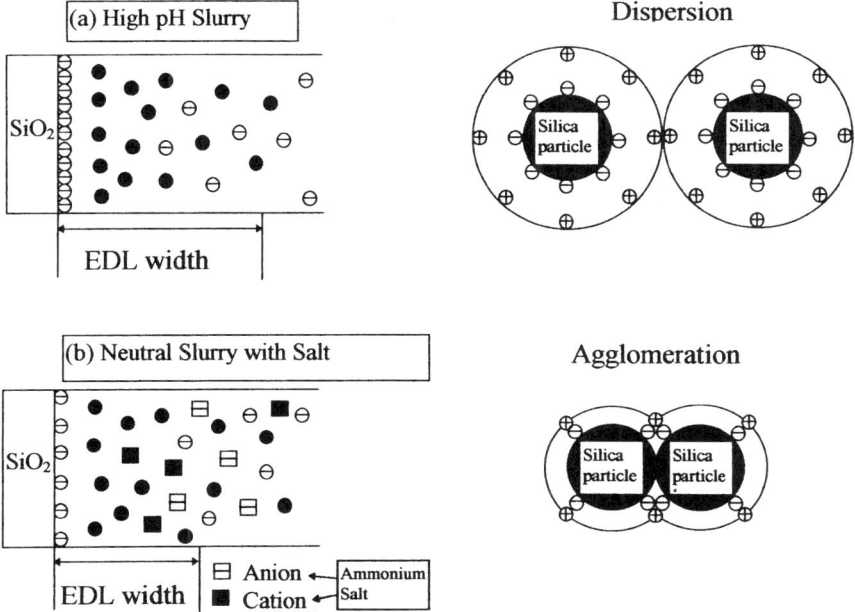

FIG. 7. Illustration of agglomeration mechanism of silica (after Hayashi et al., Ref. [6]).

the zeta potential of abrasives is a function of pH value in the solution. The concept of zeta potential is similar to the space charge potential at the interface between oxide and semiconductor. As the doping level (pH value) changes in semiconductor (slurry solution), the space charge potential (zeta potential) will change, as will the width of space charge (width of electric double layer, EDL). This change in zeta potential means that the abrasive particles may lose their ability to repel each other and will begin to agglomerate (see Fig. 7) for an illustration of the case of silica).

4. MILLING

To reduce or eliminate unwanted peaks or tails in a particle size distribution curve (see Figs. 2b and 6), milling can be applied to break apart the chains of the abrasive particles. Milling is a process for continuous dispersion and wet fine grinding of solid in a liquid phase. It is essential that besides breakage of chains no other chemical reactions occur under the influence of pressure, temperature, and milling action. A depiction of the milling equipment is shown in Fig. 8. Abrasives in liquid (usually in deionized water) can be pumped into the milling chamber. A rotor is rotating in the chamber, where agglomerates in the liquid can be broken apart.

Figures 9a and 9b show the SEM pictures of the alumina powders before and after milling. Note that the magnifications of Figs. 9a and 9b are different. In other words, milling has significantly reduced the agglomerate sizes from those in Fig. 9b to those in Fig. 9a.

Milling can be run with one or multiple passes, depending on the specifications of the particle sizes. More passes further tighten the particle size distribution.

III. Slurry Solution

The slurry solution plays a different role between oxide (as a hydrolizer) and metal (as an oxidizer) slurries. It is more complex in metal than in oxide, because traditionally, oxide slurry is used only for polishing oxide (for ILD, for example), whereas the metal slurry (for tungsten, for example) is used to polish tungsten, titanium nitride, titanium, and oxide. Accordingly, the choice of a metal slurry oxidizer must first satisfy the requirement of the selectivities between each different deposited film. Selection of solution for oxide slurry does not have such constraints.

1. For Oxide Slurry

For oxide CMP, the purpose of the solution is two fold. First, water weakens the Si—O bond in a silicon dioxide film and softens the surface as it becomes hydrated with Si—OH bonds [6, 7]. Figure 10 shows the reaction mechanism. Second, the solution is to provide a basic environment (pH > 10), which accelerates the hydration rate. An environment with high pH values will allow the polishing-induced reaction to be further accelerated because the surface $Si(OH)_x$ species will be partially dissolved into water. In the meantime, the zeta potential of silica increases with increasing pH values. At high zeta potentials silica particles will repel each other, whereby a better-suspended slurry is formed.

Commercial oxide slurries are available with different chemistries. The most common ones are the NaOH-based, the KOH-based, and the NH_4OH-based slurries. The NaOH-based slurry is the best medium for OH^- groups because the NaOH solvent is cheap and stable. However,

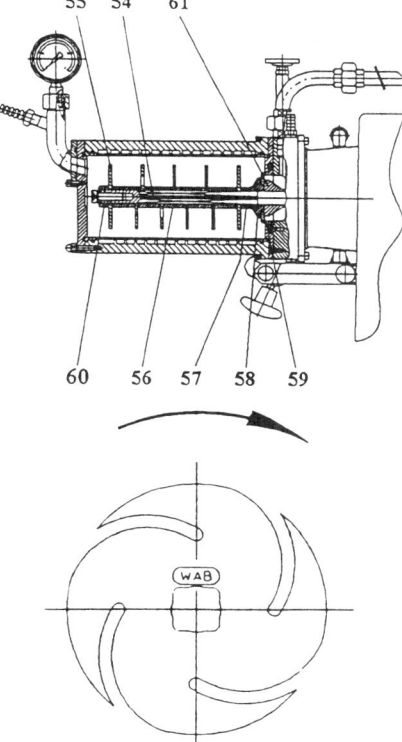

Fig. 8. Schematics of milling. The numbered parts are locking cap (60), agitator discs (55), spacers (56 and 57), gap separator (58), rotor ring (59), and standard spacer (61). (Courtesy of CB Mills, Division of Chicago Boiler Co., Gurnee, IL.)

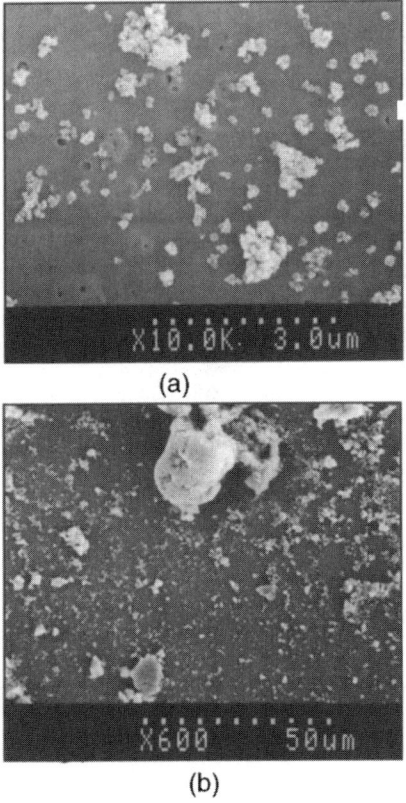

Fig. 9. SEM pictures of alumina abrasive powder (a) after and (b) before milling. (Courtesy of Baikowski International Corp., Charlotte, NC.)

because the cation Na^+ is a notorious mobile ion, which will ruin the transistors in integrated circuits if the ions migrate to the gate oxide, fab engineers are often opposed to allowing Na in any process chemicals. This is the reason that NH_4OH-based slurry has come into the market. While the NH_4OH-based slurry is mobile ion-free, it has the disadvantages of unstable colloidal suspension and cost. The KOH-based slurry is the most popular slurry, because the slurry is stable and the K^+ ion can be easily gettered by the premetal oxide such as the borophosphosilicate glass (BPSG).

$$\text{Si-O-Si} + \text{H}_2\text{O} \longleftrightarrow 2\text{SiOH}$$

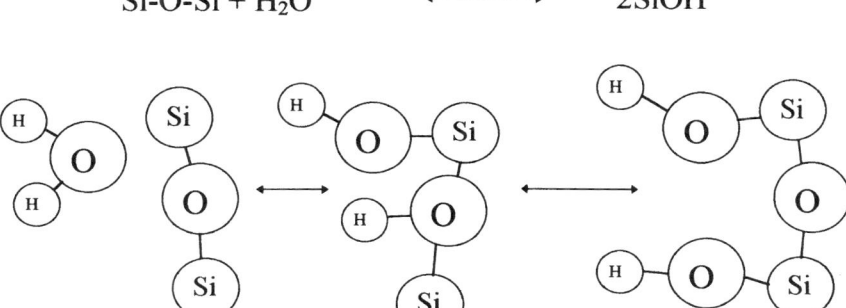

FIG. 10. Schematics of reaction mechanism between H_2O and an SiO_2 surface in glass polishing (after Hayashi et al., Ref. [6]).

2. For Metal Slurry

As mentioned, the solution (oxidizer) plays a difficult and dominating role in the metal slurry. On one hand, the selectivity between different metal layers should be close to unity. On the other hand, the selectivity between metal and oxide should be as large as possible. Further, on one hand, the polish rate for metal needs to be higher than 3000 Å/min. On the other hand, dishing or plug recession must be minimized at the metal areas on the patterned wafers.

Currently, there are four different commercial oxidizers for tungsten slurries; namely, $Fe(NO_3)_3$-based, H_2O_2-based, KIO_3-based, and H_5IO_6-based slurries. The purpose of each oxidizer is the same: converting tungsten into tungsten oxide (WO_3), which is softer than tungsten, so that the surface WO_3 can be polished mechanically. However, each oxidizer has different oxidizing mechanisms. Briefly speaking, the $Fe(NO_3)_3$-based, the KIO_3-based, and the H_5IO_6-based oxidizer will dissociate into cations and anions (with different NO_x and IO_y complexes), and the anions will oxidize tungsten. In contrast, an H_2O_2-based slurry decomposes into H_2O and dissolved O_2, and the O_2 directly reacts with tungsten. In general, the oxidizing ability is in the order of

$$H_2O_2 > Fe(NO_3)_3 > H_5IO_6 > KIO_3$$

assuming that the concentrations of each oxidizer are kept the same.

IV. Comparisons Among Slurries

Several commercially available oxide and tungsten slurries are compared in this section.

1. OXIDE SLURRIES

Fumed silica and colloidal slurry with and without filtration have been evaluated. All slurries are KOH-based. All slurries have pH values greater than 10. A same polisher was used, and a same pad was used. However, to separate the effect from post-CMP cleaning, neither scrubbing nor chemistry was applied. The oxide wafers were only water-rinsed after polishing. The results are shown in Fig. 11. The slurry with fumed silica left more particles compared to that with colloidal silica.

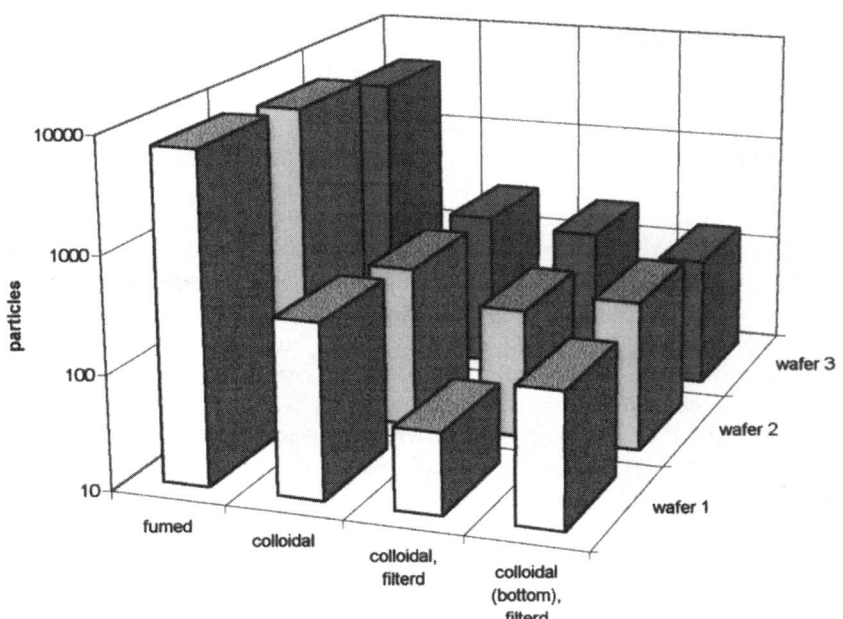

FIG. 11. Evaluation of particle post-CMP performance for commercial oxide slurries with fumed and colloidal silica with and without point-of-use filtration. The filter size is 0.3 μm. The "bottom" denotes that the slurry used is from the bottom of the drum.

Filtration can reduce the particles adhered to the wafer, but only slightly. Filtration will, however, reduce scratch defects. Slurry particles can be more effectively cleaned if more elaborate post-CMP clean chemistry is used [8].

2. TUNGSTEN SLURRY

Three commercial slurries, the H_2O_2-based, the KIO_3-based, and the $Fe(NO_3)_3$-based slurries, are evaluated. Their abrasives are SiO_2, Al_2O_3, and Al_2O_3, respectively [9]. The experiments were conducted using an orbiting polisher and a soft pad (Suba 500). The post CMP clean used was scrubbing with NH_4OH chemistry. For the H_2O_2- and the KIO_3-based slurries, one-step polishing was used. However, for the $Fe(NO_3)_3$-based slurry, an oxide buffing step followed. The results of the comparison are summarized in the following [10].

Blank/test structure wafers

Slurry	Particles	Surface roughness	Plug recess	Fe content
H_2O_2 based	42	First	200–800 Å	141E10/cm^2
KIO_3 based	4650	Third	0–100 Å	81E10/cm^2
$Fe(NO_3)_3$ based	3300	Second	100–400 Å	10101E10/cm^2

Production wafers

Slurry	Electrical	Ease for photoalignment	Yield
H_2O_2 based	Normal	Better	Slightly better
$Fe(NO_3)_3$ based	Normal	Reasonable	Reasonable

The H_2O_2-based slurry performed the best in this evaluation. However, it has problems with plug recess and field oxide erosion.

The pot life has also been compared among the H_2O_2-based, the $Fe(NO_3)_3$-based, and the H_5IO_6-based slurries after mixing with the abrasives (suspended in water). Figure 12 shows the polish rate change as a function of time after the mixing. We see that the $Fe(NO_3)_3$-based slurry can last the longest (more than 6 months) without significantly losing polish rate. The H_5IO_6-based slurry slightly loses its polish rate. However, the H_2O_2-based slurry has lost 50% polish rate in 100 h. This indicates the poor stability of the slurry. The H_2O_2 decomposes and the dissolved O_2 will evaporate in a short period of time. Therefore, to run the H_2O_2-based slurry, *in situ* mix and/or autotitration are required.

Plug recess and oxide erosion between the H_2O_2 (silica as abrasives) and the H_5IO_6 (mixture of different abrasives) based slurries are also evaluated on patterned wafers. The same polisher (rotating), recipe, and pad (IC1000/

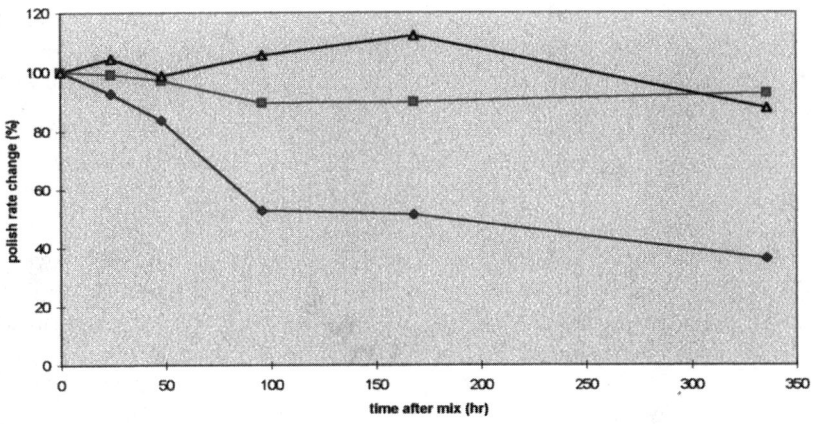

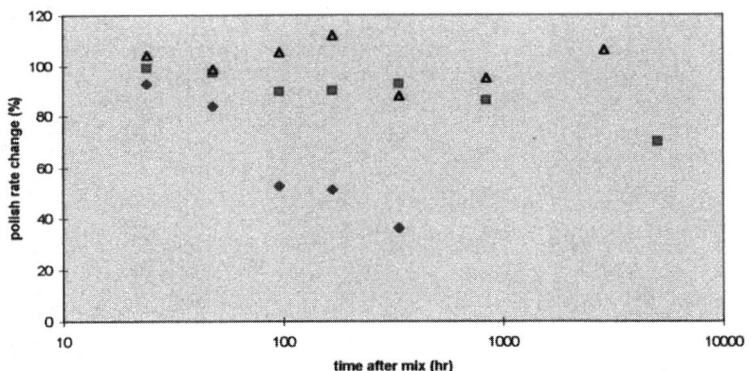

FIG. 12. Comparison of pot life for commercial available tungsten slurries with H_2O_2 (diamonds), $Fe(NO_3)_3$ (triangles), and H_5IO_6 (squares) oxidizers: (a) linear and (b) log scales.

SubaIV) are used. The results are

Patterned wafers

Slurry	Plug recess	Oxide erosion
H_2O_2 based	164–227 Å	289–713 Å
H_5IO_6 based	36–51 Å	91–551 Å

The plug size is 0.5 μm in diameter, and the oxide erosion measurement is taken on two kinds of plug arrays; 3 or 1 μm in pitch. Both arrays are 3000 μm^2 in size.

To summarize, for tungsten slurry, there is not yet an optimal formulation. Each kind of slurry has its strengths and weaknesses in terms of performance and ease of use. Obviously, more development work needs to be done in this area.

References

1. Tutorial, Planarization Processes for ULSI Fabrication to the Year 2002 (SEMICON WEST, July 16, 1997) p. 51.
2. J. C. Kotz and K. F. Purcell, *Chemistry and Chemical Reativity* (Saunders College Publishing, New York, 1987) pp. 98 and 790.
3. Marketing materials, Baikowski International Corporation, Charlotte, NC, 1996.
4. R. Iler, *The Chemistry of Silica* (John Wiley & Sons, New York, 1979) Chapter 1.
5. Marketing materials, Millipore Corporation, Bedford, MA, 1998.
6. Tutorial, *Planarization Processes for ULSI Fabrication to the Year 2002* (SEMICON WEST, July 16, 1997) pp. 52–53; Y. Hayashi et al., *Jpn. J. Appl. Phys.* **2B**, (1995), p. 1037.
7. J. M. Steigerwald, S. P. Murarka, and R. J. Gutmann, *Chemical Mechanical Planarization of Microelectronic Materials* (John Wiley & Sons, New York, 1997), p. 139.
8. More detailed information is found in Chapter 7 (post-CMP clean) of this book.
9. S. H. Li, H. Banvillet, C. Augagneur, B. Miller, M.-P. Nabot-Henaff, and K. Wooldridge, *Proceedings of CMP-MIC Conference*, Santa Clara, CA, 1998, p. 165.
10. More detailed data are also used in Tables 2, 3, and 4 of Chapter 8 (metrology chapter) of this book.

CHAPTER 6

CMP Consumables II: Pad

Lee M. Cook

RODEL INC.
NEWARK, DELAWARE

I. INTRODUCTION	155
II. CLASSES OF PADS AND THEIR MANUFACTURE	156
1. *Classes of Pads*	156
2. *Primary Manufacturing Processes*	156
III. STRUCTURE, PROPERTIES, AND THEIR RELATIONSHIP TO THE POLISHING PROCESS	162
1. *Local-Level Models for the Polishing Process*	162
2. *Die-Scale Models*	166
3. *Wafer-Scale Models*	167
4. *Impact of Structure and Properties on the Polishing Process*	169
IV. APPLICATION TO SEMICONDUCTOR PROCESSING	170
1. *Dielectric CMP*	170
2. *Metal CMP*	177
REFERENCES	180

I. Introduction

Polishing in its simplest sense is the controlled abrasion of a surface to produce a flat specular-defect-free surface. This is generally effected by rubbing the surface to be polished with a sheet of material (a pad) in conjunction with a water-based solution containing very fine particles, which are generally inorganic oxides (a slurry, covered in Chapter 5 of this work). Polishing is one of the oldest manufacturing processes still in use. The earliest documented polishing occurred more than 3000 years ago in the Phoenician and Egyptian cultures. These original processes used natural materials such as pitch, cork, or leather for pads, and dispersions of fine iron or tin oxides for slurries. The art was practiced in essentially the same fashion up until the 20th century, with only minor refinements, such as the use of cloth, both woven fabric and felt, as polishing pads [1].

Polishing pads as we now know them have existed only since the 1930s. The progenitors of today's polyurethane pads were developed in Germany prior to World War II as synthetic leather substitutes for lens finishing applications. Until the advent of chemical mechanical polishing (CMP) as a potential semiconductor process technology, virtually all commercial polishing pads were designed as analogs to earlier materials.

Successful application of CMP to device processing requires a degree of control and reproducibility not achieved in earlier polishing applications. Consequently, an enormous amount of research into the polishing process has been conducted over the last two decades, the objective of which has been to replace polishing "art" with the same degree of scientific understanding that is typical of other equally complex semiconductor processes [e.g., chemical vapor deposition (CVD), etch, etc.]. While this is by no means complete, significant progress has been made, as evidenced by the various sections in this volume.

II. Classes of Pads and Their Manufacture

1. Classes of Pads

Polishing pads used in semiconductor applications can be grouped into four main classes based on their structural characteristics. These are

Class I. Felts and polymer impregnated felts
Class II. Microporous synthetic leathers
Class III. Filled polymer films
Class IV. Unfilled textured polymer films

A tabular summary of these pad types, with subcategories, common trade name varieties, applications, and property ranges is given in Table I.

2. Primary Manufacturing Processes

As might be expected, each class of pads in Table I is associated with specific manufacturing techniques. General process flow diagrams for each of the four pad classes are outlined in Figs. 1–4. Manufacturing processes can be categorized as web or roll processes (classes I and II), batch processes (class III), and single-unit or net shape processes (classes III and IV). Each

TABLE I

SUMMARY OF POLISHING PAD CLASSES

	Class I	Class II	Class III	Class IV
Major structural characteristic	Felted fiber with polymer binder	High-porosity film on substrate	Solid urethane sheet with filler (voids, SiO_2, CeO_2, etc.)	Solid polymer sheet with surface texture
Subcategories	Spun bond nonwovens	Free-standing thin films; felt substrates	Foams; oxide filled	
Hardness	Medium to high	Low	High	Very high
Compressibility	Medium to high	High	Low	Very low
Slurry carrying capacity	Low to high	Very high	Low	Minimal
Bulk microstructure	Continuous channels between fibers	Complex foam to vertically oriented channels	Closed cell to open cell foam	None
Polymer types employed	Urethanes; polyolefins (fiber phase)	Urethanes; polyolefins (fiber phase)	Urethanes	Various
Representative commercial trade names	Pellon™, Suba™	Politex™	IC1000	OXP3000
Application(s)	Si stock polish; tungsten damascene CMP	Si final polish; metal damascene CMP; post-CMP buff	Si stock; ILD CMP; metal dual damascene	ILD CMP; shallow trench isolation; metal dual damascene

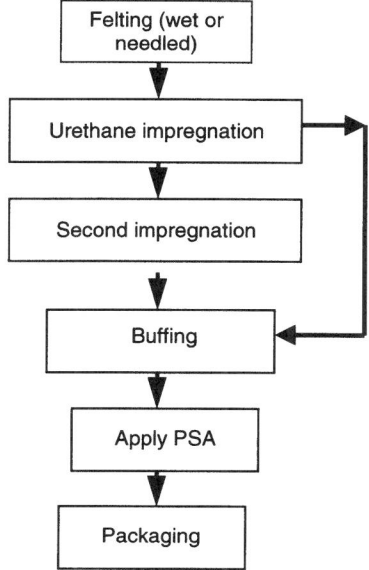

FIG. 1. Manufacturing process flow for class I pads.

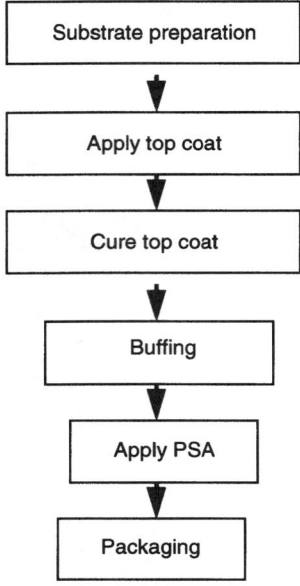

FIG. 2. Manufacturing process flow for class II pads.

type of manufacturing process has its own advantages and disadvantages. However, all current commercial manufacturing processes have similar ranges of dimensional and property control. The choice of manufacturing processes employed is largely determined by cost and the range of physical properties that are required. Because of their relative importance in CMP processing, detailed analysis of the effects of manufacturing process variables on pad performance has largely been restricted to class III pads. This is covered in more detail in the following.

All commercially available polishing pads are relatively complex composite materials, as evidenced in photomicrographic cross sections of the major pad types illustrated in Figs. 5–7. The signature structural characteristics of each class of pads (Table I) are readily apparent. The impact of manufacturing process on microstructure is sufficiently strong that the manufacturing process used to produce an unknown pad sample can be readily determined from microscopic examination.

The composite microstructure (and, therefore, its composite properties and polishing performance) is largely determined by the mode of manufacture. A summary of the major process variables associated with pad manufacturing and the anticipated effects on physical properties is given in Table II.

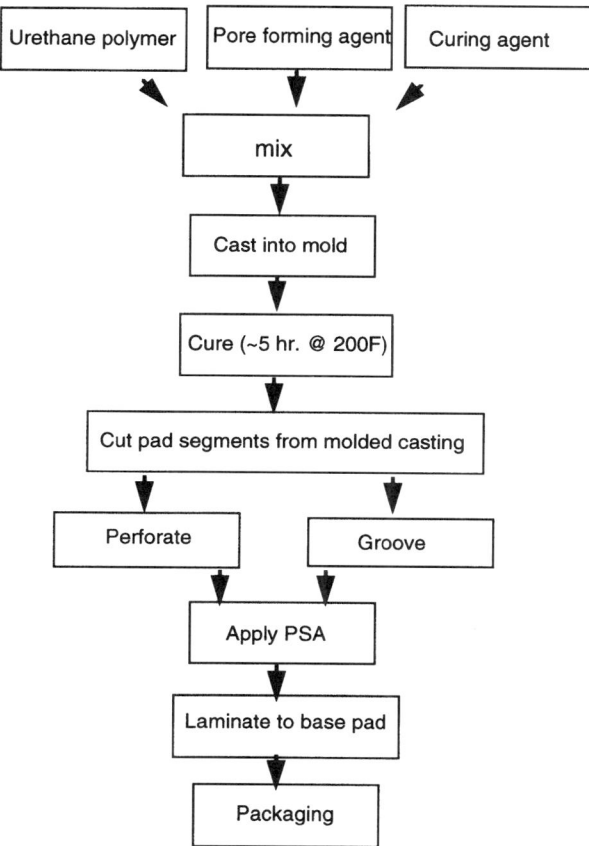

FIG. 3. Manufacturing process flow for class III pads (from Ref. [2]).

A number of common material properties emerge from Table II:

- Surface roughness and texture
- Liquid permeability
- Hardness and compressibility
- Elastic modulus
- Viscoelasticity

It is also apparent that the composite pad properties of interest are affected by a large number of process and structural variables. Interaction effects make analysis of property variability particularly difficult. As a consequence,

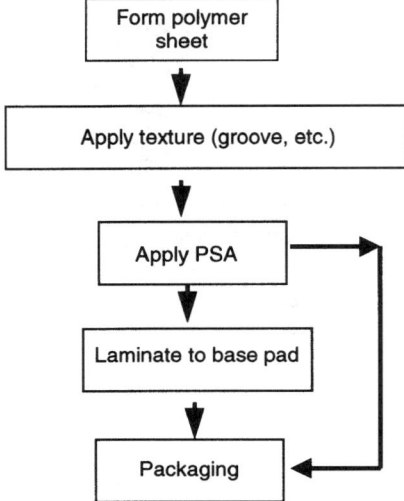

Fig. 4. Manufacturing process flow for class IV pads (from Ref. [3]).

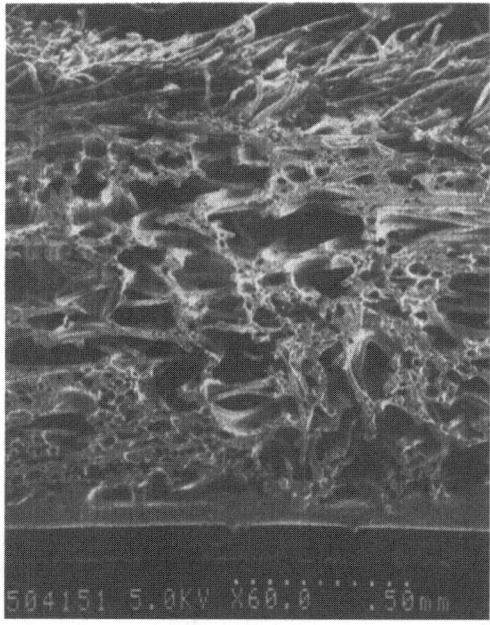

Fig. 5. Photomicrograph of cross section of Suba-4 (class I).

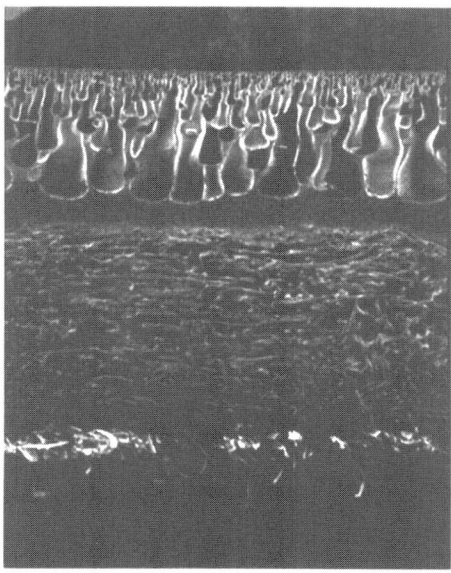

Fig. 6. Photomicrograph of cross section of Politex (class II).

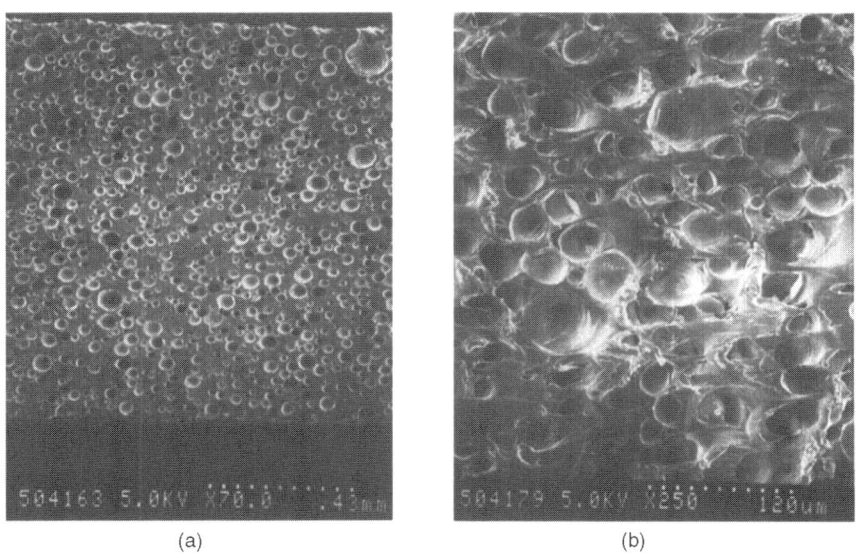

(a)   (b)

Fig. 7. Photomicrograph of cross section of IC1000 (class III): (a) side view and (b) top view.

TABLE II

MANUFACTURING PROCESS VARIABLES AND THEIR EFFECTS ON PAD PROPERTIES

Process	Variables	Effects
Felting (web)	Fiber size	Composite modulus
	Packing density	Volume fraction for impregnation polymer; composite modulus; hardness; liquid permeability
	Random vs nonrandom fiber orientation	Anisotropy of physical properties
Impregnation (web)	Polymer type	Composite modulus; hardness; viscoelasticity
	Volume fraction	Composite modulus; hardness; liquid permeability
	Microstructure	Composite modulus; hardness; liquid permeability
Coating (web)	Pore dimension	Composite modulus; permeability
	Coating height	Composite modulus; permeability
	Random vs nonrandom porosity	Anisotropy of physical properties
Buffing (web, batch)	Removal depth	Pore dimension; coating height
	Abrasive size	Surface roughness
Casting (batch, net)	Polymer type	Hardness; modulus; viscoelasticity
	Filler size	Composite modulus; liquid permeability; surface roughness
	Filler volume fraction	Composite modulus; liquid permeability; abrasivity
	Thermal history	Modulus; hardness; viscoelasticity
Texturing (web, batch, net)	Texture dimensions	Liquid permeability; pad hydrodynamics
Laminating (web, batch, net)	Member thickness	Composite modulus
	Property changes due to processing	Modulus; surface roughness; liquid permeability

the historical trend is toward production of less complex structures (e.g., class III and class IV pads) via less complex manufacturing processes to reduce the number of variables affecting properties. The significance of these pad material properties is discussed in later portions of this chapter within the context of mechanisms for the polishing process.

III. Structure, Properties, and Their Relationship to the Polishing Process

1. LOCAL-LEVEL MODELS FOR THE POLISHING PROCESS

The most generally satisfactory local-level model for CMP is an asperity contact model such as that described by Yu et al. [4]. In this model, applied load from the polisher is transmitted to the substrate by direct contact with pad asperities and by the liquid film between the contacting asperities. At low velocities, asperity contact dominates. At higher velocities, hydrodynamic pressure and liquid film thickness increase to become dominant [5]. Material removal from the substrate surface is effected by the passage of slurry particles under load [6]. The load on the slurry particles is imparted by contact with asperities on the pad surface or by impingement of particles in motion due to hydrodynamic turbulence in the liquid film. This is graphically illustrated in Fig. 8, top.

The model of Yu *et al.* [4] is of particular interest as several important aspects have subsequently been experimentally confirmed. The model assumes spherical asperities whose height z and radius β have a Gaussian

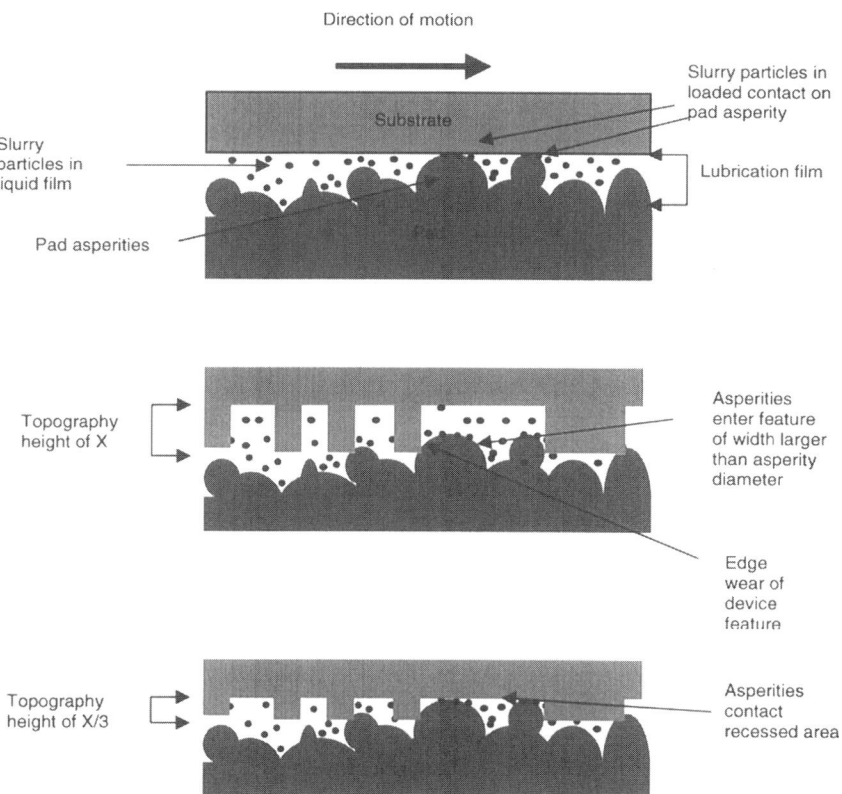

FIG. 8. Schematic diagram of (top) asperity contact during CMP, (middle) die-scale asperity contact for patterned wafer with topography height x; and (bottom) die-scale asperity contact for patterned wafer with topography height $x/3$.

distribution, and that removal during polishing is due to asperity contact rather than hydrodynamic effects. This was found to be in good agreement with measured profilometry data for the class III pad of interest (IC1000). From fits to experimental data, the mean and standard distribution of asperity density, radius, and height were calculated. Contact behavior of a single asperity was calculated for Hertzian contact:

$$\text{Contact area } a = \pi\beta(z - d) \quad (1)$$

where $z - d$ is the difference between the asperity height z and the lubrication film thickness d,

$$\text{Load } l = 4/3 E' \beta^{1/2}(z - d)^{3/2} \quad (2)$$

where E' is the composite Young's modulus of the asperity.

Total contact area A_{con} and load L over the overall pad area A are then

$$A_{\text{con}} = nA \int_d^\infty \int_0^\infty a \Phi_\beta \Phi_z \, d\beta \, dz \quad (3)$$

$$L = nA \int_d^\infty \int_0^\infty l \Phi_\beta \Phi_z \, d\beta \, dz \quad (4)$$

where n is the asperity density per unit area. The contact pressure $P_{\text{con}} = L/A_{\text{con}}$ is considerably greater than the nominal applied pressure $P = L/A$, as $A_{\text{con}} \ll A$. For varying P, P_{con} remains nearly constant while A_{con} increases, implying that the pressure effect on polishing rate is primarily due to changes in the area of contact rather than by changes in local contact pressure.

The relative effect of asperity contact vs hydrodynamic effects on rate has been examined experimentally [7–10]. Levert *et al.* [7] used capacitance probe techniques to measure slurry film thickness during CMP. For IC1000, a film thickness of $\sim 25 \, \mu\text{m}$ was observed for contact velocities typical of CMP. This film thickness is comparable to the peak to valley roughness of IC1000 pad surfaces measured by Yu *et al.* [4] and others [10]. Liquid film thickness increased with increasing linear velocity. The film thickness and kinetics were closely similar to the smooth plane surfaces (acrylic sheet on steel) used for reference purposes. Experimental data gave a good fit to a traditional hydrodynamic thrust bearing model:

$$h = Hl(W\eta ub/w)^{1/2} \quad (5)$$

where h is the minimum fluid film thickness, H is the dimensionless fluid film thickness, l is sample length, W is the dimensionless load, η is the fluid viscosity, u is the contact velocity, b the sample width, and w the load.

This study was later extended to include the effects of contact velocity on coefficient of friction and polishing rate for the system IC1000–slurry–SiO_2 [8]. Significant observations were that below a minimum applied pressure, hydrodynamic separation of the pad and wafer was observed, with extremely low coefficients of friction and negligible removal. In the pressure regime normally employed for CMP, significantly thinner lubrication layers and high coefficients of friction were observed. For worn (glazed) IC1000 pads a critical velocity was observed, above which the coefficient of friction fell rapidly and lubrication film thickness increased appreciably. This was attributed to the onset of hydrodynamic pad–substrate separation due to the low asperity concentration. No such effect was observed for conditioned pads.

Bhushan et al. [9] independently confirmed the lack of polishing activity due to hydrodynamic lubrication. The depth of the wafer carrier in CMP was adjusted so that samples either projected above the surface of the carrier (the normal case), were essentially coplanar with the carrier, or were recessed below the plane of the carrier. This produced wafer–pad lubrication film thicknesses of controlled dimension. For the case of a wafer recess of 75 μm ($\sim 3\times$ the lubrication film thickness reported in Ref. [7]), removal rate was negligibly small.

Coppeta et al. [10] made slurry film measurements during using laser-induced fluorescence. By addition of a fluorescent dye to the polishing slurry film thickness was experimentally from the fluorescence intensity of the lubrication film as measured through a transparent substrate. Film thickness measurements were in good agreement with those of Levert et al. [7, 8]. This technique can also be used to study slurry transport across the wafer surface, diameter variation in lubrication film thickness, and slurry mixing effects [11].

The dominant role of asperity contact is also apparent from analysis of the texture of polished surfaces. Figure 9 illustrates a typical post CMP surface as examined via atomic force microscopy (AFM). Surface texture is composed of innumerable randomly oriented nanogrooves of a width and depth consistent with traveling Hertzian loaded contact [12]; particle bombardment during turbulent liquid flow produces profoundly different texture. All classes of semiconductor materials examined show similar textures, indicating the general nature of the process. From the data to date, it appears that asperity contact is the dominant wear mechanism in CMP.

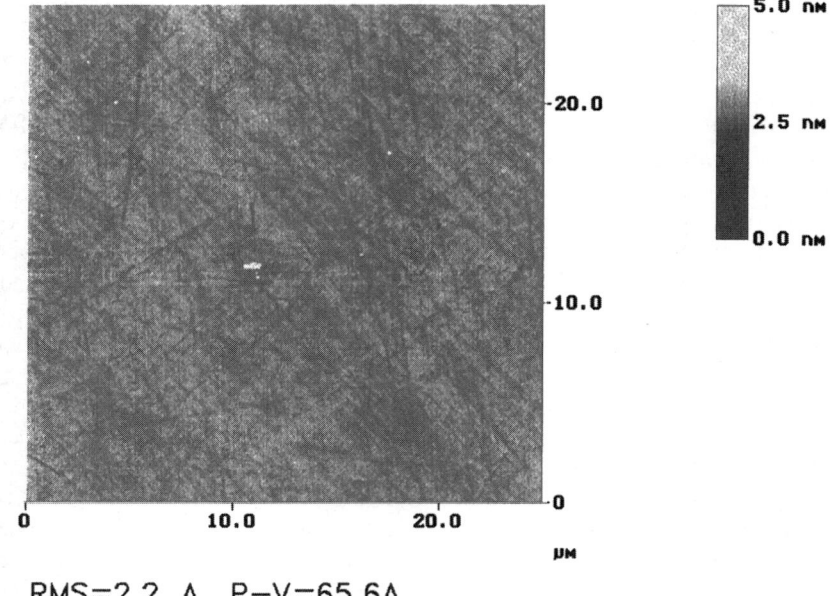

RMS=2.2 A, P–V=65.6A

FIG. 9. AFM image of post CMP surface showing nanoscratches produced by slurry particle contact.

2. DIE-SCALE MODELS

The primary purpose of CMP is to remove unwanted topography from semiconductor device structures to produce a planar die surface (planarization). To achieve this end, the ideal CMP process must exhibit a high degree of spatial selectivity (i.e., a high ratio of polishing of projecting vs recessed surface features) over a wide range of sizes and densities of topography at the die level. Consequently, a great deal of investigation into planarization process has been conducted.

The seminal model for planarization is that of Warnock [12] who developed a phenomenological model centered on pad properties. Three adjustable parameter were used: a horizontal length scale corresponding to pad deformation; a vertical length scale, corresponding to pad roughness; and a kinetic component related to the contact velocity. A good fit to experimental polishing behavior for a range of feature widths and densities was obtained using a horizontal length scale of 4.2 μm, a vertical length scale of 20 nm, and a kinetic factor of 11:1 (ratio of the horizontal to vertical polish rate at a feature edge). Although the physical significance of these values is not readily apparent, both the horizontal and vertical length scales are well within the measured surface roughness of the pad and are therefore most likely reflective of the role of pad asperities rather than bulk mechanical properties of the pad.

This conclusion is reinforced by Yu et al. [4] in the portion of their model that addressed planarization behavior. They argued that spatial selectivity is determined by the dimensions and behavior of the pad asperities. From geometry, a pad asperity of radius β cannot enter a trench of width w and depth h if w is below some critical value w_{cr}:

$$w_{cr} = 2(2\beta h - h^2)^{0.5} \tag{6}$$

For the polishing behavior of trench features, removal of material at the top and bottom of the trench is proportional to the fraction of the area in asperity contact. From Eq. (3), the reduction in step height dh/dt could then be expressed as

$$dh/dt = K_2(A_+ - A_-)v = K_2 nvA \int_d^{d+h} \int_0^\infty a\Phi_\beta \Phi_z \, d\beta \, dz \tag{7}$$

where K_2 is a constant reflective of the chemical and abrasive nature of the slurry, v is the contact velocity, and A_+ and A_- are the asperity contact areas on the projecting and recessed features, respectively.

The model was fit to experimental planarization data with good success. Divergence from experimental results was explained by viscoelastic deformation of pad asperities. Incomplete elastic recovery reduces penetration of the trench and improves selectivity. This may be due to velocity effects (i.e., the interaction times are below the relaxation time for the material).

A schematic diagram of die-scale asperity interaction is given in the middle and bottom graphs in Fig. 8.

3. WAFER-SCALE MODELS

Efficient die-scale planarization must extend across the entire wafer for rational use of CMP in a production environment. CMP rate nonuniformity has been a significant and persistent problem since its first use, and has been the subject of much investigation.

TABLE III
Root Causes for Wafer Scale Rate Variation

Root cause	Effect	Sources
Change in asperity concentration	Change in rate; sensitivity to hydroplaning	Viscoelastic deformation (burnishing); nonuniform distribution of pad asperities; pad texture
Change in pressure	Change in rate	Change in applied wafer pressure (carrier related) pad texture (grooves, perforations, texture)
Change in slurry particle concentration in lubrication film	Change in rate	Slurry flow; slurry nonuniformity; pad texture and porosity
Change in pad to wafer fit	Change in lubrication film thickness; change in contact pressure	Pad or wafer deformation; changes in pad shape due to deformation or wear
Change in lubrication film thickness	Change in asperity contact area	Change in pad to part fit; slurry flow
Change in contact velocity	Change in rate	Orbital mechanics (i.e., nonsynchronous rotation)

From the local-scale models already presented, rate variation on the wafer scale can be attributed to the root causes listed in Table III. Wafer-scale-rate nonuniformity in CMP is generally observed in three modes: (1) randomly distributed nonuniformity, (2) radial nonuniformity (i.e., rate varies from the center to edge of the wafer in a regular fashion), and (3) wafer edge effects. For well-regulated processes, the edge effect is the most pronounced. It has an undesirable impact on die yield if the nonuniformity extends beyond the edge exclusion zone typical of the coating and photolithographic processes.

Baker [13] presented a model for the edge effect that showed good agreement to experimental results. The system modeled was a composite pad consisting of an IC1000 upper layer and a Suba IV base layer, a typical configuration for oxide CMP. Contact pressure variation at the wafer edge was modeled using a differential element elastic plate approach and radial variation in rate was held proportional to contact pressure. Incorporation of measured pad physical properties gave good agreement with experimentally measured edge nonuniformities. The location of the maximum nonuniformity from the edge X_{peak} was given by:

$$X_{peak} = 2.53/B \qquad (8)$$

where

$$B = (k/D)^{1/4} \qquad (9)$$

$$K = 1/Ct \qquad (10)$$

where C is the base pad compressibility, t is the base pad thickness, and

$$D = Eh^3/12(1 - v^2) \qquad (11)$$

where E is Young's modulus, h is the thickness, and v is Poisson's ratio of the top pad. No factors other than pad mechanical properties were required to yield the effect. One of the model conclusions was that the use of retaining rings on the wafer carrier that were coplanar with the wafer surface would largely eliminate the edge effect. This has been subsequently implemented by a number of polishing machine manufacturers with good success.

A number of models for wafer-scale nonuniformity have been published. While of significance for understanding the CMP process, they have limited relevance for this present chapter as, for the most part, pad effects are neglected. The reader is referred to the primary references listed here by focus area:

Pressure distribution — Tseng et al. [14]
Velocity distribution and kinematic effects — Hocheng [15]
Frictional effects (shear and Von Mises stress effects) — Murthy [16] and Guo et al. [17]

Of these models, Tseng et al. [14] is of relevance as it, like Baker [13], uses a flexural plate model for pressure distribution across the wafer to accurately fit data for radial nonuniformity.

4. IMPACT OF STRUCTURE AND PROPERTIES ON THE POLISHING PROCESS

The expected impact of the material properties reviewed in Table II on the polishing mechanisms reviewed in the previous section is summarized in Table IV. As might be expected from previous discussion on structure vs properties, a high degree of interaction between properties and process effects is evident. Publicly available evidence to support materials properties effects on the CMP process is relatively limited. This is reviewed in Section IV.

TABLE IV

SUMMARY OF EXPECTED PAD MATERIAL PROPERTY EFFECTS ON THE CMP PROCESS

Material property	Impact on polishing mechanism	CMP process impact
Pad roughness	Asperity concentration	Rate
	Asperity dimensions	Planarization efficiency
Pad texture (e.g., grooving)	Thickness of lubrication film	Rate; nonuniformity
Liquid permeability	Slurry concentration	Rate
	Thickness of lubrication film	Rate; nonuniformity
	Slurry particle–reaction product buildup	Rate; defectivity
Hardness (asperity)	Contact pressure	Roughness; defectivity
Hardness (bulk)	Asperity hardness	Roughness; defectivity
	Bulk modulus	Planarization length; nonuniformity
Compressibility (asperity)	Asperity concentration; contact pressure	Rate; roughness; defectivity
Compressibility (bulk)	Fit to wafer	Nonuniformity
Modulus (asperity)	Asperity concentration; contact pressure	Rate; roughness; defectivity
Modulus (bulk)	Fit to die	Planarization length
	Fit to wafer	Nonuniformity
Viscoelasticity (asperity)	Asperity radius	Planarization efficiency
Viscoelasticity (bulk)	Fit to wafer	Nonuniformity

IV. Application to Semiconductor Processing

1. Dielectric CMP

The most widely used pad for dielectric CMP is IC1000, a class III material [2]. A micrograph of IC1000 pad microstructure is given in Fig. 7. A summary of material property data taken from a large volume analysis [18] is given in Table V.

One of the most widely observed characteristics of IC1000 during oxide CMP is the need to abrade the pad surface to maintain a constant removal rate (a process typically termed conditioning). The need for abrasive conditioning is not restricted to IC1000 but is a general characteristic of most known type III and IV pads. As reviewed by Bajaj et al. [19], the rate of decay observed when abrasive conditioning is stopped during oxide CMP fits the equation

$$R_t = R_i - m * \ln(t) \qquad (12)$$

TABLE V
Property Summary for IC1000

Property	Average	Standard deviation (SD)	SD/Ave (%)	Maximum	Minimum	Range
Thickness (mil)	0.0500	0.0007	1.5	0.0519	0.0476	0.0042
Density (g/cm^3)	0.748	0.051	6.8	0.827	0.624	0.203
Hardness (Shore D)	52.2	2.5	4.7	55.6	46.5	9.1
Static Comp. (%)	2.1	0.4	17.7	3.4	1.2	2.2
Rebound (%)	73.3	4.0	5.5	88.3	64.4	23.9
Prop. Limit (psi)	1317	185	14.0	1617	909	707
Ten. Str. (psi)	3132	406	13.0	3776	2220	1556
Elong. to break (%)	176	21	12.1	218	130	88
Storage modulus (Pa)	3.06E+08	4.16E+07	13.6	3.64E+08	2.04E+08	1.60E+08
Loss modulus (Pa)	2.79E+07	4.52E+06	16.2	3.44E+07	1.71E+07	1.73E+07
Tan delta	0.091	0.004	4.7	0.101	0.082	0.019

where R_t is the instantaneous removal rate at time t after conditioning is stopped, R_i is the removal rate with conditioning, and m is the decay rate. In this study, the authors demonstrated that both R_i and m were dependent on the applied stress at the pad–wafer interface. Data for four different types of pads was compared, and R_i was found to be inversely proportional to pad density, while m was inversely proportional to the shear modulus of the pad. The authors explained the decay in terms of a combination of the smoothing away of pad asperities under the shear force induced by polishing coupled with reduction in slurry carrying capability.

While the results and conclusions are consistent with the asperity contact model discussed earlier, the data does not unambiguously demonstrate the connection to asperity deformation. One of the complicating assumptions in Ref. [14] was that the shear modulus used in the comparison was a composite modulus calculated from the bulk material properties of each component in a two-pad stack. If asperity deformation is a dominant factor, a more appropriate value is the shear modulus of the contacting member.

Additional insight was obtained from conditioning studies conducted on class IV pads. A pad of the type described in Ref. [3] was prepared from a solid urethane sheet and used in an oxide CMP process. Pad material properties and process conditions employed are given in Tables VI and VII, respectively.

The polishing rate decay kinetics observed in this experiment are graphically illustrated in Fig. 10. Data fit Eq. (12), yielding $R_i = 1524$ Å/min and

TABLE VI
Physical Properties Summary for Class IV Pad

Property	Average	Standard deviation (SD)	SD/Ave (%)	Maximum	Minimum	Range
Thickness (mil)	0.0505	0.0007	1.4	0.0532	0.0487	0.0045
Density (g/cm^3)	1.182	0.002	0.2	1.186	1.175	0.011
Hardness (Shore D)	73.1	1.1	1.5	75.5	70.1	5.4
Static Comp. (%)	1.0	0.2	17.6	1.5	0.6	1.0
Rebound (%)	84.4	4.8	5.7	92.2	71.0	21.3
Shear modulus (Pa)	3.8E+08					
Ten. Str. (psi)	10883	361	3.3	11774	9853	1921
Elong. to break (%)	335	19	5.7	382	300	82
Storage modulus (Pa)	8.47E+08	6.08E+07	7.2	1.03E+09	7.42E+08	2.89E+08
Loss modulus (Pa)	8.68E+07	3.87E+06	4.5	9.59E+07	7.79E+07	1.80E+07
Tan delta	0.103	0.005	4.6	0.110	0.093	0.018

TABLE VII
Experimental Conditions for Conditioning Study

Variable	Setting
Polisher	Westech 372
Downforce	9 psi
Platen speed	20 rpm
Carrier speed	42 rpm
Conditioner type	Westech RPC1, 10.5 in., 68 μm diamond
Conditioning platen speed	10 rpm
Conditioner wheel speed	25 rpm
Conditioner downforce	0.4 psi
Slurry–flow	ILD1300 at 150 ml/min

$m = 446$ Å/min. These values are well within the range reported in Ref. [14] and are consistent with predictions for the shear modulus of the pad material.

During the course of the experiment, pad surface roughness was measured by contact profilometry. Removal rate was found to be directly proportional to pad roughness (Fig. 11). Microscopic examination of the pad surface clearly showed the progressive smoothing away of the upper pad asperities (Figs. 12a–12d). No evidence of shear deformation of the asperities was

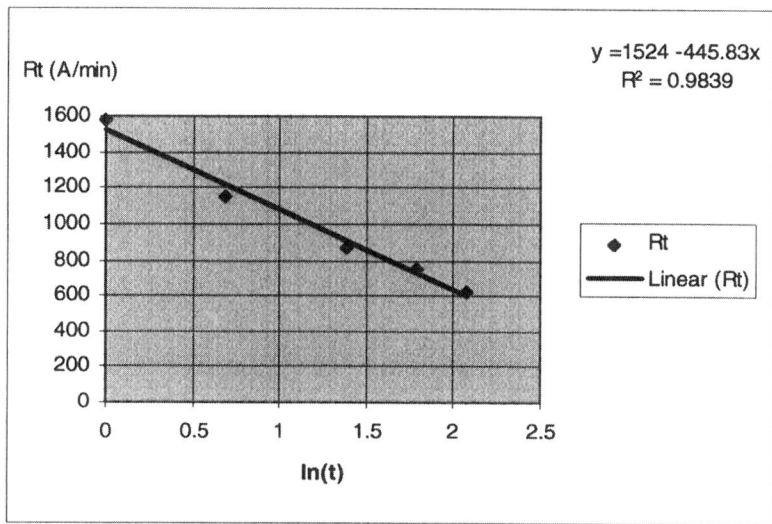

FIG. 10. Change in rate vs time after abrasion conditioning is stopped for a class IV pad.

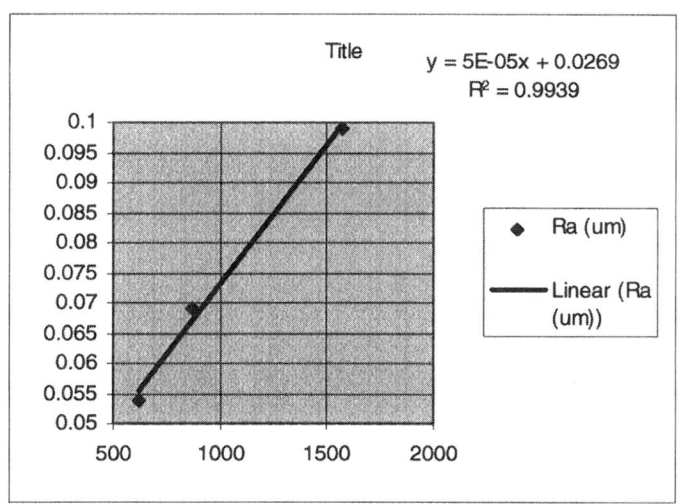

FIG. 11. Pad surface roughness vs rate.

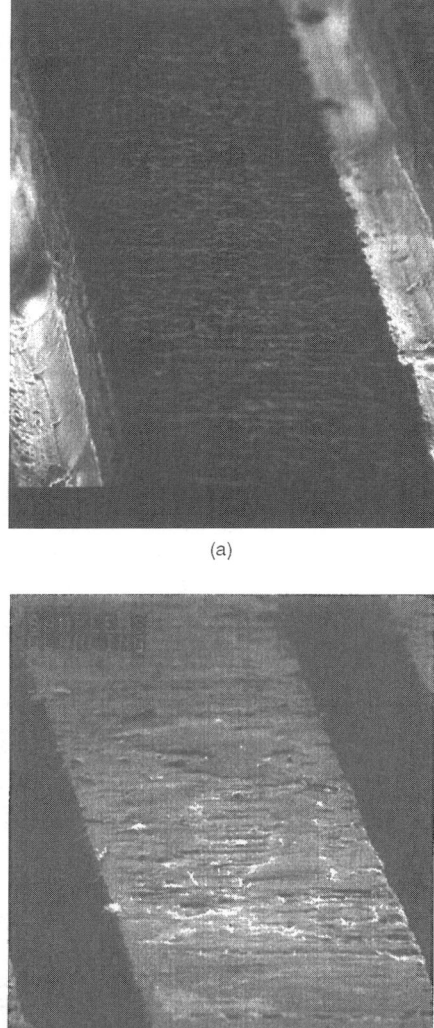

FIG. 12. (a) Class IV pad surface after abrasive conditioning and pad surface after (b) 2 min, (c) 4 min, (d) 6 min, and (e) 8 min of CPM without conditioning.

6 CMP CONSUMABLES II: PAD 175

(c)

(d)

FIG. 12. Continued.

(e)

Fig. 12. Continued.

observed; instead pad asperities were simply polished away, resulting in the formation of smooth mesa structures. The effect on rate is consistent with expectations from the asperity contact model reviewed here; that is, the number of asperities is reduced due to burnishing, and the radius of the asperities is greatly increased (i.e., approximately the dimensions of the mesas). A similar study conducted on IC1000 [20] did not show a clear relationship between pad roughness, although, qualitatively, high-roughness pad surfaces gave higher rates than low-roughness surfaces. This lack of correlation is attributed to the effects of the porosity of IC1000, which produces significantly higher surface roughness, obscuring the effects of pad burnishing.

Linear profilometry is not the most appropriate measurement technique for pad asperities. A more appropriate metric is the bearing area (i.e., the fraction of the nominal surface area in contact with a reference plane). Hetherington *et al.* reported more details on the structure and composition of mesa structures on the surface of IC1000 following the cessation of conditioning in the oxide CMP process [21]. Fourier transform infrared (FTIR) and laser Raman analysis of the mesa areas in worn pads showed no significant change in material composition relative to unworn material. Bearing areas were calculated for the pad samples via confocal microscopy. Good correlation was obtained between bearing area and removal rate (Fig. 13). These data are consistent with the class IV results already reviewed.

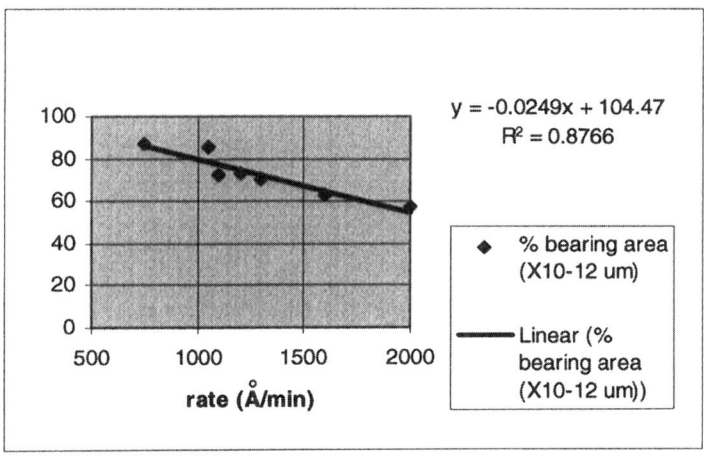

FIG. 13. Removal rate vs bearing ratio (from Ref. [21]).

One of the consequences of the rate decay behavior already reviewed is that the measured polishing rate will vary with the duration of polishing time if the pad is not conditioned during use (i.e., postpolish conditioning is employed). Moreover, if the pad asperities worn away during the polishing cycle are not completely regenerated by the postpolish conditioning process used, the rate for the subsequent polish step will vary. This is a major source of process variability. Effects of the conditioning process on rate stability for IC1000 were reviewed by Baker [22]. He proposed a quantitative index of performance, the degree of conditioning (DOC), which is proportional to the amount of pad thickness removed during the conditioning step. The DOC was shown to affect the change in rate vs polish time: high DOC significantly reduced the rate of decay. Of greater significance, DOC was shown to affect the rate stability of the interlevel dielectric (ILD) process. Underconditioning—a low initial DOC during the break-in step followed by low DOC during the polishing process—results in a downward drift in removal rate to some terminal rate (Fig. 14). Overconditioning—when the DOC of the break in step is low and the DOC during polishing is high—yields the reverse behavior.

2. METAL CMP

Despite its importance as a process step, there have been relatively few public reviews of pad effects in metal CMP. The pronounced effect of the choice of pad type on metal CMP process yields has undoubtedly inhibited

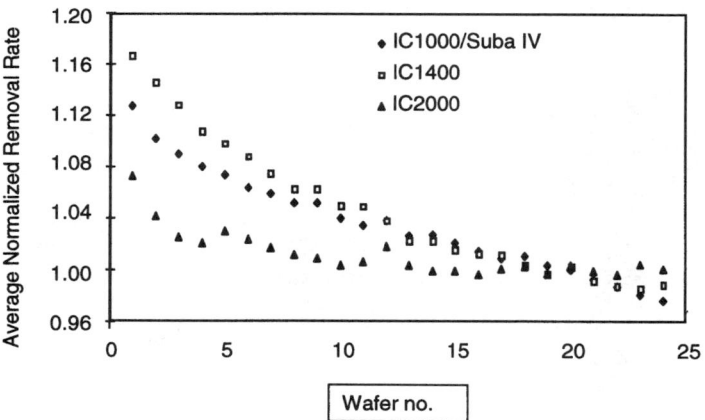

Fig. 14. Blanket ILD baseline indicating process drift.

public review of pad property–performance. Historically, class I, II, and III pads have been employed in production processes. A comparison summary for these materials in metal CMP is given in Table VIII. Overall, class III pads (e.g., IC1400) are preferred. Deficiencies reported in surface roughness and defect densities can be largely corrected through process optimization and choice of slurry. As in the case of oxide CMP, multistep processes have been the norm, with a secondary buff step, usually with a class II pad, employed to reduce defects to acceptable levels. This is a less acceptable alternative in dual-damascene processes, where excessive feature erosion may occur during the buffing step. This has become a strong driver for the development of low-defectivity single-step metal slurries.

One of the most significant differences between metal CMP and dielectric CMP is the effects of pad conditioning on rate for class III pads. Unlike dielectric CMP, where abrasive pad conditioning is required to maintain rate, CMP of the soft metals (e.g., Cu and Al) does not require abrasive conditioning other than an initial conditioning step to generate a starting concentration of pad asperities. Tungsten CMP shows quite different behavior, depending on the oxidant employed in the slurry. For oxidants such as KIO_4, which produce WO_3 as CMP wear by-products, tungsten removal rate increases linearly with time when conditioning is not employed. Use of abrasive conditioning is required to produce stable removal rates. This effect is attributed to buildup of solid WO_3 deposits on the pad surface, which act as abrasives and contribute to the substrate removal. In contrast, hydogen peroxide, which produces soluble peroxytungstate wear by-products, does not produce this effect.

TABLE VIII

PAD PERFORMANCE DIFFERENCES IN METAL CMP

Pad type	Class I	Class II	Class III
Representative pads employed in metal CMP	Suba™ 500; Pellon™ PAN-W	Politex™	IC1000; IC1400
Relative postpolish roughness	Medium	Low	High
Relative defect density (with single-step process)	High	Very low	Medium
Ability to remove metal topography	Medium	Low	High
Relative dishing and erosion	Medium	High	Low
Conditioning required	No	No	W: yes; Cu, Al: no

One of the earliest published reviews of polishing pads and their process effects on tungsten CMP was given by Shih et al. [23]. A comparison study of four polishing pads was conducted using common slurry, process, and wafer structures. The pads evaluated were IC1400 (class III), Politex™, Meritex™, and URII pads (all class II). A single-step CMP process was used with constant slurry and process conditions to assess pad effects on oxide erosion, defectivity, and barrier layer removal rate. A summary of differences in pad properties together with average oxide erosion and defect density is given in Table IX. All four pads studies were composite pads consisting of a top and bottom layer. IC1400, URII, and Politex™ had varying hardnesses of the top pad component with common base layer properties. Politex™ and Meritex™ had identical top layer components, with Meritex™ having a much higher modulus base layer. Oxide erosion and defectivity showed trends that are consistent with trends expected from the models previously reviewed in this chapter (i.e., defect density increased directly with the hardness of the top pad layer while oxide erosion decreased directly with the hardness of the top pad layer). Variation in the bottom layer modulus had no significant effect on either oxide erosion or defectivity. Changes in the hardness of the top pad layer are expected to produce higher contact pressures during CMP, and therefore, higher surface roughness and defectivity. At the same time, as hardness and modulus are generally related, planarization efficiency should increase with increasing hardness. Changes in base pad modulus should only affect the planarization efficiency of very large structures and have no effect on contact pressures.

A similar but more detailed comparison of IC1400 and Politex™ pads in a tungsten CMP process was given by Stein et al. [24]. Overall, the process differences observed were the same as in the previous study cited (i.e., the

TABLE IX
Pad Properties vs Performance in Tungsten CMP from Reference [23]

	Low modulus bottom pad	High modulus bottom pad
High hardness top pad	IC1400 Oxide erosion: 220 Å Defect density: 700	
Medium hardness top pad	UR II Oxide erosion: 400 Å Defect density: 180	
Low hardness top pad	Politex™ Oxide erosion: 820 Å Defect density: 110	Meritex™ Oxide erosion: 770 Å Defect density: 80

Data is taken from average values at 8 psi down force. Erosion data is wafer flat. Defect data is unclustered defects $>0.3\,\mu$m.

softer pad produces lower surface roughness and defectivity, but unacceptably high feature erosion). For the consumables set used, PETEOS removal rate was found to be essentially independent of process conditions, while tungsten rate increased linearly with increasing pressure and velocity. Consequently, selectivity was directly proportional to tungsten rate. However, PTEOS roughness, as expected, varied directly with pad hardness.

References

1. Cornish, D. C. (1961). *The Mechanism of Glass Polishing. A History and Bibliography.* British Scientific Instrument Research Association Research Report R. 267.
2. Reinhardt, H., Roberts, J., McClain, H., Budinger, W., Jensen, E. (1996). Polymeric polishing pad containing hollow polymeric microelements. U.S. Patent 5,578,362, Nov. 26, 1996.
3. Cook, L., Roberts, J., Jenkins, C., Pillai, R. (1996). Polishing pads and methods for their use. U.S. Patent 5,489,233, Feb. 6, 1996.
4. Yu, T.-K., Yu, C. C., Orlowski, M: (1993). A statistical polishing pad model for chemical-mechanical polishing. *Tech. Digest, Int. Electron. Devices Meeting.* Washington, DC, Dec. 5–8.
5. Runnels, S., Eyman, M. (1994). Tribology analysis of chemical-mechanical polishing. *J. Electrochem. Soc.* vol. 141, pp. 1698–1701.
6. Cook, L. (1990). Chemical processes in glass polishing. *J. Non-Cryst. Solids.* vol. 120, pp. 152–171.
7. Levert, J., Mess, F., Grote, L., Dmytrychenko, M., Cook, L., Danyluk, S. (1995). Slurry film thickness measurements in float and semipermeable and permeable polishing pad geometries. *Proc. Int. Tribology Conf.*, Yokahama.

8. Levert, J., Baker, R., Mess, F., Salant, R., Danyluk, S. (1997). Mechanisms of chemical-mechanical polishing of SiO_2 dielectric on integrated circuits. *STLE '97 WTC- London/ STLE Tribology Transactions.*
9. Bhushan, M., Rouse, R., Lukens, J. (1995). Chemical-mechanical polishing in semidirect contact mode. *J. Electrochem. Soc.* vol. 142, pp. 3845–3851.
10. Coppeta, J., Rogers, C., Philipossian, A., Kaufman, F. (1997). Characterizing slurry flow during CMP using laser induced fluorescence. *Proc. 2nd Int. CMP for ULSI Multilevel Interconnect. Conf.,* Santa Clara.
11. Coppeta, J., Racz, L., Philipossian, A., Kaufman, F., Rogers, C. (1998). Pad effects on slurry transport beneath a wafer during polishing. *Proc. 3rd Int. CMP for ULSI Multilevel Interconnect. Conf.,* Santa Clara.
12. Warnock, J. (1991). A two-dimensional process model fror chemimechanical polish planarization. *J. Electrochem. Soc.* vol 138, pp. 2398–2402.
13. Baker, A. (1997). The origin of the edge effect in chemical mechanical planarization. *Electrochem. Soc. Proc.* 96–122, p. 228.
14. Tseng, W.-T., Kang, L.-C., Pan, W.-C., Chin, J.-H., Cheng, P.-Y. (1998). Distribution of pressure and its effects on the removal rate during chemical-mechanical polishing process. *Proc. 3rd Int. CMP for ULSI Multilevel Interconnect. Conf.,* Santa Clara, pp. 87–94.
15. Hocheng, H. (1997). A linematic analysis of CMP based on velocity model. *Proc. 2nd Int. CMP for ULSI Multilevel Interconnect. Conf.,* Santa Clara, pp. 277–280.
16. Murthy, A. (1997). Non-uniformity in CMP process: an effect of stress. *Proc. 2nd Int. CMP for ULSI Multilevel Interconnect. Conf.,* Santa Clara, pp. 281–284.
17. Guo, Y., Tang, J., Dornfield, D. (1998). A finite element model for wafer material removal rate and non-uniformity in chemical mechanical polishing process. *Proc. 3rd Int. CMP for ULSI Multilevel Interconnect. Conf.,* Santa Clara, pp. 113–118.
18. Fury, M., James, D. (1996). Relationships between physical properties and polishing performance of planarization pads. *SPIE Microelectronic Manufacturing Symposium. Austin Oct. 16, 1996.* (unpublished).
19. Bajaj, R., Desai, M., Jairath, R., Stell, M., Tolles, R. (1994). Effect of polishing pad material properties on chemical mechanical polishing (CMP) processes. *Proc. MRS Spring Meeting.*
20. Stein, D., Hetherington, D., Dugger, M., Stout, T. (1996). *J. Electron. Mater.* vol. 25, p. 1623.
21. Hetherington, D., Achuthan, K. (1998). Tribological evaluation of CMP pads. *Proc. CMP Technology for ULSI Interconnection,* SEMI, pp. L-3–L-20.
22. Baker, A. (1997). Conditioning for rate stability. *Proc. 2nd Int. CMP for ULSI Multilevel Interconnect. Conf.,* Santa Clara, pp. 339–342.
23. Shih, Y.-C., Sethuraman, A., Wang, H.-M., Lavoie, R., Cook, L. (1997). Polishing pads and process effects on tungsten CMP. *Proc. 2nd Int. CMP for ULSI Multilevel Interconnect. Conf.,* Santa Clara, pp. 237–240.
24. Stein, D., Hetherington, D., Cecchi, J. (1998). Tungsten CMP performance of Politex and IC1400 pads using an iodate/alumina based slurry. *Proc. 3rd Int. CMP for ULSI Multilevel Interconnect. Conf.,* Santa Clara, pp. 161–164.

Note: Politex™, Suba™, and Meritex™ are trademarks for pad products produced by Rodel Inc. Pellon™ is a trademark for polishing pads produced by Freudenburg Corp.

CHAPTER 7

Post-CMP Clean

François Tardif

LETI GRENOBLE, FRANCE

I. INTRODUCTION 183
II. SURFACE CONFIGURATIONS AFTER CMP PROCESSES 184
III. CLEANING REQUIREMENTS AFTER CMP PROCESSES 184
    1. *Particle Effects* 184
    2. *Metallic Contamination Effects* 185
    3. *Damaged-Layer Effects* 186
IV. CORROSION EFFECTS 186
    1. *Electrochemical Corrosion* 186
    2. *Photoassisted Corrosion* 190
V. SLURRY REMOVAL 193
    1. *Particle-Removal Mechanisms* 193
    2. *Scrubber Cleanings* 202
    3. *Wet Cleanings* 204
VI. METALLIC CONTAMINATION REMOVAL 206
    1. *Metallic Contamination during the Full CMP Step* 206
    2. *Metallic Contamination Cleaning* 207
VII. DAMAGED LAYER REMOVAL 208
    1. *Practical Determination of the Damaged Layer* 209
    2. *Elimination of the Damaged Layer* 209
VIII. FINAL PASSIVATION 210
IX. EXAMPLES OF PRACTICAL POST-CMP CLEANING PROCESSES 210
    1. *STI and Silicon Oxide CMP* 211
    2. *Tungsten CMP* 212
    3. *Copper CMP* 212
X. CONCLUSION 213
    REFERENCES 213

I. Introduction

With the decrease of device dimensions, planarization of both front and back end layers by chemical mechanical polishing (CMP) now seems an absolute must for technologies smaller than 0.5 μm. Unfortunately, CMP processes leave a large amount of particles, parasitic metallic contamination, and a damaged layer at the top surface. Thus, a powerful new cleaning process must be added after each CMP step, which was unnecessary in the past.

II. Surface Configurations after CMP Processes

Today's CMP applications concern both front-end steps such as shallow trench isolation (STI) and inter metal dielectric (IMD) in the back end of the line.

Figure 1 shows the surface configuration after STI CMP. Both the nitride stop layer and the isolation oxide contaminated with SiO_2 slurries, particles, and metals must be cleaned.

As shown in Fig. 2 two different back end surface configurations can be found after CMP. For technologies down to about 0.25 μm, the metal layers are embedded in silicon oxide. Insofar as the integrity of this oxide is sufficient, any aggressive conventional cleaning processes can be used. For more advanced technologies where the damascene structure is adopted, outcropping metals such as aluminum, copper, tungsten, titanium, titanium nitride, tantalum, and tantalum nitride may be present with the silicon oxide. In this case, the difficulty lies in cleaning up all these different materials without damaging the conductive lines or the diffusion barriers.

III. Cleaning Requirements after CMP Processes

1. PARTICLE EFFECTS

The more obvious and understandable issue concerns the very high level of residual particles left by CMP. These particles essentially originate from the used slurries (SiO_2, Al_2O_3, CeO_2) but also from the polished surface materials and to a lesser extent from the polishing equipment environment. The typical particle levels encountered depend greatly on the type of CMP

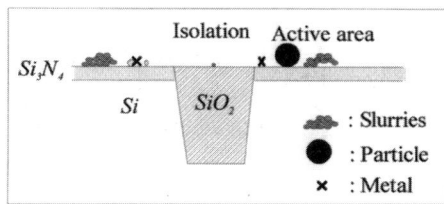

FIG. 1. Schematic illustration of the surface configuration after STI CMP. The nitride stop layer is contaminated with SiO_2 slurries, particles, and metals. This layer is removed by etching after cleaning.

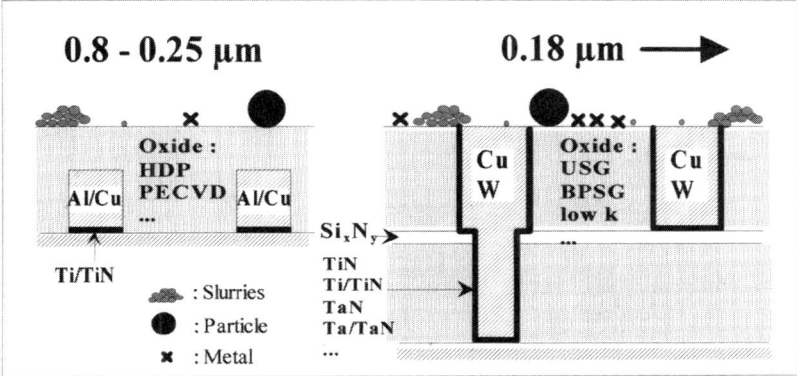

FIG. 2. Schematic illustrations of two back-end surface configurations after CMP. For the conventional technology, a simple oxide surface must be cleaned up, whereas for damascene structures, both outcropping metals and insulator must be processed together. The difficulty is to preserve the metal lines and the diffusion barriers.

and the polishing conditions: from several hundreds for oxide CMP to several 10^4 particles by wafer when alumina slurries are used. These particles can be physically adsorbed at the surface or even in the worst case partially embedded in the top layer due to the mechanical pressure exerted by the pad.

For advanced 0.18-μm technologies the commonly measured particles at 0.2 μm are very close to the linewidth and thus potentially very dangerous. The SIA roadmap suggests that back-end processes for 0.18-μm technologies should contribute no more than 50 adders at 0.09 μm for a 200-mm wafer.

2. METALLIC CONTAMINATION EFFECTS

CMP processes also leave a metallic contamination typically in the 10^{11}–10^{12} at/cm^2 range. These contaminants arise from the outcropping metals, the slurries, and the mechanical environment of the polishers. In front-end applications (STI), these levels are prohibited because they are not compatible with the following hot processes. In the case of back-end steps, these parasitic metals must be removed as well, even if this seems more paradoxical with the use of metallization steps. Indeed a large amount of charges at the interconnection level or the presence of mobile ions such as sodium or potassium can induce disturbances during the electrical information transfer. Furthermore, a superficial conductive metallic contamination can generate shorts between two adjacent lines by percolation conduction mechanism. And last but not least, fast diffusers such as copper can reach the active area from the backside surface during the following thermal processes even if performed at relatively low temperature (450°C). The SIA roadmap suggests for 0.18-μm technologies that critical metals have to be reduced to below 4×10^9 at/cm^2 for front-end applications and to below 5×10^{11} at/cm^2 for back-end applications.

3. Damaged-Layer Effects

A damaged layer between 1 to 10 nm thick according to the material and the polishing conditions is generated by the CMP process. This layer would seem to have to be removed as it presents poorly defined physical properties, for example, in terms of contamination, internal stress, insulating characteristics, and the like. Nevertheless the detrimental effects of this layer still have to be clearly demonstrated.

Finally, the challenge of post-CMP cleaning consists of removing particles, metallic contamination, and the damaged polished layer without damaging the insulator, the metallic plug, or lines and the diffusion barriers when present at the top surface. Furthermore post-CMP cleaning processes must avoid the enhancement of microdefects present in the insulator such as vertical cracks, microscratches, surface voids, ripouts, and so on.

IV. Corrosion Effects

When the metallic lines or plugs outcrop at the surface, post-CMP cleaning processes must be specifically designed to preserve them from any corrosion effects. The same nonaggressive chemistries also have to be used for embedded lines when the insulator integrity is not absolutely guaranteed. Conventional RCA[1] based chemistries are absolutely forbidden in this case because they contain the very high oxidant specie H_2O_2.

1. Electrochemical Corrosion

Metal corrosion is due to the oxidation mechanism:

$$M \rightarrow M^{n+} + n.e^-$$

The ionic species generated by this reaction can be solvated or complexed by the solution or can form oxides — M_xO_y — which in certain cases protect the metal surface by stopping the oxidation mechanism.

a. Thermodynamic Aspect

The use of conventional electrochemistry laws and knowledge of the stability of the generated oxides allows the behavior of metals in solutions to be predicted. Pourbaix diagrams [2] present the voltage–pH equilibria for all metals in aqueous solutions at 25°C without the presence of any complexing agents. The oxidation potentials and pH of the solution define the "corrosion" or "immunity" areas from an electrochemistry point of view. The existence and stability of passivation oxides for each considered element define the "passivation" areas. Figure 3 gives the Poubaix diagrams of the usual back-end metals: aluminum, tungsten, copper, titanium, and tantalum.

Extrapolation to HF- or Ammonia-based chemistries is not directly possible. In each of these cases, the equilibria must be recomputed taking

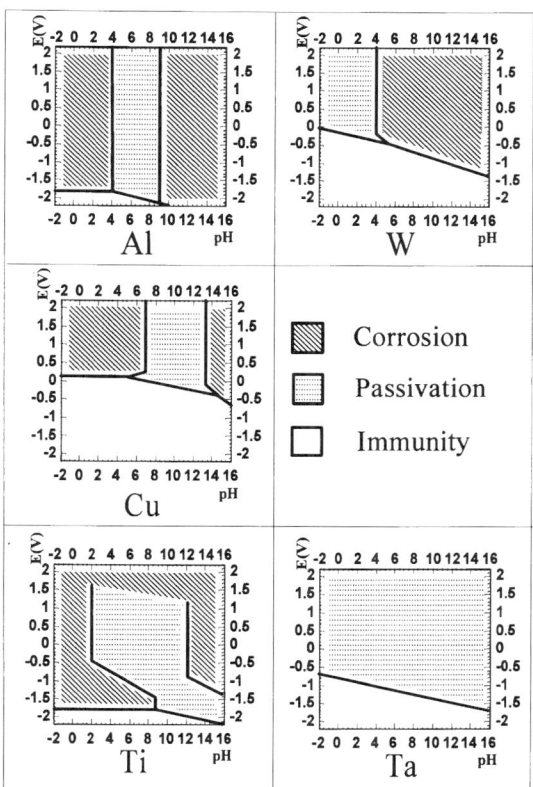

FIG. 3. Pourbaix diagrams of the metals usually used in back-end processes. The results are valid only in aqueous solutions at 25°C without any complexing agent.

into consideration the complexing effects that can reduce the stability areas to their simplest expression. Nevertheless, the Pourbaix diagrams give precious guidelines to predict the general trends.

Even if oxidant species are not intentionally introduced in the cleaning chemistries the latter are all oxidant due to the omnipresence of dissolved oxygen. The only way to annul this effect is to introduce a reducing agent. From the same reference [2] the behavior of the metals used in IMD processes can be predicted:

- *Aluminum* is a strong reducing agent ($E_{0Al/Al^{3+}} = -1.66$ V), that is, it can be very easily oxidized in cleaning solutions. This behavior could represent an additional limitation of the use of aluminum in the case of damascene structures for which post-CMP cleaning is required. Nevertheless aluminum oxide (Al_2O_3) constitutes a very well known passivating layer but is only stable in relatively neutral pH (4–8.5).
- *Tungsten* is a weak reducing agent ($E_{0W/WO_3} = -0.09$ V), which can be oxidized into WO_2, W_2O_5, or WO_3. Passivation is achieved in aqueous solutions by WO_3 which is stable at pH lower than 4 except in the presence of HF, which forms fluorinated tungsten complexes. Nevertherless corrosion is very low in diluted ammonia.
- *Copper* is the only noble metal to be found among back-end materials ($E_{0Cu/Cu^{2+}} = 0.34$ V). Its two oxides—Cu_2O and CuO—are usually present together. They are not able to really constitute a good passivation layer due to their high porosity. Cu_2O is dissolved by diluted HF, H_2SO_4, and H_3PO_4. CuO is dissolved by all diluted acids. These two oxides are stable between pH 7 and 12.8. Ammonia and copper form very stable complexes, which drastically reduce the passivation area until it almost disappears even without any oxidant.
- *Titanium* is a strong reducing agent ($E_{0Ti/Ti^{2+}} = -1.63$ V). It can be oxidized into TiO_3, TiO_2, and TiO but this reaction is rapidly stopped by the exceptional integrity of TiO_2 which is the most common oxide form present in aqueous solutions. The passivating properties of TiO_2 are even higher than those of Al_2O_3. Only very oxidant solutions such as concentrated H_2O_2 can in some cases corrode titanium. The stability of TiO_2 is very high whatever the pH.
- *Tantalum* is a reducing agent as well ($E_{0Ta/Ta_2O_5} = -0.75$ V). It presents many oxidized forms such as Ta_2O, TaO, TaO_2, Ta_2O_5, and Ta_2O_7. Ta_2O_5 is mainly present in aqueous solutions. Its stability is high whatever the pH. The behavior of tantalum is therefore comparable to that of a noble metal without any corrosion in air or in aqueous solutions. Only very concentrated HF can dissolve Ta_2O_5.

Titanium nitride and tantalum nitride present different resistances to the chemical reagents compared to their metals (these data are not available in the literature) but their oxidized species are the same.

The behavior of these metals, in fact, must be carefully checked in the actual device configurations on account of the simultaneous presence of two different metals — conductor plus diffusion barrier — which induces different electrochemical potentials. As an example, the difference of behavior of copper in HF-based chemistry (pH = 7) with or without titanium nitride is illustrated in Fig. 4. According to the Pourbaix diagram, at this pH the copper is theoretically passivated by CuO as verified by the slight increase of copper layer thickness. But in presence of titanium nitride copper is severely corroded.

b. *Kinetic Aspect*

The kinetic aspect must be investigated as well because corrosion may be thermodynamically possible but is so slow that it can be ignored in practice.

We arbitrarily considered a maximum acceptable material removal during the cleaning process of 5 nm, which corresponds for an industrial 10-min process time to a 0.5 nm/min etching rate. Different oxidant-free mixtures able to remove particles by underetching mechanisms and covering the whole pH range were adjusted to remove — in 10 min — 5 nm of PECD

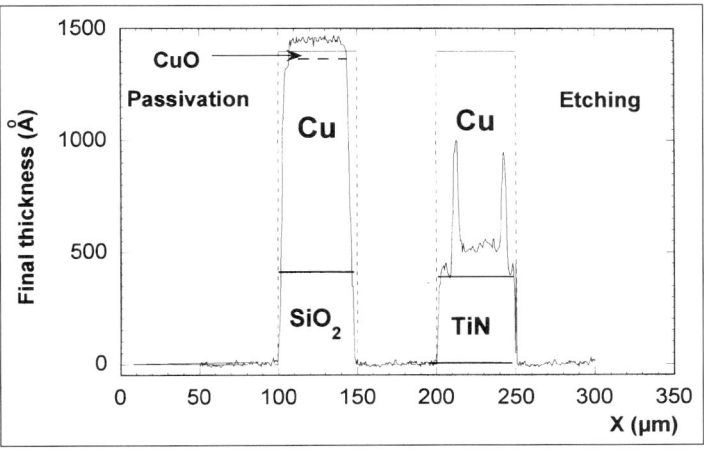

FIG. 4. Profilometer results illustrating the difference of copper etching in the presence or not of titanium nitride, after 1.5 h in 0.1% HF + FNH_4 at pH = 7.

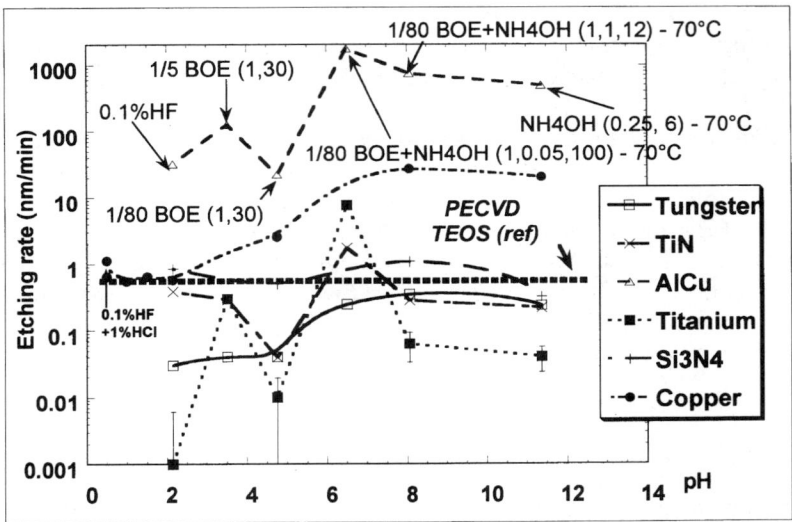

FIG. 5. Etching rate of some back-end materials in different nonoxidant cleaning mixtures adjusted to etch 5 nm of PECVD TEOS oxide in 10 min.

TEOS oxide selected as the back-end reference material. Figure 5 indicates the etching rates of the different back-end materials in these mixtures.

The etching rate of PECVD silicon nitride is comparable to PECVD TEOS oxide. Ti, TiN, and W present an acceptable etching rate whatever the pH. In the presence of copper, only HF-based chemistries can be used. For Al/Cu none of the tested mixtures are suitable.

2. PHOTOASSISTED CORROSION

The effects of the light from the clean room during wet etching of IC layers has been clearly identified in the past [3, 4]. This phenomenon can be due to the presence of p–n junctions, which can act as photogenerators. In the case of interconnections, some outcropping lines or plugs can be connected to the two different parts of a junction. Dipping these structures in a cleaning solution closes the electrical circuit, which can lead to catastrophic corrosion effects.

As represented in Fig. 6, the electron–hole pairs photogenerated in the junction are separated by the presence of the junction electrical field. These carriers, in excess with respect to the thermodynamic equilibrium, induce a

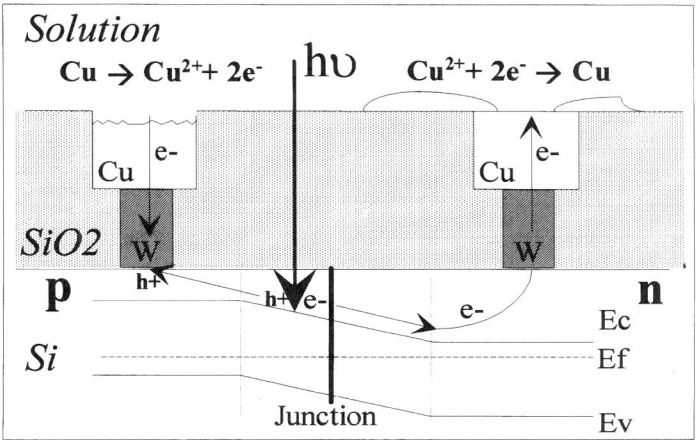

FIG. 6. Schematic illustration of a photoassisted corrosion phenomenon during the post CMP cleaning of damascene interconnections. At the electrode connected to the *p*-side of the junction, the metal is corroded by the oxidation reaction: $M \rightarrow M^{n+} + n.e^{-}$, while the produced soluble M^{n+} species diffuse to the other electrode where the opposite reaction can occur.

potential difference between the two sides of the junction. This potential increases with the light intensity and in open-loop conditions can theoretically reach a maximum value corresponding to the junction height (typically between 0.3 and 0.6 V in silicon). At this potential, the electrical field in the junction is annulled, which interrupts the electron hole separation mechanism. In short circuit conditions, photogenerated carriers are drained out in the external circuit. The current is roughly proportional to the light intensity.

In solution, photovoltage and photocurrent between electrodes are controlled by the different redox couples present in solution such as foreign ions, but also dissolved oxygen and water.

At the electrode connected to the *p*-side of the junction, the metal is corroded by the oxidation reaction — $M \rightarrow M^{n+} + n.e^{-}$ — while the produced soluble M^{n+} species can diffuse to the other electrode where the opposite reaction can occur — reduction of M^{n+} — into its metallic form.

As seen in Fig. 7, the metal can be even fully removed in 1 min at the line connected to the *p*-side of the junction — one can see the plugs — and the redeposition of the metal can be observed on the lines connected to the *n*-side of the same junction. The better method of reducing this phenomenon consists of adding corrosion inhibitors and reducing the light as far as possible using for example close chamber cleaning equipments.

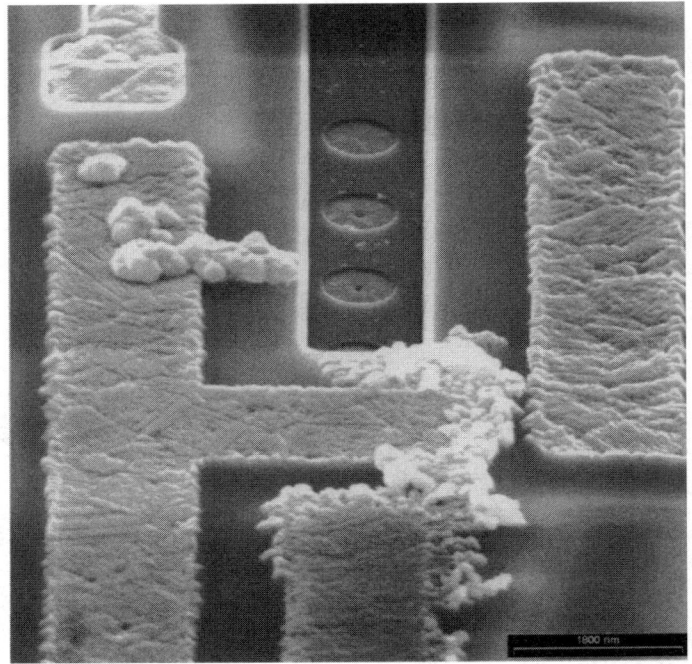

FIG. 7. Scanning electron microscope (SEM) picture after a Cu post-CMP cleaning process (dHF).

The corrosion inhibitor can be a complexing agent that stops the metal redeposition reaction (reduction) by eliminating free metal ions from the solution. Theoretically, this would consecutively stop the associated oxidation reaction. Due to parasitic reduction reactions, however, the metal oxidation can continue, even enhanced by the complexing agent effect.

The corrosion inhibitor can also be a redox couple presenting a reversible and fast electrochemical behavior that is able to react in place of the metal. This is obtained when its redox couple potential is lower than that of the considered metal. The reversible behavior allows the continuous regeneration of the corrosion inhibitor. These reducing agents are often organic compounds soluble in aqueous solutions. A nonexhaustive list is given in Ref. [5].

The deposition of a passivation layer such as described in Section VIII can, in certain conditions, represent an interesting solution.

V. Slurry Removal

As CMP uses slurries it can be easily understood why this process is one of the most polluting steps in IC fabrication in terms of particles. Nevertheless, the global strategy to reduce particles must begin by trying to limit the particle level left by CMP [6]. In spite of the continuous efforts made in this area, the residual high particle level after CMP still constitutes the main challenge of post-CMP cleaning.

1. Particle-Removal Mechanisms

a. Forces Involved

The main forces liable to be exerted on fine particles are calculated in Table I. We can see from this table that the two main parameters that drive the particle adhesion–removal mechanisms are the van der Waals and electrostatic forces.

The van der Waals forces [7, 8], which link the particles to the substrates, can easily be calculated in the case of spherical particles:

$$F = \frac{A \cdot R}{6 \cdot x^2} \qquad (1)$$

where

A = Hamaker constant depending on the nature of the particle, the substrate, and the solution

TABLE I

Calculation of the Forces Susceptible to be Exerted on a Fine Particle Deposited on a Substrate (Case of Spherical Silicon Nitride Particles in Water at Room Temperature, Substrate Potential: $-300\,\text{V}$)

Nature of forces	Particle dimension dependencies	Forces, order of magnitude (N)		
		$R = 0.1\,\mu\text{m}$	$R = 1\,\mu\text{m}$	$R = 10\,\mu\text{m}$
van der Waals	R	10^{-7}	10^{-6}	10^{-5}
Electostatic	—	10^{-7}	10^{-6}	10^{-5}
Surface tension	R	10^{-8}	10^{-7}	10^{-6}
Drag	R	10^{-9}	10^{-8}	10^{-7}
Gravitation	R^3	10^{-16}	10^{-13}	10^{-10}
Archimede	R^3	10^{-17}	10^{-14}	10^{-11}

R = Particle radius
x = Distance between the particle and the substrate

The distance between the particle and the substrate is never null as very strong repulsion forces occur at short distances between the electronic clouds of the atoms constituting the particle and the substrate.

The van der Waals forces can be much higher in the actual case of nonspherical particles, which usually present more than one contact point with the substrate.

The electrical interactions can be easily calculated when the particle is close to the substrate and the ionic strength is low. In this case, we can simplify the problem to a conventional punctual particle with a charge q placed in the electrical field generated by the substrate in solution* [9]. Using the Gouy–Chapman model for the calculation of the electrical field generated by the substrate leads to the following equation:

$$F_{EL} = q * \frac{-4kTY_0\kappa}{ze(\exp(\kappa x) - Y_0^2 \exp(-\kappa x))} \quad (2)$$

where

$$Y_0 = \frac{\exp(ze\Psi_0/2kT) - 1}{\exp(ze\Psi_0/2kT) + 1}$$

k = Boltzmann's constant

T = Absolute temperature

κ^{-1} = Double-layer thickness

e = Elementary charge

z = Electrolyte valency

Ψ_0 = Wafer surface potential

x = Particle–wafer distance

Figure 8 shows the variations of both van der Waals and electrostatic forces as a function of the particle–substrate distance. Close to the wafer surface the van der Waals forces are predominant, then the electrostatic forces can remove or reattract the particles according to the relative sign of the particle and substrate charges.

*In fact, the proper calculation has to take into account the charges contained in the double layer [16], q must therefore be modified. This concept is introduced later in this chapter.

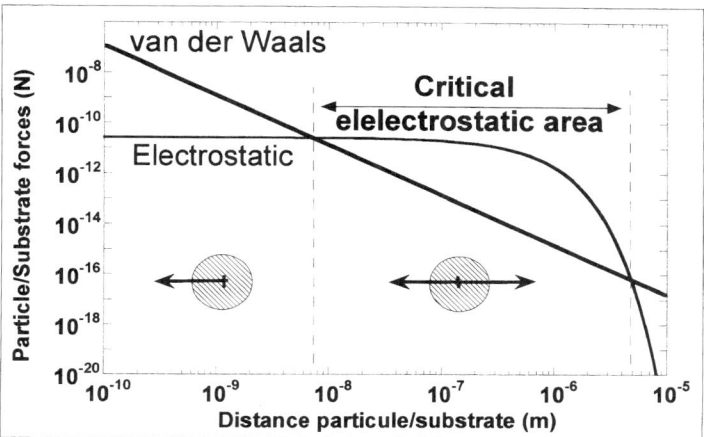

FIG. 8. Van der Waals and electrostatic interactions between a 0.2-μm Al_2O_3 particle and a SiO_2 substrate at -300 V (ionic strength: 10^{-7} mol/liter). Close to the substrate, van der Waals forces are predominant, then the electrostatic forces can remove or reattract the particle (from Ref. [16]).

Finally, to remove particles, the van der Waals forces first must be overcome to separate the particle from the substrate using mechanical effects such as scrubbing or by chemically etching the particle and/or the substrate to purely and simply eliminate the two surfaces in contact. As seen in Table I, harsh accelerations or high-pressure sprays, for example, are not able to remove the fine particles. Then the electrostatic interaction must be turned into favorable conditions to avoid particle readhesion.

b. *Separation from the Substrate*

The easiest way to produce a mechanical effect consists in using a brush that is actually brought into intimate contact of the substrate. Other techniques have been recently proposed such as laser flash [10] and shot-peening with argon or ice microballs [11] but they are still not mature enough to be developed in this chapter.

In the case of the underetching removal process, one of the main parameters is the etching thickness. On silicon, a 2-nm etching is necessary to remove the particles [12]. This distance corresponds to a theoretical decrease of the van der Waal's interactions of about three orders of

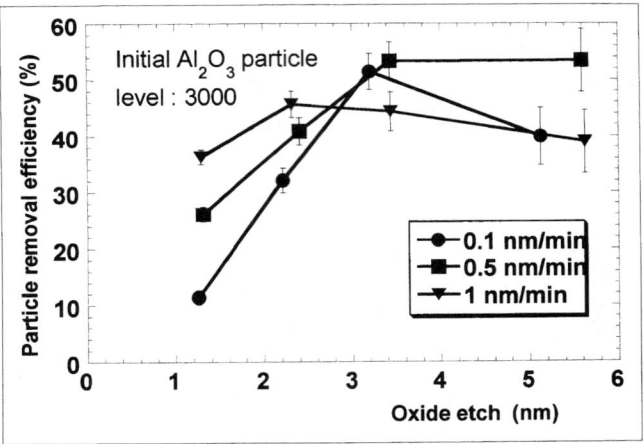

FIG. 9. Removal efficiency of alumina particles ($>0.3\,\mu$m) in HF-based chemistries; 4 to 5 nm of underetching whatever the etching rate are necessary in the case of silicon oxide.

magnitude [13]. According to Fig. 9, a 4- to 5-nm underetching seems to be more appropriate for oxide cleaning. Furthermore the optimal removal efficiency does not seem to depend on the etching rate, unlike what could be expected from the dynamic aspect of the redeposition process. As seen in Fig. 10, a 3- to 4-nm underetching is necessary in the case of silicon nitride.

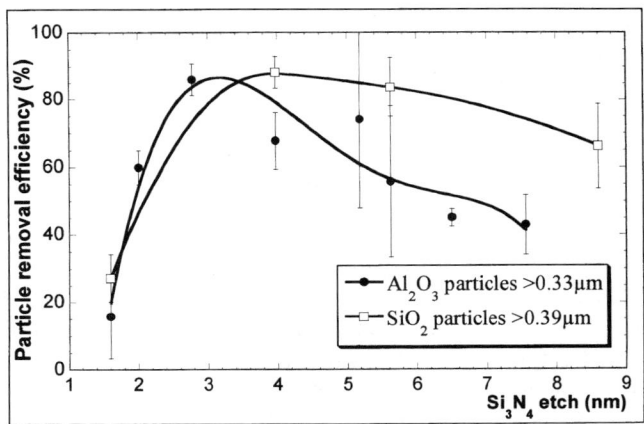

FIG. 10. Silicon oxide and alumina particles ($>0.3\,\mu$m) removed in NH_4OH-H_2O [0.25, 6] at 65°C; 3 to 4 nm of underetching are necessary in the case of silicon nitride.

c. *Prevention of Electrostatic Readhesion*

The second step consists of preventing the readhesion of the just-liberated particles by annihilating the electrostatic attraction forces between the substrate (scrubbing brushes when used) and the particles or even better by obtaining a repulsion force.

The charges are usually mainly located at the particle or substrate surface. Their origin is generally mainly due to the chemical terminations of these surfaces. These terminations are in equilibrium with the solution and can therefore be modified by the pH or by some species, to a certain extent in the same way as ion exchange resins.

Unlike what happens in air, the surfaces of charged particles or substrates dipped in aqueous media are immediately surrounded by a layer containing an equivalent but opposite charge of ions from the solution. Even though there is actually no concentration discontinuity in this layer, the double-layer theory distinguishes two areas: a first compact layer of adsorbed counterions localized close to the surface — referenced as Stern layer — and then a larger surrounding layer (see Fig. 11). Even in deionized (DI) water, where the amount of ions is limited, there are still sufficient H^+ or OH^- species to counteract the particle charge.

The size of this double layer depends on the ionic strength of the solution: the higher the ionic strength, the thinner the double layer. This behavior can be compared to what takes place in semiconductor space charge regions, which are thinner the higher the doping (see Fig. 12). From the Poisson–Boltzmann equation and using the Debye–Huckel approximation leads to the emergence of a quantity κ^{-1}, which has a unit of length and can be assimilated to the double-layer thickness [15]:

$$\kappa^{-1} = \left(\frac{2000 \cdot e^2 \cdot N_A \cdot I}{\varepsilon \cdot K \cdot T} \right)^{-1/2} \quad (3)$$

where

N_A = Avogadro's number

I = Ionic strength

ε = Permittivity of the solution

The net charge of the particles are not directly available in practice. Furthermore we prefer to speak in terms of surface potential generated by these charges because this parameter depends only on the nature of the

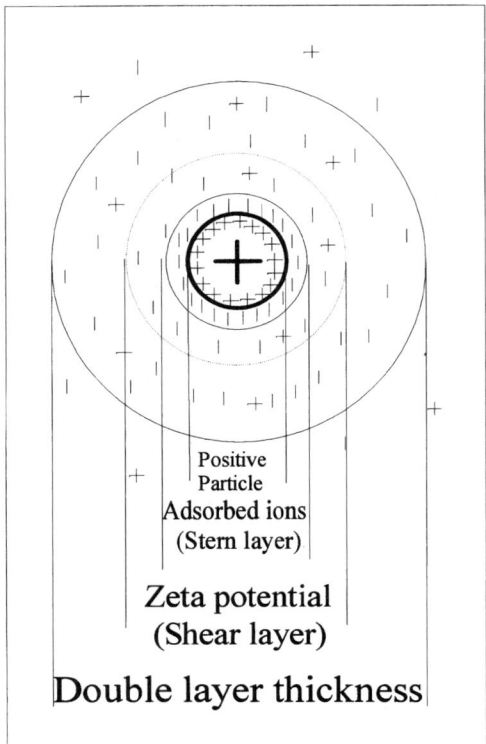

FIG. 11. Electrostatic double layer around the charged surface of a particle [14].

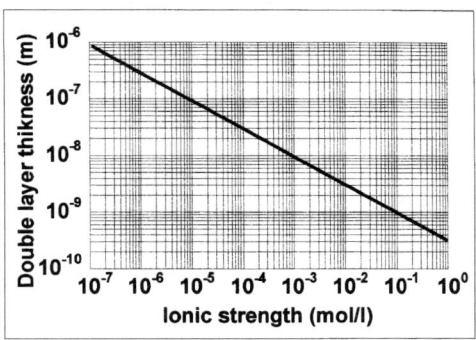

FIG. 12. Evolution of the double-layer thickness with the ionic strength of the solution for a 1:1 electrolyte.

material (not on the size of the particles). This surface potential is characterized in the studied solution by the measurement of the mobility of the particles under an electrical field (electrophoresis). When the particles move relative to the solution under the electrical field only a part of the double-layer thickness is swept by the particle. This layer is called the shear layer. The surface potential obtained by electrophoresis is therefore measured at this level: this potential is called the zeta potential.

As the charge of the particles is in equilibrium with the other species in solution, the zeta potential depends on the pH as well. Figure 13 shows the zeta potential of some materials concerned by the CMP process, where SiO_2 stands for both the substrate and the fumed glass slurries; PVA is the material used for the scrubber brushes (poly vinyl alcohol); Si_3N_4 is used as a polish stop layer; and Al_2O_3 and CeO_2 represent the alumina and ceria slurries, respectively.

The zeta potential is also modified by the ionic strength. When ionic strength increases, the absolute value of Zeta potential reduces. This observed phenomenon can be explained by both the presence of more counterions in the shear layer due to the decreasing double-layer thickness and to the increasing counterion adsorption into the stern layer.

As seen in Fig. 13, the favorable pH are below 1.8 — area A — and above 9 — area B — where both the substrates and all the different particles present the same electrical sign (repulsion). From the electrostatic interaction point

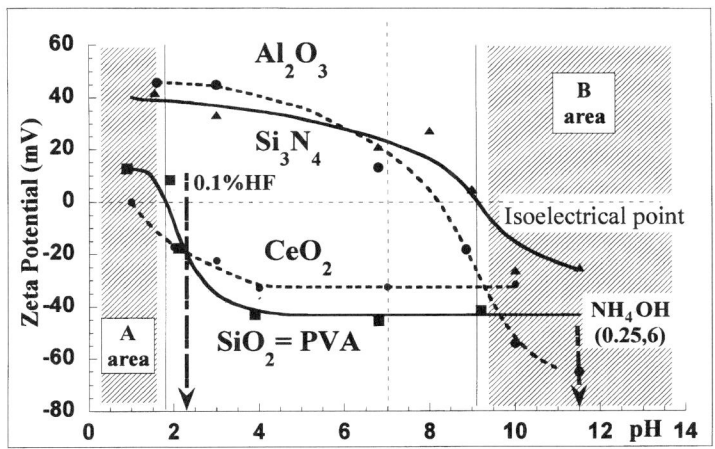

FIG. 13. Zeta potential of the main back-end materials as a function of the pH (ionic strength = 10^{-1} mol/liter). PVA is the polymer used in the scrubber brushes; its zeta potential is very similar to that of silicon oxide.

of view, the more difficult cases are alumina slurries deposited on silicon oxide and silicon oxide or ceria slurries deposited on a silicon nitride layer. This is verified on the curve in Fig. 14, where the removal efficiency of alumina particles on PECVD TEOS oxide is optimal in the two just mentioned favorable areas. The better removal efficiency obtained in the *B* area can be explained by the higher absolute values of the zeta potentials.

Another solution consists in using surfactants that stick on the particle and substrate surfaces and therefore modify their apparent charges. The results of Fig. 15 were obtained under the same experimental conditions as those reported in Fig. 14. Some anionic surfactants such as TA sulfate injected in 0.1% HF at the critical miscellanies concentration (CMC) level achieve the same good performances as those obtained in alkaline media.

The ionic strength of the solution also plays an important role [16]. As represented in Fig. 16, the electrostatic interactions between substrate and particles are eliminated at a smaller distance in the case of a thin double layer (high ionic strength), which leads to a better removal efficiency. This feature highlights the limitation of the use of diluted chemistries.

Remark. It is important to note the fundamental difference between slurries and the conventionally measured particles: the elementary particles that constitute the commonly used slurries are more than one order of

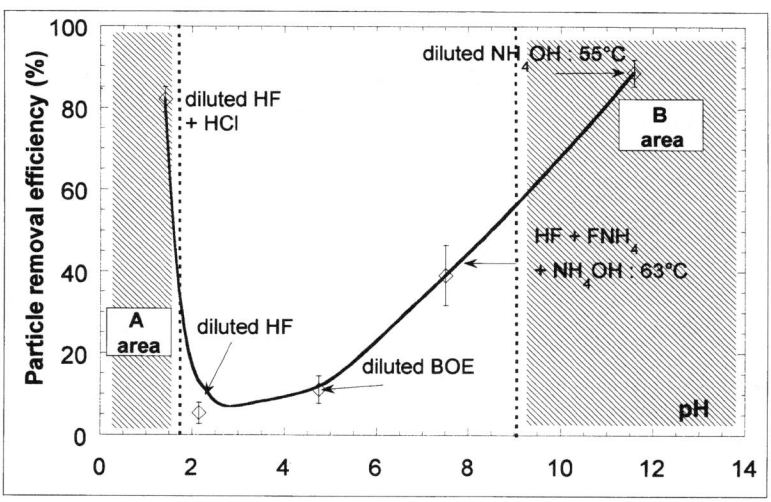

FIG. 14. Evolution of alumina particle removal efficiency from PECVD TEOS oxide as a function of pH. As expected, better results are obtained in areas *A* and *B*.

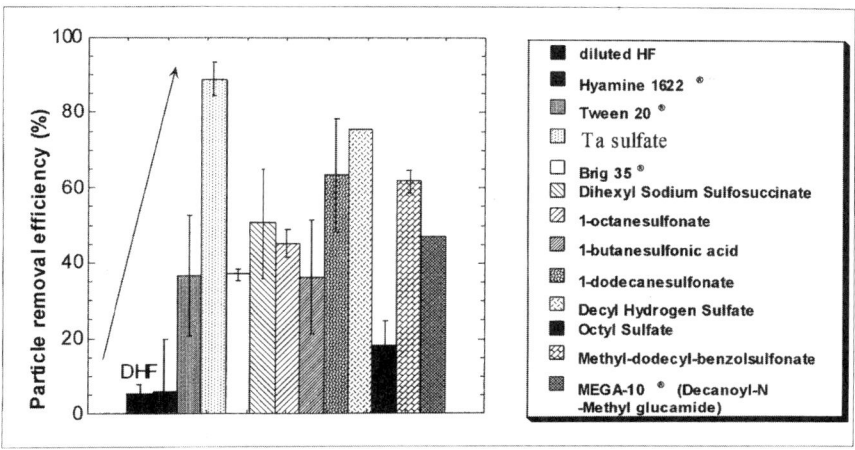

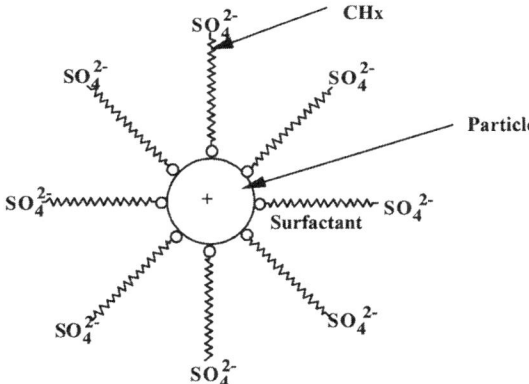

FIG. 15. Performances of anionic and nonionic surfactants in 0.1% HF to remove positive Al$_2$O$_3$ particles deposited on a negative PECVD oxide. Schematic view of the interaction of anionic surfactant–Al$_2$O$_3$ particles.

magnitude smaller. The slurries that are measurable today by light scattering (TENCOR) are therefore only the biggest conglomerates. Figure 17 shows the difference between the two types of alumina particles: the supple aspect of the slurries conglomerates enables them to maximize the contact surface with the substrate enhancing the van der Waals forces and leading to flat particle geometries. This phenomenon is further enhanced by the

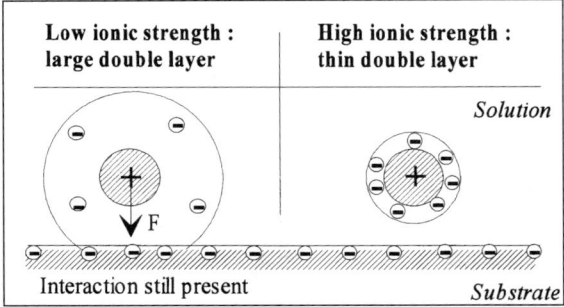

Fig. 16. Influence of the ionic strength in the prevention of readhesion phenomenon in the case of a positive particle on a negative substrate (alumina slurries on silicon oxide layer): a high-ionic-strength limit the double layer thickness. Particle and substrate are therefore electrically masked at a closer particle–substrate distance. (The double layer of the substrate is not represented here.)

pressure exerted by the pad. These smooth defects are intuitively more difficult to catch with the scrubber brush and their masking aspect can slow down the under-etching mechanisms. The different cleaning processes therefore must be validated with actual slurries left by the actual CMP process. This is what was done for the following tests described in this chapter.

2. SCRUBBER CLEANINGS

The old scrubber technique is in fact very attractive for post-CMP cleaning as the same mechanical effect is active for all the materials present at the surface (insulators, metal barriers). Doubled-sided scrubbers for cleaning the frontside and the backside of the wafer and lateral brushes to take care of the wafer side are now proposed on the market. Furthermore, the implementation of megasonic sprays in the scrubber can sometimes help for difficult cases. The major limitation is in terms of cost of ownership (COO) as a single-wafer process is involved. Indeed according to Witt et al. [17] who used the standard SEMATECH COO model, brush cleaning is more than three times more expensive than wet cleaning, which was confirmed by other economic studies [18].

Scrubber optimization is performed by adjusting the brushes and wafer rotation speeds, the DI water flow, and the brush height. The brush must be compressed 2 to 3 mm onto the wafer surface to come in direct contact

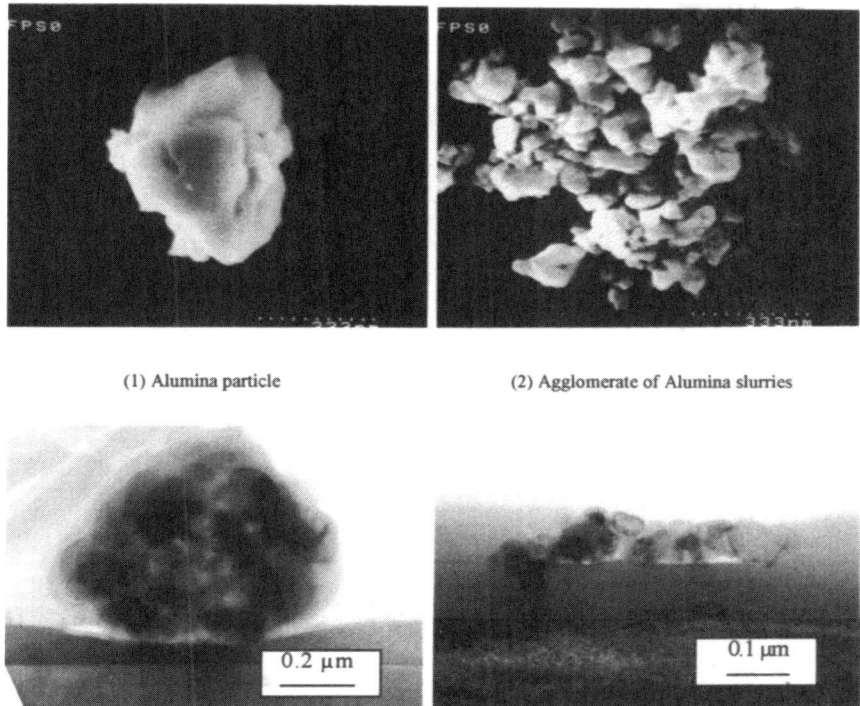

FIG. 17. Observation of alumina particles and slurries (Upper: SEM top views, tilt is 40°; lower: cross sections obtained by TEM).

with the wafer, which represents the only way to remove the fine particles due to the weakness of drag forces (see Table 1).

In the case of tungsten or copper CMP where alumina slurries are used, the pH of the solution must be greater than 9 or lower than 2 to avoid adhesion of the slurries in the porous structure of the brush (back to Fig. 13). This phenomenon, called the "loading effect," increases the final particle levels on the wafers and therefore drastically reduces the brush lifetime. This effect can be greatly attenuated by injection of 0.5 to 2% ammonia, for example.

A development reported by Zhao *et al.* combines mechanical brush scrubbing with *in situ* oxide underetching mechanism in a single tool [19]. The HF diluted at 0.5 to 1% injected in the second brush of the scrubber optimizes slurry removal.

3. WET CLEANINGS

In conventional wet benches or spray tools, the wafers are processed in batches of 25 or more, which leads to lower COO. This is the main advantage of wet cleanings. As demonstrated earlier [18], the slurries from the market unfortunately contain foreign particles as well. Therefore wet processes must be able to remove both positive and negative particle types to reach a low final level.

a. Slurry Removal in Alkaline Media

Both silicon oxide and alumina slurries can be efficiently removed on PECVD TEOS oxide or silicon nitride substrates in a conventional SC1 or in a SC1 without any water peroxide in the case of outcropping tungsten (see Fig. 5). When water peroxide is not present to continuously regrow a protective oxide layer, OH^- species can etch the silicon. In the latter case, the backside of the wafer must therefore be protected with a nitride or oxide layer to avoid a severe silicon roughening effect. Nevertheless to achieve the same particle removal efficiency obtained with a scrubber, power megasonics also have to be used (see Fig. 18).

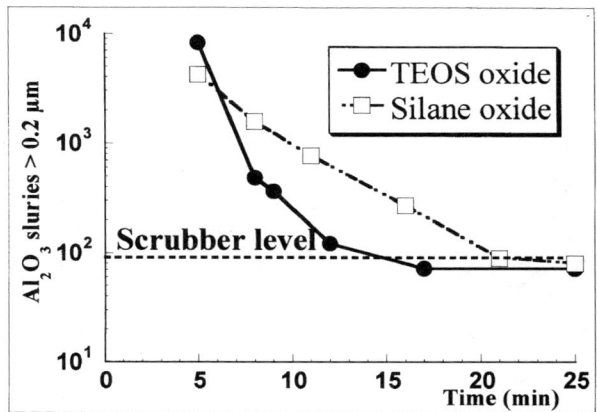

FIG. 18. PECVD TEOS oxide: slurry removal efficiency in ammonia-based chemistries (0.25, 6) performed at 55°C during 10 min. Diluted ammonia with megasonics gives results as good as the scrubber.

b. Slurry Removal in Acid Media

HF-based chemistry is particularly interesting due to its compatibility with all back-end metals and barriers. Unfortunately as the absolute values of the zeta potential in the A area of Fig. 13 are lower than in alkaline media, the removal mechanism is even more difficult. Indeed as seen in Fig. 19, the particle removal efficiency in the HF–HCl mixture is almost zero for actual alumina slurries. Very-high-power megasonics performed in a specific HF-compatible bath are absolutely necessary to obtain the same good residual particle level as with the scrubber.

When using HF-based chemistries, the processing time must be limited to reduce the enhancements of the substrate defects which then appears as particles when reaching a certain size (see Fig. 20).

Note that, as expected, the underetching step must be enhanced in the case of actual slurries as compared to particles.

Finally, although the more direct way to remove slurries is still to use a scrubber, wet processes represent a cheaper alternative and achieve comparable residual particle levels on silicon oxides and silicon

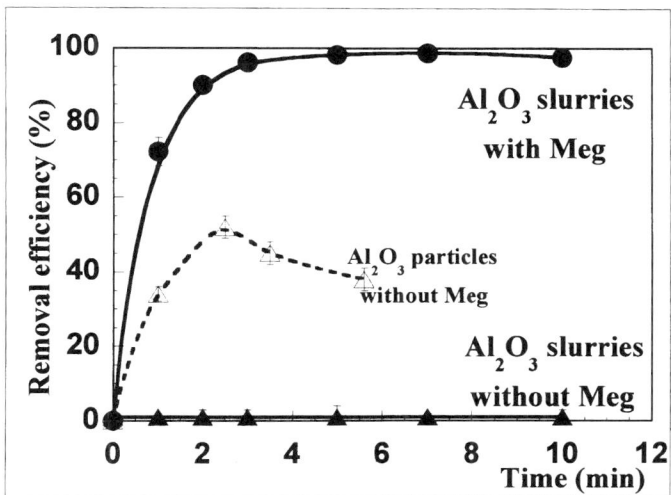

FIG. 19. Alumina slurry removal efficiency in HF mixtures on PECVD TEOS oxide. Behavioral differences between alumina particles and slurries and efficiency of a specific HF- compatible very-high-power megasonics tank. (Direct Coupling system from SubMicron Inc.)

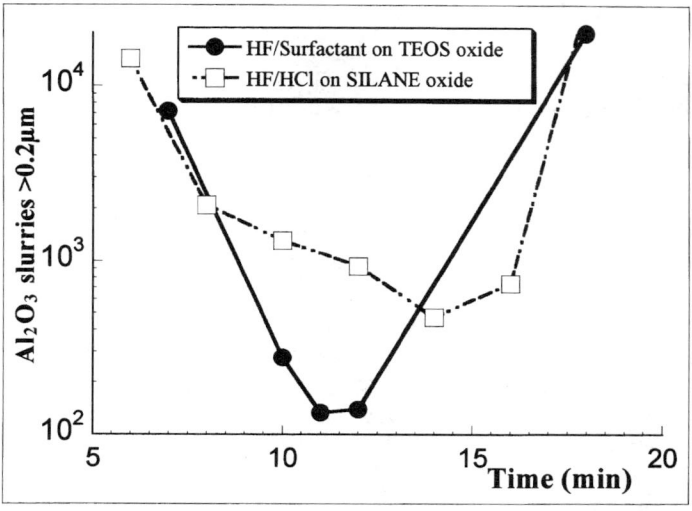

Fig. 20. Example of light point defect increases when using HF-based chemistries due to the revelation of different oxide defects. The processing time must be reduced in this case.

nitrides since strong megasonics are used [20]. Nevertheless when using wet cleanings on outcropping metals, the particle removal efficiency must be verified on the metals as well. This can be achieved by electrical defectivity tests because of the lack of sensitivity of light scattering methods on highly reflecting metals.

VI. Metallic Contamination Removal

1. Metallic Contamination during the Full CMP Step

The metallic contamination left by the polishing equipment as measured by total x-ray fluorescence (TXRF) is usually in the 10^{11} to 10^{12} at./cm^2 range. Furthermore metallic contamination can be deposited in the neutral (DI water) or basic (ammonia etc.) chemistries used during the following scrubber or wet cleaning steps [21]. In the case of IMD, with outcropping metals the scrubber brush is able to transport traces of the metal lines and barriers on to the adjacent insulator. As shown in Fig. 21, copper and to a lesser extent tungsten are transported when using 1% ammonia. In the case of copper, this behavior can lead to 100% electrical shorts. Zhao et al. [22]

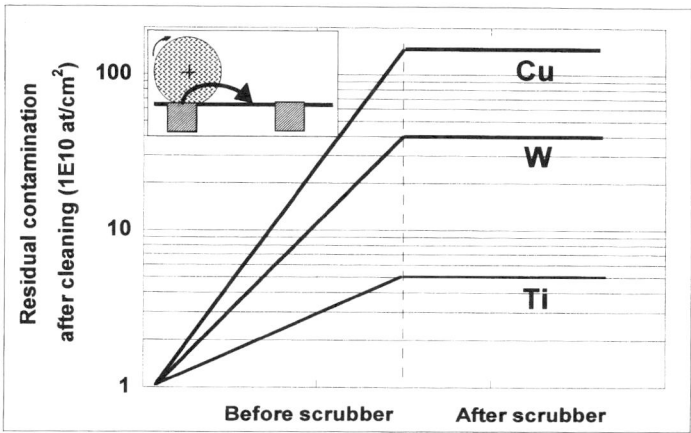

FIG. 21. Metal transportation during the scrubbing process.

demonstrated that this transportation occurs on oxidized copper layer by a brush loading phenomenon.

2. METALLIC CONTAMINATION CLEANING

Metallic contamination is mainly present on the wafer surface as adsorbed ions, oxides, hydroxides and salts. These species are generally instantaneously dissolved in acidic media, even in weak acids such as HF or citric acid. Furthermore HF-based chemistries are able to remove the contamination diffused into silicon oxides or nitrides by lift-off mechanism [23] even faster than a conventional SC_2 [1].

The residual noble metals present in their metallic form are usually oxidized by the dissolved oxygen naturally contained in solutions ($E_{0\,O_2/H_2O} = +1.23$ V in acid media) if no other oxidant species are intentionally added (case of IMD). As shown in Fig. 22, copper can be satisfactorily removed and therefore cleaned in HF mixtures since the dissolved oxygen concentration is sufficient. Of course, this cleaning effect does not occur when using a reducing agent as a corrosion inhibitor.

Unfortunately this mechanism can be drastically limited on substrates presenting even a few not fully oxidized silicon atoms, which can be oxidized by the noble metals in solution. These substrates are quite impossible to clean. They are therefore forbidden in the case of copper metallization. This

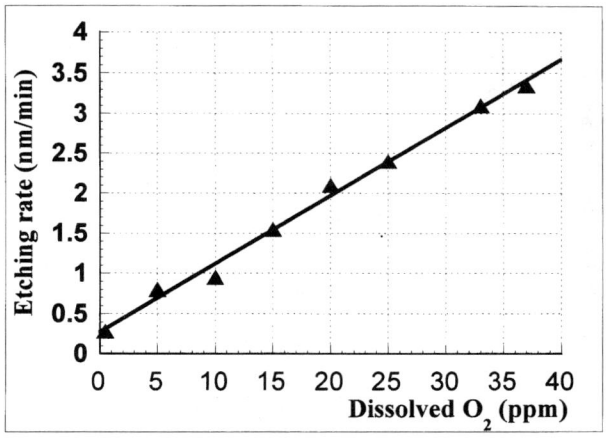

FIG. 22. Etching of copper in 0.1% HF by the dissolved oxygen. This moderated reaction enables us to clean the copper contamination with an acceptable consumption of the conductor lines.

is the case, for example, for SiON, unannealed PECVD TEOS oxide, and the like.

In the case of wet processes, if an alkaline step is used to remove the slurries, then an acid step is necessary to remove the metals.

When using a scrubber, an additional acid treatment must be implemented, preferably afterward. If an HF bath is used prior to the scrubber to improve global particle removal [24], the metallic cleanliness obtained can be preserved if a complexing agent is used in the scrubber process — EDTA in water or choline instead of ammonia. The acid treatment can be introduced in the second scrubber station: citric acid or HF. In the latter case, the scrubber has to be especially designed for corrosion and security reasons.

In conclusion, note that an acid step is always required at the end of the post-CMP cleaning process to remove the residual metallic contamination.

VII. Damaged-Layer Removal

The damaged layer left by the CMP on the insulator depends to a large extent to the polished material and CMP conditions. It can practically reach a thickness comprise between 1 and 10 nm. The thickness of this electrically not well-defined layer therefore must be known precisely for each different CMP condition.

1. PRACTICAL DETERMINATION OF THE DAMAGED LAYER

The damaged layer can be determined by measuring its chemical resistance by a succession of short dips in diluted etching mixtures. When the etching rate reaches a constant value (corresponding to the bulk material) the damaged layer is removed. Figure 23 shows that both 0.1% HF and hot ammonia lead to the same thickness.

Optical methods such as x-ray specular reflectivity and specular ellipsometry give very similar thicknesses to the etching method [24]. This validates to a certain extent the just described procedure, which is much easier to use.

A not well-defined layer, which probably exists on the outcropping metal lines, may also have to be removed.

2. ELIMINATION OF THE DAMAGED LAYER

In the case of the scrubberless approach, the damaged layer is in practice generally removed during the underetch particle removal process (ammonia

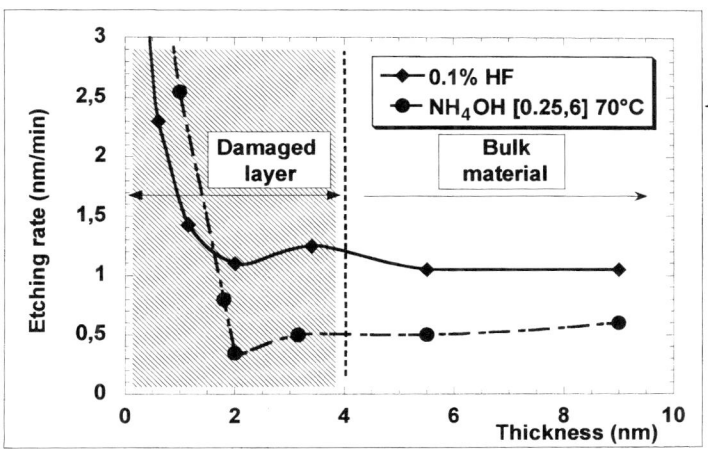

FIG. 23. Example of damaged layer estimation in PECVD TEOS oxide as measured by ellipsometry using 0.1% HF and [0.25, 6] NH_4OH 70°C.

or HF). If the damaged layer is particularly thick, the particle removal process can be prolonged.

When using a scrubber without any etching step (HF), an additional etching bath can be implemented in order to remove the damaged layer.

We must ensure that the etching steps performed during the cleaning processes are sufficient to remove the damaged layer as well.

VIII. Final Passivation

In the case of aluminum, tantalum, titanium, and tungsten, the oxygen and the humidity of the air form a protective oxide layer that passivates the surfaces. This oxide is generally removed just before the next process step. In the case of copper, Cu_2O and CuO are not water- and airtight. The oxidation reaction can therefore continue, leading to a severe corrosion of the copper lines. This behavior is further enhanced if the lines are connected to the two parts of a junction (see Subsection 2 in Section IV). As suggested by V. Brusic *et al.* [25], corrosion inhibitors such as benzotriazole (BTA) can be introduced in the final cleaning step. For over half a century BTA has been used against atmospheric copper corrosion. (1H-BTA) reacts on Cu, Cu_2O, and CuO to form a stable layer of 0.5 to 4 nm of (Cu–BTA), which protects the surface from any further corrosion. The compatibility of this layer with the following processes must be closely verified. The wafers can be stored under nitrogen as well, for example in INCAM type containers [26].

IX. Examples of Practical Post-CMP Cleaning Processes

As CMP materials are constantly progressing, it would be delusive to propose definitive practical recipes. However, taking into account all the elements previously set out, some cleaning processes using a scrubber or wet chemistries adapted to some typical CMP situations can be proposed for example purposes (see Table II).

- The simple scrubber process is very efficient to eliminate slurries but does not remove the metallic contamination or the damaged layer. The simplest additional process is to use an HF-based step which removes both of them. The use of an HF-compatible scrubber saves an additional wet bench with a dryer and a wafer transfer. The chemistries used must avoid loading effects.

TABLE II

EXAMPLES OF POST-CMP CLEANINGS PERFORMED BY SCRUBBING OR BY WET PROCESSES

Application	Configuration		Scrubber		Wet process	
	Wafer	Slurries	Station 1	Station 2	Step 1	Step 2
STI	Si_3N_4	SiO_2	Water	HF	SC1 Meg.	HF dip
Oxide	SiO_2	SiO_2	Water	HF	SC1 Meg.	HF dip
Tungsten	SiO_2	Al_2O_3	NH_4OH Meg.	HF +surfactant	Hot dNH_3 Power Meg.	HF/HCl dip
Copper	SiO_2	Al_2O_3	Water +surfactant +corrosion inhibitor Meg.	HF +surfactant +corrosion inhibitor	HF +surfactant +corrosion inhibitor Power Meg.	

- Diluted HF-based chemistries are potentially very interesting in the case of scrubberless processes as they are theoretically able to remove the slurries, metallic contamination, and the damaged layer in a single step. Furthermore, they are compatible with the presence of all the metal lines and barriers in the case of IMD CMP. The only limitation of HF-based chemistries is in terms of defect enhancement of some materials. Limiting the underetching process time induces the use of expensive powerful megasonics especially designed for this application. This is why ammonia-based chemistries are still interesting if they are compatible with the materials present at both surfaces of the wafer. In this case, a short 0.1% HF dip is required as a final step to remove the metallic contamination or better in some cases a 0.1% HF—1% HCl dip is required to prevent any particle readhesion.

Scrubber and wet benches are sometimes used one after the other in the same post-CMP cleaning, but this probably does not represent the most suitable solution. In the near future, the more rational way will be to integrate a scrubber or megasonic bath in the CMP tool to avoid handling wet wafers.

1. STI AND SILICON OXIDE CMP

- SiO_2 slurries are quite easy to remove and do not produce any brush-loading effect since the commercial slurry used does not contain too many foreign particles. The simplest scrubber process consists in a

conventional first stage in water to remove the silicon oxide slurries followed by a second HF scrubbing to remove the metallic contamination and the damaged layer. This process could be performed in a single HF step but single-brush stations usually result in less repeatable performances.
- On a wet bench, a conventional SC1 with megasonics to remove the particles followed by an HF dip to remove both the damaged layer and metallic contamination gives satisfactory results.

2. TUNGSTEN CMP

- At the first station, where alumina slurries are removed, ammonia must be added to avoid a loading effect. The introduction of megasonics at this stage can in certain cases provide a slight advantage. The second HF stage then removes both the damaged layer and the metallic contamination. An anionic surfactant may be added if required to enhance the brush lifetime.
- The efficient scrubberless alternative consists in using hot diluted ammonia in a specific bath with very high megasonic power. In this case, the backside surface must be protected with an oxide or nitride layer to prevent a severe silicon roughening effect from occurring. Then an HF–HCl dip enables the metallic contamination and damaged layer to be removed. HCl turns the respective zeta potentials into favorable conditions that limit the particle redeposition.

3. COPPER CMP

- Although ammonia at room temperature only slightly etches copper, here it is preferred to reduce the loading effect by an anionic surfactant. Then the damaged layer and the metallic contamination are removed at the second station by HF. If the metallic copper transported by the brush is not removed in this step, an additional HF dip with a high dissolved oxygen content is required. Nonreducing corrosion inhibitor agents must be added at both of the stations.
- Only the new generation of very high power megasonic tanks compatible with diluted HF can be used in this case. As HCl is not acceptable with copper, annihilation of the attractive particle–substrate forces can be achieved by using an anionic surfactant at it miscelanies concentration. A nonreducing corrosion inhibitor agent has to be added as well.

X. Conclusion

Understanding post-CMP cleaning processes requires some basic knowledge of electrochemistry and colloid sciences. We hope that this chapter gives a sufficiently thorough outline of the theoretical knowledge and methodologies for engineers to be able to adapt the cleaning chemistries and tools to the varying natures of polished surfaces and CMP processes. Developments in the post-CMP cleaning field must be continuously in progress to follow the fast-moving changes in the polished materials. New techniques will probably emerge to complete the advantages offered by today's two approaches: scrubber and wet processes. This represents a very interesting challenge for IC engineering in the forthcoming years.

Acknowledgments

The author would like to acknowledge P. Patruno, J. Torres, J. Palleau, L. Liuzu, T. Lardin, I. Constant, L. Mouche, H. Bernard, F. Mondon, C. Tabone, and A. Wright for their active contributions.

References

1. W. Kern and D. Poutinen, *RCA Rev.* 31, p. 187 (1970).
2. *Atlas d'équilibres électrochimiques à 25°C*, M. Pourbaix, ed., Gauthier-Villars & Cie, Paris (1963).
3. H. Nielsen and D. Hackleman, *J. Electrochem. Soc.*, vol. 130, no. 3, March (1983).
4. R. E. Hines, *Micro. Reliab.*, vol. 11, no. 6, p. 537, December (1972).
5. *Handbook of Chemistry and Physics*, CRC Press, 71st ed., pp. 8–24 (1990).
6. L. Zhang and S. Raghavan, *Mater. Res. Soc. Symp. Proc.*, vol. 477 (1997).
7. P. C. Hiemenz, *Principles of Colloid and Surface Chemistry*, Dekker Press, New York, 2nd ed. (1986).
8. M. B. Ranade, *Aerosol Sci. Technol.* vol. 7, p. 161 (1987).
9. L. Mouche, F. Tardif, and J. Derrien, *J. Electrochem. Soc.*, vol. 141, no. 6, June (1994).
10. A. C. Tam, W. P. Leug, W. Zapka, and W. Ziemlich, *J. Appl. Phys.* Vol. 71, p. 3515 (1992).
11. M. Takenaka, Y. Sato, A. Ishihama, and K. Sakiyama, *Mater. Res. Soc. Symp. Proc.* vol. 386, p. 121 (1995).
12. M. Meuris, M. Heyns, P. Mertens, S. Verhaverbeke and A. Plilipiosian, *Microcontam. Rev.*, May 92 (1992).
13. L. Mouche thesis, Faculté des Sciences de Luminy, Marseille, France (1994).
14. R. P. Donovan, T. Yamamoto, and R. Periasamy, *Mater. Res. Soc. Proc.* vol. 315, p. 3 (1993).
15. D. J. Shaw, *Introduction to Colloid and Surface Chemistry*, Butterworths (1966).
16. I. Constant, F. Tardif, R. J.-M. Pelleng, A. Delville, ECS Fall Meeting, Hawaii (1999), to be published.

17. K. Witt, M. Jolley, and P. Burke, CMP-MIC Conference, Santa Clara (1997).
18. F. Tardif, I. Constant, T. Lardin, O. Demoliens, M. Fayolle, and Y. Gobil, MAMS '97, Villard de Lans, France (1997).
19. E. Y. Zhao, R. Emanmi, I. Malik, K. Mishra, W. C. Krusell, J. de Larios, and D. J. Hymes, *Mater. Res. Soc. Symp. Proc.*, vol. 477 (1997).
20. B. Morrison, M. B. Olesen, and G. Liebetreu, CMP-MIC Conference, Santa Clara (1997).
21. L. Mouche, F. Tardif, and J. Derrien, *J. Electrochem. Soc.*, vol. 142, no. 7, July (1995).
22. E. Zhao, L. Zhang, H. Li, D. Hymes, J. de Larios, and W. Krussel, CMP-MIC Conference, p. 359 (1998).
23. F. Tardif, T. Lardin, C. Paillet, J. P. Joly, A. Fleury, P. Patruno, D. Levy, and K. Barla, ECS Fall meeting (1995).
24. I. Constant, S. Marthon, T. Lardin, C. David, M.N. Jacquemond, and F. Tardif, ECS fall meeting, Paris (1997).
25. V. Brusic, M. A. Frich, B. N. Eldridge, P. P. Novak, F. B. Kaufman, B. M. Rush, and G. S. Frankel, *J. Electrochem. Soc.*, vol. 138, no. 8, August (1991).
26. C. Doche and M. Morin, IES Symposium, Orlando FL, May (1996).

CHAPTER 8

CMP Metrology

Shin Hwa Li, Tara Chhatpar, and Frederic Robert

STMICROELECTRONICS
PHOENIX, ARIZONA

I.	INTRODUCTION	215
II.	REFLECTOMETRY	216
	1. *Substrate Modeling*	218
	2. *Measurement Patterns and Points*	220
	3. *Measurement of Patterned Wafers*	224
	4. *Integration Issues*	225
III.	DEFECTIVITY MONITORING	226
	1. *Laser Scanning Method*	228
	2. *Digital Image Comparison Method*	228
IV.	NONCONTACT CAPACITIVE MEASUREMENT	229
	1. *Flatness*	231
	2. *Bow and Warp*	232
V.	TOTAL X-RAY FLUORESCENCE	233
VI.	STYLUS PROFILOMETRY (FORCE MEASUREMENT)	236
VII.	ATOMIC FORCE MICROSCOPY	236
VIII.	FOUR-POINT PROBE	241
	REFERENCES	243

I. Introduction

Most metrology tools in modern integrated circuit fabrication plants were not designed for CMP applications, but rather for the other modules: thin films, diffusion, photolithography, and etching. As the CMP processes have become more and more important in silicon processing for sub-0.35-μm design rules since 1995, the CMP module has been forced to adopt almost all of the existing tools, often facing incompatibilities, misinterpretation, and subcapabilities in doing so. For example, when a tool measures the thickness and thickness uniformity of a film across a wafer, typically a pattern with the measurement points radially located (called a polar map) is used, which is appropriate and sufficient for thin films. In chemical vapor deposition (CVD) or physical vapor deposition (PVD), the film thickness deposited does not vary drastically from the center to the edge. Therefore, a polar map can better represent a circular wafer statistically. In CMP, however, the polish nonuniformity behavior tends to be center slow, and edge extremely fast, because CMP is a process in which the wafer carrier retaining ring and the polish pad play significant roles. It is more reasonable to use a contour map (a square lattice of measurement sites), because the nonuniformity in CMP will be incorrectly represented using the polar map. A difference of 2–10% in nonuniformity may be found between the measurements using the contour and polar maps.

In this chapter, many such CMP-related metrology issues are discussed for the most-often-used tools.

II. Reflectometry

In an oxide CMP process, the film thickness and nonuniformity are two of the most important process control parameters [1]. The most commonly used method to measure these parameters is reflectometry. When white light illuminates a wafer, the light interacts with the topmost films. The light can be absorbed, refracted, and/or reflected. In this particular method, a significant portion of the light must be reflected from the sample. With fixed incident angle, polarization, and wavelength of the incoming light [2], the reflected light can undergo constructive or destructive interference depending on the thicknesses and refractive indices of the films that beam penetrates (see Fig. 1). The resultant reflected light is collected using a photomultiplier tube, and the reflected intensity (reflectivity) is plotted against the wavelength (see Fig. 2). The system also calculates a theoretical spectral response curve based on the nominal thickness entered and the light dispersion model representing the films and the substrate. The equations of the dispersion model of each film can be expressed as

$$n = n_1 + \frac{n_2}{\lambda^2} + \frac{n_3}{\lambda^4}$$

$$k = k_1 + \frac{k_2}{\lambda^2} + \frac{k_3}{\lambda^4}$$

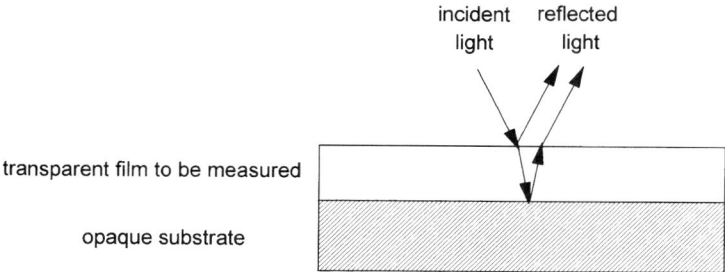

FIG. 1. Schematic of light refracted and reflected in reflectometry.

where n is the refractive index of the film, k is the Cauchy coefficient (also called extinction coefficient), and λ is the wavelength. The film thickness is determined by mathematical iteration of the theoretical curve, using various thickness values (or whichever quantity is to be measured) until the theoretical and measured curves match [3].

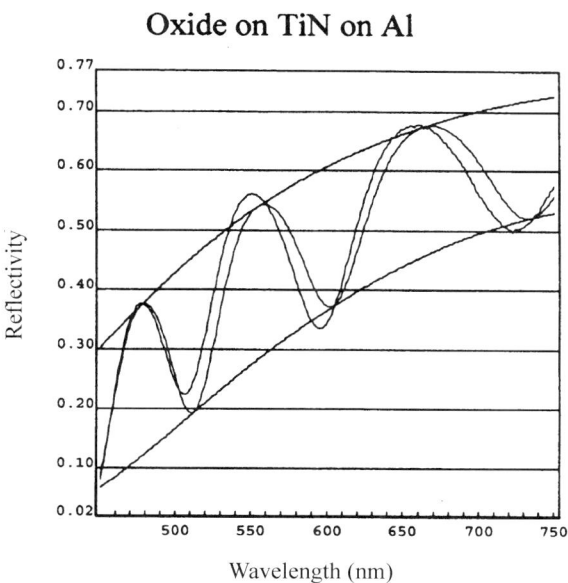

FIG. 2. Empirical and theoretical response spectra of a TEOS–TiN–Al stack using the reflectometry. The TiN and TEOS thicknesses are 250 and 10,000 Å, respectively. The empirical and theoretical curves are matched within a tolerance.

In general, reflectometry has advantages over other dielectric-film-thickness measurement methods, such as the ellipsometry, in that it has much less possibility of order skipping that leads to incorrect data, and it is significantly faster. The origin of the so-called order skipping is multivalued solutions. When the theoretical and empirical spectral response curves are being matched, a value of thickness that enables the match is not a singular solution, but a set of different values that occur cyclically (with 2000- to 3000-Å increments for the case of oxide thickness on silicon substrate, for example). Depending on the nominal or initial values chosen, the determination of the final value can be different. Noise in the recorded data, insufficient discrimination in the cycle-selection algorithm, and interference introduced by other effects in the film stack can all contribute to an incorrect final value [4].

In reflectometry, the wavelength range typically used is that of visible light, approximately 4000 to 8000 Å. When ultraviolet (UV) light is incorporated into the measurement, the range extends to include wavelengths of 2000–4000 Å. The UV light can help to increase the resolution, to determine thickness less than 250 Å. In the case of premetal dielectric (PMD) or interlayer dielectric (ILD) CMP processing, the thicknesses to be measured far exceed 250 Å, so the UV range is unnecessary. For a shallow trench isolation (STI) process, on the other hand, where oxide is polished away either completely or to where only a few hundred angstroms remain, the UV range is necessary.

To enable reflectometry to provide accurate, reproducible, and efficient measurements, several factors must be considered. Choice of the substrate material, substrate modeling, number of measurements per wafer, choice of the measurement patterns, and the setup of the pattern recognition program are all critical to the measurement process, as discussed in the following.

1. SUBSTRATE MODELING

In reflectometry, the light passes through the films to be measured. Beneath the transparent films, there must be an opaque substrate through which light does not pass. The substrate characteristics must be modeled correctly to calculate the thicknesses of the films above. In silicon processing, theoretically, any of the commonly used metal materials, such as the titanium nitride (TiN), aluminum (Al), and tungsten (W), can be used as substrates. However, in reality, whereas a PMD oxide can be measured on the polysilicon material used in poly interconnections, an ILD oxide can not be measured directly on TiN, because the TiN layer used is too thin to be opaque. TiN is semitransparent if its thickness is less than 1000 Å. A thin

TiN film of approximately 300 Å is typically used in the back-end interconnect process, as both the cap layer for the aluminum metal deposition sequence and an antireflective coating for the subsequent photolithography step. Since this TiN cannot be a substrate for oxide thickness measurement in ILD, the aluminum beneath the TiN must be used as the substrate. In other words, the TiN is a component of the film to be measured. Thus, its refractive index or thickness must be known to determine the unknown oxide thickness. However, the refractive index of TiN is not constant, but varies with thickness. As a result, the TiN thickness must be precisely controlled to enable the validity of the substrate modeling.

Also shown in Fig. 2 are the measured spectrum of a TEOS (tetra-ethyl-ortho-silicate)–TiN–Al stack, overlaid with the theoretical spectrum. The theoretical spectrum was based on such substrate modeling treatment. Given this empirical spectrum, both the TiN and TEOS thicknesses can be simulated and determined simultaneously. Alternatively, the TiN thickness can be fixed with only the TEOS needing to be measured. The latter method is better than the former because the more unknown film thicknesses there are, the more error is introduced. This can be demonstrated by the data in Table I, where the measurement results of the oxide thicknesses on TiN–Al substrates on the same wafer, for TiN thickness fixed and varying, are compared.

We see from the thickness nonuniformity that if the TiN thickness is fixed, the oxide thickness has a lower standard deviation, indicating a tighter distribution of the measurement results. Since the measurement is on the same wafer, the difference suggests that the effect of fixed TiN thickness to help improve the repeatability of the measurement. (Of course, the TiN film must be uniform for this to be true.) In short, if the control of the TiN

TABLE I

COMPARISON OF OXIDE THICKNESS MEASUREMENT ON TiN/AL SUBSTRATE WITH THE TiN THICKNESS FIXED AND AS A VARIABLE IN THE RESPONSE SPECTRA SIMULATION

	Oxide thickness (Å)/standard deviation			
	Trial 1	Trial 2	Trial 3	Mean
TEOS measured TiN fixed Al substrate	8485.67/5.25%	8485.68/5.25%	8485.76/5.25%	8485.70/5.25%
TEOS measured TiN measured Al substrate	8464.96/6.15%	8465.07/6.16%	8465.49/6.14%	8465.17/6.15%

thickness and uniformity is stable in a wafer fabrication plant, it is preferable to fix the TiN thickness in the substrate modeling so that only the TEOS thickness on top of the stack is a variable.

Another aspect that must be accounted for in substrate modeling is scattering. Some substrates chosen have very rough surfaces and tend to scatter the light. Common examples of such nonspecular surfaces are PVD aluminum and CVD tungsten [3]. Scattering due to a rough surface interferes with the normal interaction of the light with the film materials and thicknesses. To reduce this kind of noise, special-modeling algorithms must be added. In addition, for the trench oxide thickness measurements in STI, it is a concern that the substrate for the films (silicon) may be damaged by plasma in the dry etch process used to form the trench. Even though the substrate is silicon, which is well known and for which most metrology tools have built-in algorithms, the substrates must be reevaluated and remodeled as "plasma-damaged" silicon.

2. Measurement Patterns and Points

As mentioned in the chapter's introduction, the measurement pattern will affect the results. Figure 3 shows four types of patterns, namely, (1) diameter scan, (2) polar map, (3) contour map, and (4) 9-point contour map. For the 49-point polar map there are 24 points at the edge (49%). For the 49-point diameter map, there are 2 points at the edge (4%), and for the 21-point contour map there are 8 points at the edge (38%). Since CMP has a very strong edge effect in polish rate, this makes the measurement pattern an important factor in the measurement results. In this section, the results using a diameter scan and a contour map with the same or fewer data points are compared to the conventional type of mapping (49-point polar map, all with 6-mm edge exclusion). In addition, a 9-point contour map with no edge point is also compared [5] (sites also shown in Fig.3). The reflectometric measurement results for 47 wafers are illustrated in Figs. 4a, 4b, 5a, and 5b. The wafers have nominally 9000-Å TEOS oxide after approximately 2000 Å of oxide are removed. Figures 4a and 4b are the raw data, and Figs. 5a and 5b are the curve fitting using a linear regression method. In all figures, the bisecting line (slope equals one) represents the result using the 49-point polar map. We observe, from Fig. 5a, that for the mean value the 21-point contour pattern is nearly equivalent to that of the 49-point polar map. However, for the diameter scan and the 9-point contour map, the results tend to be smaller than those for the 49-point polar map because fewer edge data points are used. On the other hand, for the value of the standard deviation in percentage (actually standard deviation divided by mean, as

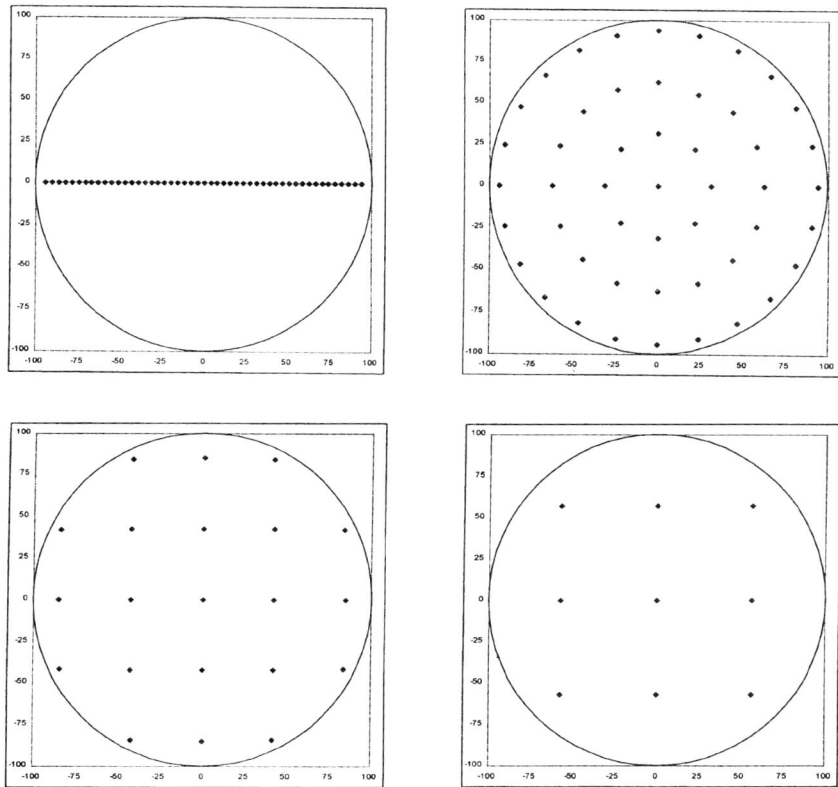

FIG. 3. Different types of measurement patterns and points: (a) 49-point diameter scan, upper left; (b) 49-point polar map, upper right; (c) 21-point contour map, lower left; and (c) 9-point contour map, lower right.

seen in Fig. 5b), all the results from other measurement patterns give lower numbers than that of the 49-point polar map, if the 49-point polar map result is higher than 3.5%. This shows that if a CMP tool does not have good uniformity performance, a polar map measurement will exaggerate the results.

Since reflectometry is a major metrology tool in CMP processes, another important issue is the number of measurement points on the wafer that are required to determine the film thickness and uniformity without sacrificing cycle time. Table II presents a comparison of number of data points vs the measurement efficiency and accuracy using the polar map pattern on the

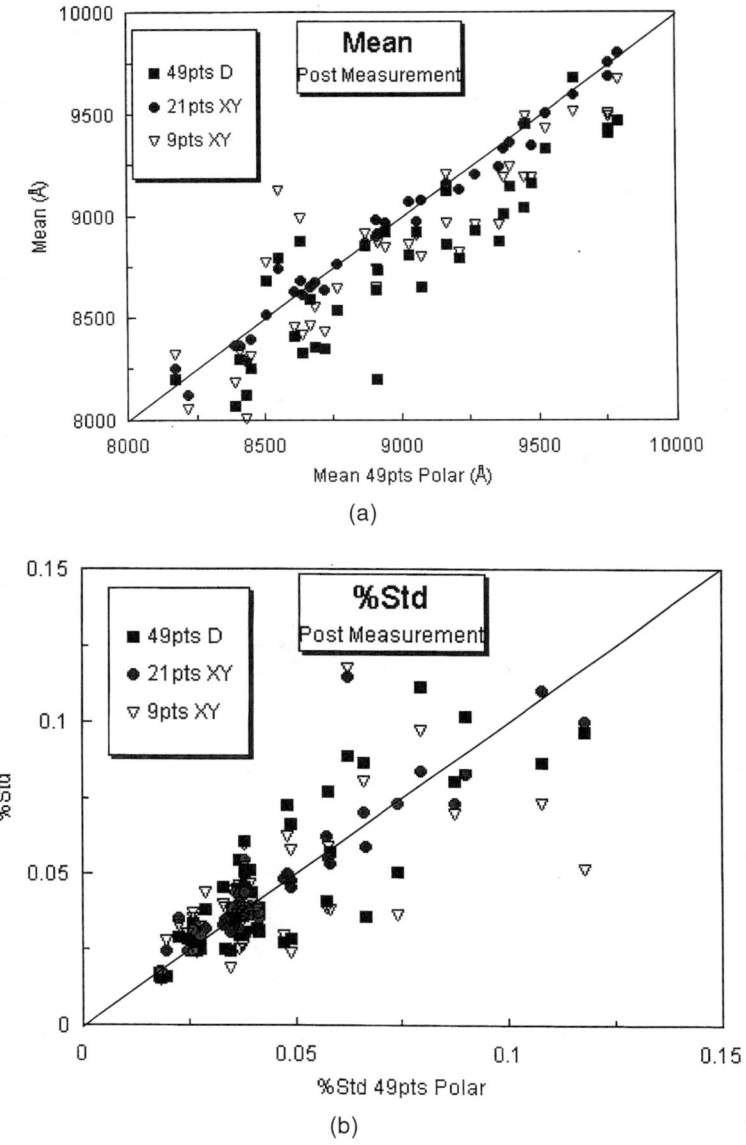

Fig. 4. Comparison of the raw data from 49-point diameter, 21-point contour, and 9-point contour measurement patterns to 49-point polar pattern data, showing (a) mean and (b) the nonuniformity (standard deviation divided by mean in percentage) using reflectometry.

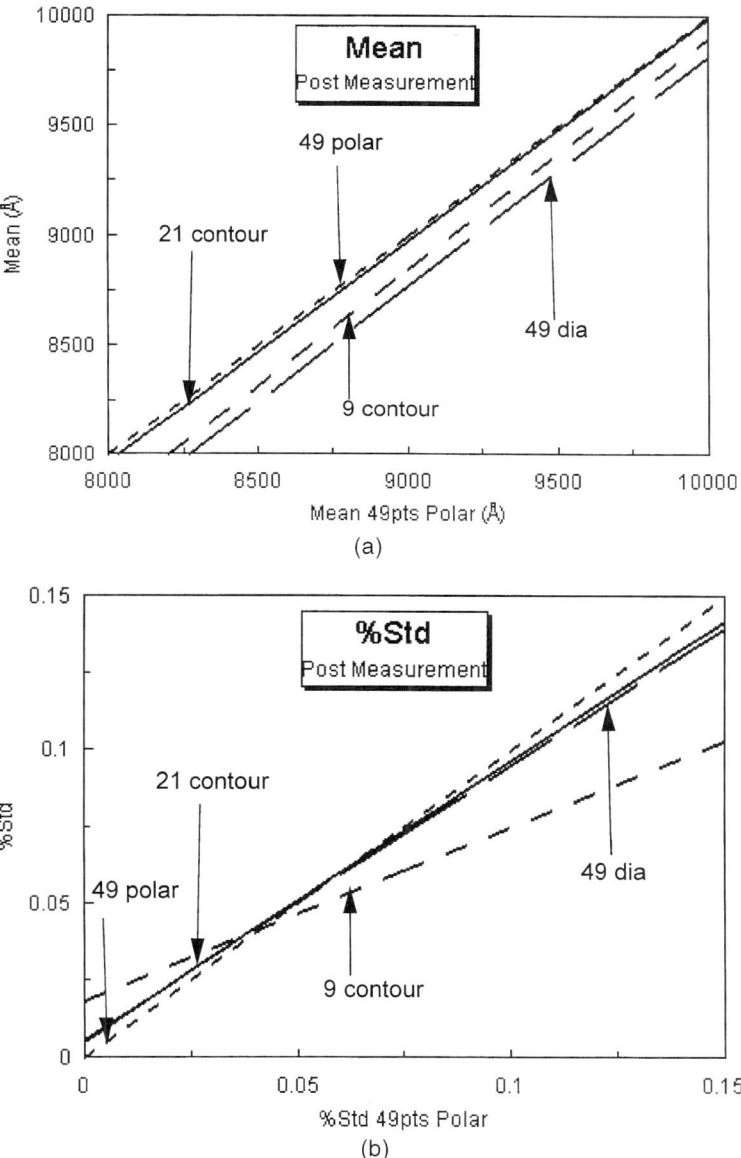

FIG. 5. Comparison of linear regression curve fitting for the data of Figs. 4a and 4b showing (a) mean and (b) the nonuniformity.

TABLE II

NUMBER OF MEASUREMENT POINTS VERSUS NONUNIFORMITY ACCURACY AND TIME CONSUMED FOR THE SAME WAFER

Number of measurements (points)	Time (min) to measure	Nonuniformity across the wafer (%)
5	2:05	6.49
9	2:45	5.09
15	3:35	5.25
30	4:20	4.34
49	7:25	4.78

same wafer. We see that the accuracy is saturated as the measurement sites are increased from 30 to 49 points. However, the time used for 30 points is only 60% of that needed to measure 49 points.

3. MEASUREMENT OF PATTERNED WAFERS

In addition to the measurement of monitor wafers, and more importantly, a reflectometry tool must be able to measure the patterned wafer in a manufacturing environment. The tool must be capable of measuring identical sites of different dies in a wafer. In general, to measure a patterned wafer, the alignment features of the pattern must be taught and recorded into software so that the tool can look for the identical places. In many cases, the equipment has pattern recognition software built in. Sometimes these features may be difficult to identify after the wafer has been polished, however, because occasionally polish nonuniformity may result in pattern color variation, which can alter the pattern contrast. Also, after a lamp is changed, the image brightness and contrast may change. In these cases, the recorded alignment features work for some dies but fail for others. Some tools have the added capability to enhance alignment features, whereby several pixels in the recorded pattern are combined together making lines thicker and more defined. This makes the memorized pattern more identifiable, independent of the contrast and brightness variation [4].

To understand the impact of a CMP process on a certain product with a unique integrated circuit pattern, it is desirable to measure areas with different feature sizes and shapes. Since CMP polish rate may be affected by pattern density, areas encompassing various features should be included in the measurement program. The within-die thickness nonuniformity will indicate the planarization capability of a CMP process.

4. INTEGRATION ISSUES

To measure the final oxide thickness in the PMD process, the measurement sites can be set up over field oxide or over the polysilicon interconnections (see Fig. 6). However, since there are fewer variables in measuring over the field oxide, and the field oxide process is relatively well established, the PMD thickness is more accurately measured over the field oxide. If the measurement is over the polysilicon, the resultant PMD thickness measured can be affected by the deposited polysilicon thickness, the polysilicon doping, and the field oxide thickness variation, while if the measurement is over the field, the PMD thickness is affected only by the field oxide thickness.

For the ILD process, it is less feasible to continue measuring over field oxide, because multiple oxide layers are present (also see Fig. 6). The true thickness and thickness uniformity of the particular level of an ILD process can be confounded by the presence of the underlying previous layers. Thus, after the first metal process, it is preferable to measure the oxide thickness directly on a metal line.

One of the metrology issues with the STI process is that the process utilizes three layers of different materials: (1) thin thermal oxide (less than 200 Å), (2) nitride (approximately 1500 Å), and (3) the TEOS oxide above the active regions (see Fig. 7). Ideally, the CMP process polishes the TEOS oxide and stops at the nitride. In reality, after the polish, either a very thin residual TEOS oxide is present or the TEOS is completely gone and the nitride thickness is being measured. This poses some problems in the setup

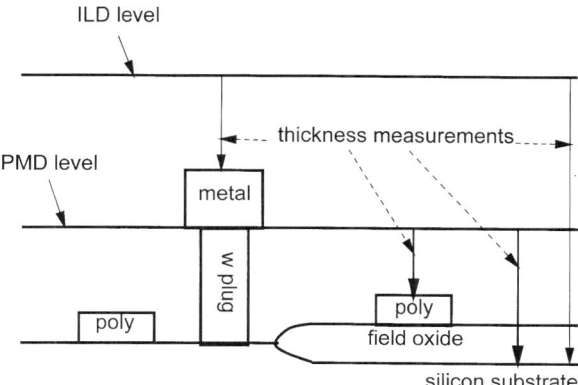

FIG. 6. Schematic of a cross section of a typical multilevel CMOS (complimentary metal oxide semiconductor) integrated circuit structure.

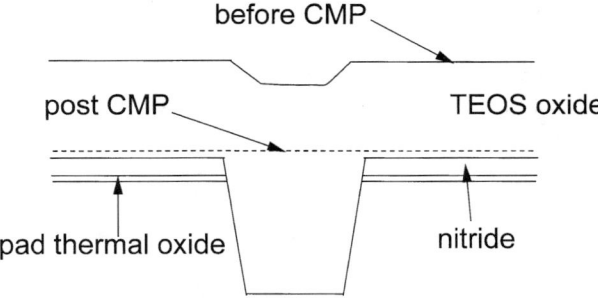

FIG. 7. Schematic of a cross section of a typical STI structure at the trench areas.

of measurement programs; it is difficult to set up a program to satisfy two different kinds of film stacks, one with two layers and the other with three layers. In addition, there is an issue with the measurement in the trench areas: the oxide thickness measured may be affected by the trench depth, which has been defined by the preceding trench etch process. If the trench depth is too deep or too shallow, the oxide thickness measured after CMP may incorrectly indicate that the oxide is under- or overpolished, respectively.

III. Defectivity Monitoring

As new technologies make CMP indispensable, new problems arise in the area of defectivity. Compared to the established planarization processes of TEOS deposition, etchback, and the like, where knowledge of the formation of defects is relatively well established, the CMP processes have introduced a whole new category of potential defect types. The physical contact of the CMP pad and slurry with the wafers are new variables to deal with. For example, if the slurry is not etched or rinsed off completely in the post-CMP cleaning process, slurry particles can be left on the surface as defects. Figure 8 shows a scanning electron microscope (SEM) micrograph of typical slurry particles. Furthermore, if a slurry drum is not stored in a proper environment with its allowed ranges of temperature and humidity, the slurry abrasives (silica for the oxide process, for example) tend to agglomerate to form large particles. These large particles can cause scratches on the oxide surface. A typical CMP scratch is indicated in Fig. 9.

Defect detection on a wafer after CMP can be accomplished by a laser scanning technique and/or by a digital image comparison technique. Both types of tools are widely used in the industry. However, neither is sufficient

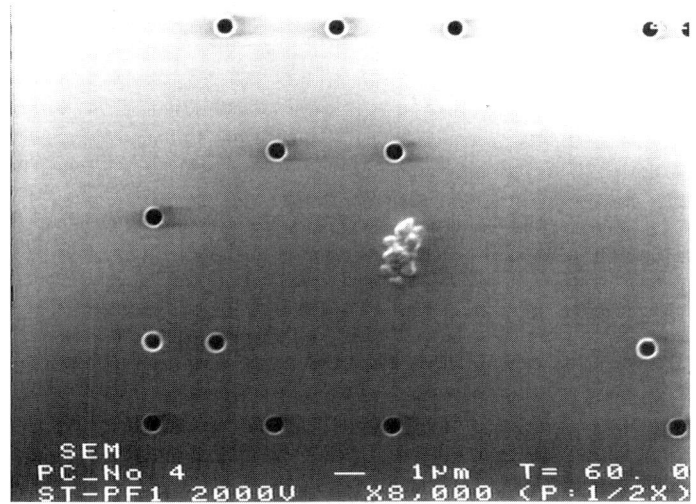

FIG. 8. SEM micrograph of a typical oxide CMP slurry residue after CMP.

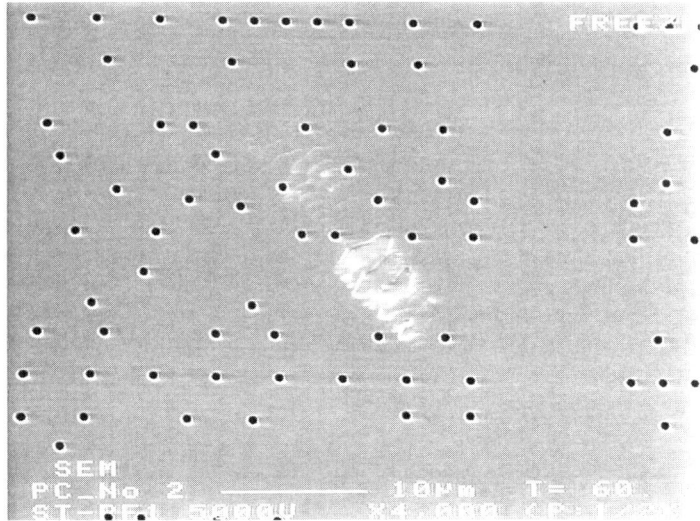

FIG. 9. SEM micrograph of a typical CMP scratch with the chatter signature.

to overcome the challenges of CMP processes. The problems in post-CMP inspection include oxide color variation due to CMP polish nonuniformity, pattern variations at and below the surface, and grainy substrates, such as metal and polysilicon, which reduce the signal-to-noise ratio [6]. Maximizing signal while consistently suppressing noise is the major challenge of the next generation of defect inspection tools. Currently, both laser scattering tools (primarily for blank monitor wafers) and digital image comparison tools (primarily for patterned wafers) provide complementary applications for CMP. It is probable that both technologies will continue to grow to be further utilized [7].

1. Laser Scanning Method

A common technique for measuring defects on unpatterned wafers employs a laser to scan across the entire wafer. If a defect is present, light is scattered away at the point of incidence. A photomultiplier tube collects the scattered light, whose magnitude is proportional to the size of the particle [8]. Laser scanning can also be used to detect defects on patterned wafers using darkfield light scattering [7]. Scattered laser light is detected using a charge-coupled diode (CCD) camera. Through an image processor and the intensity calculations, defects can be detected and quantified on the wafer. Sensitivity settings, such as laser angle, repetitive pattern filtering (also called Fourier masking), and polarization, are adjusted to maximize the number of defects detected and to minimize false identifications [6].

Occasionally, complications arise because the threshold and gain used are unable to distinguish signal from noise. Very small defects may not be detected because they do not scatter enough light. Another limitation is that the technique is applicable only for detecting nonplanar defects, which scatter light. Defects of a particulate nature such as slurry residue in Fig. 8, for example, can be easily detected, but microscratches like those in Fig. 9 are more difficult to detect because they tend to be more planar to the detector. Small-laser-angle scattering may help slightly, but is still not entirely satisfactory. In any case, the advantage of laser scanning over digital image comparison is that it is very fast and suitable for inspecting a high volume of wafers in a manufacturing environment [7]. Its disadvantage is that the tool may not pick up defects smaller than 0.5 μm.

2. Digital Image Comparison Method

Another method of defect detection is digital image comparison. The tool works by comparing a pixel of one die to that of the preceding and succeeding die. If there is a difference in contrast value, that coordinate is flagged as a defect. Inspection of post-CMP wafers using this type of tool can be very difficult, due to the inevitable color variations caused by thickness variation across the wafer [9]. If the detection program is set to be too sensitive, these color variations are translated into different contrast values, and are regarded as defects. When sensitivity is lowered, the inspection tools have difficulty detecting subtle defects such as CMP scratches directly after the polishing process [9]. To our advantage, such small CMP-related defects can be easily detected after the TiN barrier film is deposited over them, because the TiN layer serves as an antireflective coating that eliminates the light reflection or interaction from below. This benefit, however, comes with the drawback that these defects are caught several processing steps (photolithography, contact–via, Ti–TiN deposition) after CMP, rather than immediately following the polish. Accordingly, there is a great risk of misprocessing several lots before the CMP scratching problems are flagged.

Advances have been made to reduce the color variation problems following CMP in this kind of tool. New, complex computer algorithms have been introduced. Figures 10a and 10b present the maps of a wafer scanned by the digital image comparison method using the standard algorithm versus the same wafer scanned using a new algorithm. Unwanted false defects displayed in Fig. 10a are shown successfully filtered in Fig. 10b.

IV. Noncontact Capacitive Measurement

A noncontact capacitive probe can be used to measure wafer total thickness, flatness, bow, and warp. This tool was very popular in the early CMP process development days. Since CMP is a planarization process, it is intuitively understandable that the initial wafer flatness, warp, and bow will affect the polishing performance [11,12]. The gauging system is based on a simple concept: measuring the capacitance of an air gap. A wafer is put between two opposing-capacitance probes (see Fig. 11), above and below the wafer. At each measurement site, the probe measures a pair of signals corresponding to the distance from a probe to the wafer's front side (d_{upper}) and to the backside (d_{lower}), respectively. The thickness of the wafer (t) is equal to the difference between the size of the probe gap and the sum of the probe distance signals ($d_{\text{upper}} + d_{\text{lower}}$) and is independent of the wafer position between the probes [10].

$$t = d_{\text{total}} - d_{\text{upper}} - d_{\text{lower}}$$

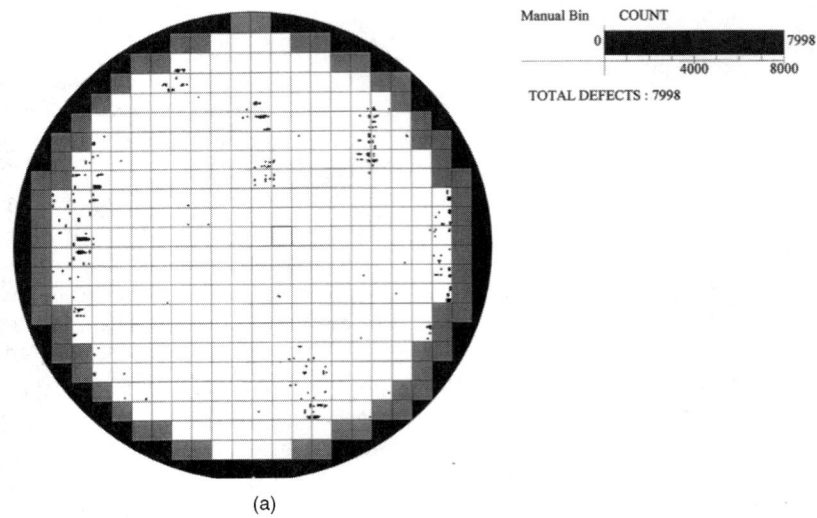

(a)

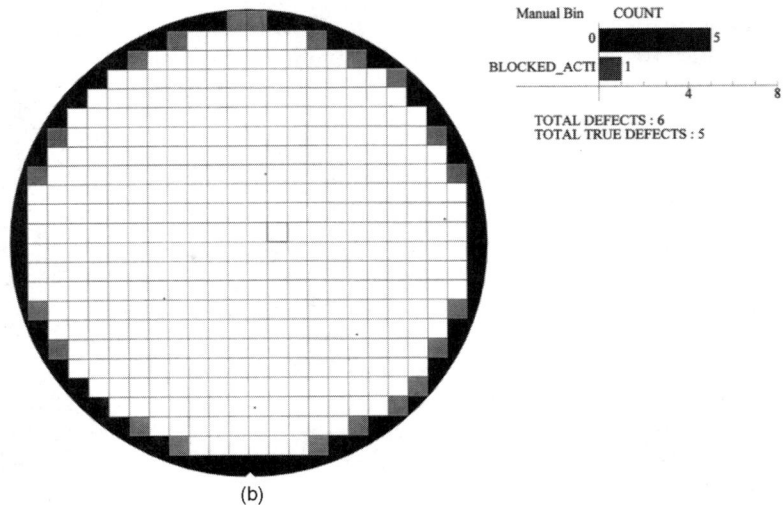

(b)

FIG. 10. Defectivity observed by the digital image comparison method: (a) wafer map showing color variation problem after CMP, and (b) the same wafer using a modified computer algorithm.

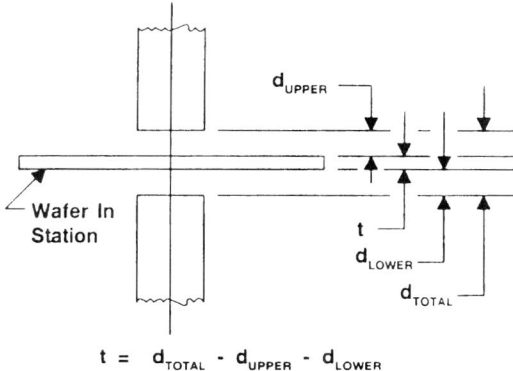

FIG. 11. Schematic of the noncontact capacitance measurement setup. (Courtesy of ADE Corporation, Westwood, MA.)

Therefore, the thickness of a wafer can be deduced. Beyond this, and more importantly, the tool enables the roughness, referencing to the upper and lower probes, to be measured, which can be converted to the so-called flatness, bow, and warp. Flatness, warp, and bow are defined in the following sections.

1. FLATNESS

If a wafer has both rough front and back sides, it is impossible to define "flatness." To define "flatness," an "ideal flatness" of a wafer backside must be assumed. In other words, a wafer is assumed to be placed on a perfectly flat vacuum chuck, as described in Fig. 12. The roughness shows only on the front side of the wafer. From the variation of the thickness, an imaginary plane can be obtained based on the modeling algorithm. The distance from

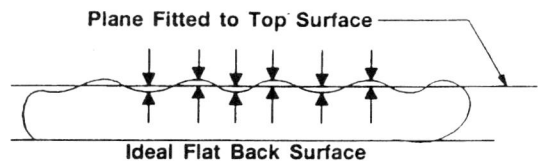

FIG. 12. Schematic of a wafer put on a perfectly flat chuck. (Courtesy of ADE Corporation, Westwood, MA.)

each individual measurement site to the plane is defined as a "focal plane deviation" (FPD), which is the characteristic of the "flatness." Another useful parameter is called "total thickness variation" (TTV), which represents the distance between the minimum and the maximum thickness measured [13]. Usually, as long as the initial TTV is less than 5 μm, the post-CMP polish uniformity will not be significantly affected [11].

2. Bow and Warp

By measuring the points on both the front side and backside surfaces, without putting the wafer on a perfect plane, the system can determine two things: (1) an imaginary reference plane (can be global and local) and (2) the wafer's median, nonplanar surface, which is the midpoints between the front side and backside rough surfaces. Note that at this point, no "ideal flatness" is used for the backside, as depicted by Fig. 13. Bow is a measure of concave or convex deformation of the median surface at the wafer center, which is calculated by measuring the deviation of the centerpoint of the median surface relative to the global reference plane. Warp is the difference between the maximum and minimum deviations of the median surface relative to the local reference plane [13]. Typically, the bow and warp must be controlled less than 40 μm for CMP process in order not to cause problems in nonuniformity.

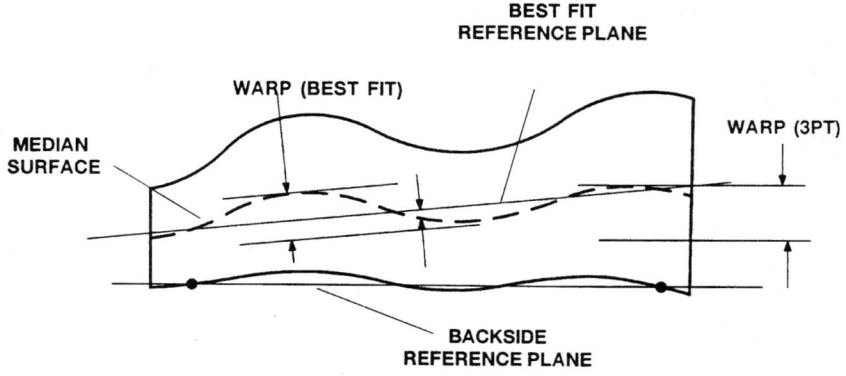

FIG. 13. Schematic of a wafer shape measured using a noncontact capacitance probe. The warp and bow are exaggerated. (Courtesy of ADE Corporation, Westwood, MA.)

V. Total X-Ray Fluorescence

Total x-ray fluorescence (TXRF) is a commonly used technique in CMP for postclean evaluation to determine contamination. In this technique, the x-ray source is generated using high-energy electrons hitting either tungsten (W) or molybdenum (Mo) targets. The x rays from different targets have different spectra of wavelengths [14]. Filters, such as nickel (Ni) or zirconium (Zr), with different thicknesses, can be used to control the needed x-ray wavelengths and intensities. The x rays are then applied to the materials to be analyzed. The x-ray photons knock out the inner shell electrons in the atoms of an element. This causes the outer shell electrons to fall into the vacancy in the inner shells, emitting photons. The phenomenon is illustrated in Fig. 14 for light and heavy elements. Because different elements have different electron shell configurations and energies, the photons emitted from various elements will have their own characteristics. From these emission spectra, the elements in a sample can be both qualitatively and quantitatively determined. Another variable in the x-ray analysis is the angle of the incident beam. As seen in Fig. 15, if x rays are shining on the sample at a vertical position, the x rays will penetrate deeply, approximately 1000–10,000 Å, depending on the x-ray intensity and the materials. If the x rays are coming from a small angle, the penetration will be less than 100 Å, which is useful for surface analysis, particularly of the metal contamination

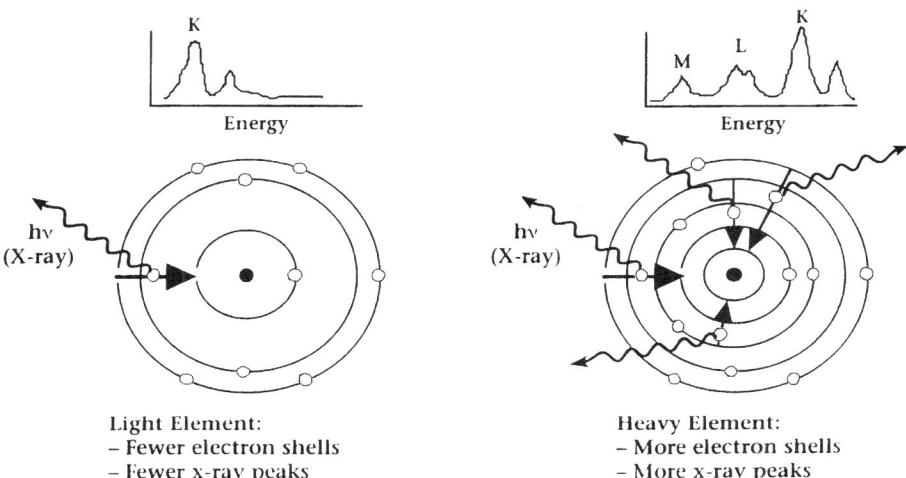

FIG. 14. Schematic of the x-ray fluorescence from light and heavy elements. (Courtesy of Charles Evans and Associates, Redwood City, CA.)

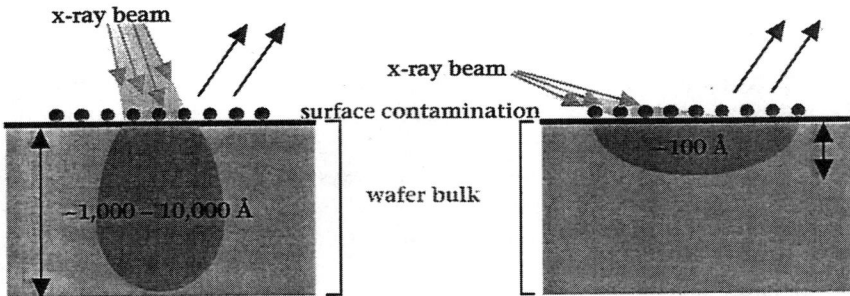

FIG. 15. Schematic of the large-angle and small-angle of x-ray fluorescence. (Courtesy of Charles Evans and Associates, Redwood City, CA.)

post-CMP process. Usually, the angle of incidence is kept below the critical angle at which total reflection occurs. The critical angle is given by

$$\phi_c = 3.72 \times 10^{-11} \frac{\sqrt{n_e}}{E}$$

where n_e is the electron density of the solid and E is the energy of the x-ray quanta [15, 16].

TXRF can be used to determine the contamination present on, in, and below the wafer surface (see Fig. 16). Changing the angle of incidence, while remaining below the critical angle of incidence, reveals the nature of the

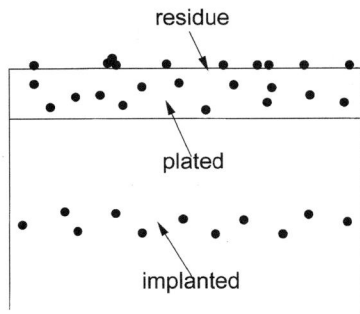

FIG. 16. Illustration of metal contaminants on (particulate), in (plated), and below (bulk) a surface.

measured surface contamination. As shown in Fig. 17, the measured signal from particulate contamination does not change when varying the angle, while the measured signal from contamination plated on the surface increases with increasing angle of incidence. Both curves decrease after the critical angle is surpassed because more x rays penetrate deep. The curve for the bulk contamination continues to increase with increasing angle of incidence.

A typical use of TXRF in CMP development is the study of post-W CMP surface metal contamination. Table III shows the surface metal residual atoms measured by TXRF after W CMP using different slurries (from vendors A, B, and C) and with or without an oxide buff process [17]. The control was a TEOS wafer without any CMP processing. We clearly see that slurry C has unacceptably high Fe contamination both with and without the buffer process. This probably is due to the fact that slurry C is $Fe(NO_3)_3$ based while slurries A and B are not.

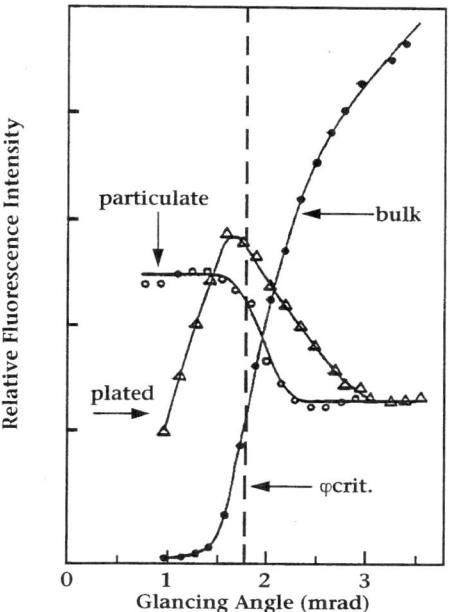

FIG. 17. Glancing angle dependence of relative fluorescence intensity of TXRF for each of the contamination situations. Here $\phi_{crit.}$ is the critical angle. (Courtesy of Charles Evans and Associates, Redwood City, CA.)

TABLE III

TXRF Metal Surface Concentrations for Wafers Polished with Various Slurries, in Dimensions of $E10/cm^2$

Slurry	K	Ca	Fe	Cu	Zn	Ti	Ag	Co
A	8.6	11	141	0.13	5.5	119	135	0
B	10.3	41.7	81	1.1	4	114	121	0
C no buff	10.4	4.6	3,324	0.3	1	3,469	148	12
C with buff	122	65	10,101	0.4	11	123	68	36
Control	2	0.7	0.7	0.5	1	0.2	7.8	0.2

VI. Stylus Profilometry (Force Measurement)

Stylus profilometry is a very simple and powerful tool in CMP. Profilometry can be used to determine the surface planarity change before and after CMP. Basically, in this technique, a stylus scans across a pattern feature in contact with a wafer, while the Z motion (height) of the stylus is monitored. This Z motion signal reflects the surface topography scanned.

The setup of a profilometer is very straightforward. The only parameters in the measurement are the stylus down force; scan rate; and choice of trench scan, mesa scan, or both. The technique can cover a large scan length, as long as several inches. However, depending on the down force and the scan rate, the resolution can vary. In general, the results are reliable if the range in surface topography is less than 1000 Å. If the step height in the topography is less than 500 Å, the resolution becomes questionable. The uncertainty comes from the surface roughness, which may vary from a few dozen to a few thousands angstroms. In addition, the stylus size and shape also play a role in the measurement accuracy. As demonstrated by Fig. 18, as a stylus scans through a trench, the actual signal observed can be distorted by the presence of the stylus itself. As the trench size decreases, the fidelity of the surface profile obtained degrades. Further, if the surface is significantly rough, the signal becomes totally meaningless. In these situations, atomic force microscopy (AFM) may be required to obtain surface topography; this technique is discussed in the next section.

VII. Atomic Force Microscopy

AFM is an advanced tool that is ideal for examination of microscopic surface topography. The main advantages of AFM over profilometry are its

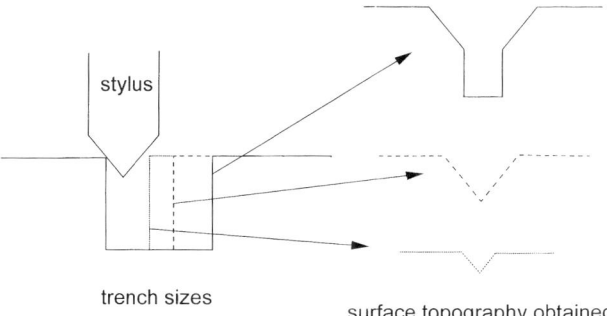

FIG. 18. Illustration showing how the surface topography obtained using profilometry is affected by the trench size and stylus shapes. As the trench size decreases to as small as a few microns, the scan may be confounded by surface roughness.

high spatial resolution, enabling it to function as an imaging tool, and the ultralow force exerted at the surface. An additional benefit is that, unlike scanning electron microscopy, which can image only on conducting materials, AFM can image on nonconducting materials, a prerequisite for the study of the dielectric films that are common in silicon processing. In AFM, a fine sharp probe that is attached to a flexible lever (cantilever) is scanned over a sample of interest. Features on the sample surface interact with the cantilever probe, due to interatomic (Lennard–Jones) potentials, and cause its position or motion to change.

AFM can be run in three different modes: contact, noncontact, and tapping mode. When AFM is in the contact mode (similar to stylus profilometry), the most common problem encountered is that under ambient conditions, sample surfaces are covered by a layer of adsorbed gases consisting primarily of water vapor and nitrogen. In addition, a dielectric film can trap electrostatic charge, which can contribute to additional attractive forces between probe and sample. These problems may cause friction in probing, which will destroy the sample or distort the resulting data.

An alternative technique is noncontact AFM [18]. Figure 19 illustrates the concept. The tip oscillates above the surface, and the modulation in amplitude, phase, or frequency of the oscillating cantilever in response to force gradients from the sample can be measured to indicate the surface topography. Even without contact, the amplitude, phase, or frequency can be affected by the van der Waals forces of the sample within a nanometer range, which is the theoretical resolution; however, this effect can be easily blocked by the fluid contaminant layer, which is substantially thicker than

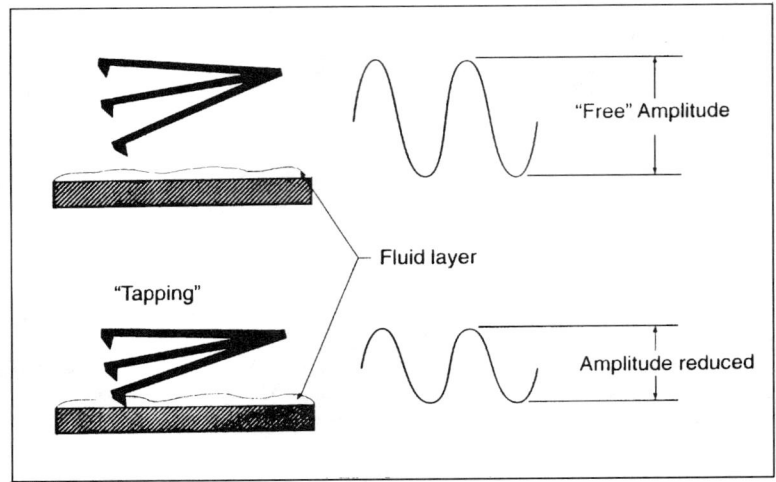

FIG. 19. Differences between noncontact and tapping mode AFM. The signal for the former is dependent on the change in oscillation due to the force gradient, while the latter is dependent on the oscillation change due to the contact. (Courtesy of Digital Instruments, Veeco Metrology Group, Santa Barbara, CA.)

nanometers. While the noncontact approach can avoid the scratch problem of the contact mode, this method cannot truly improve the resolution.

The most useful configuration of AFM for CMP purposes is its tapping mode. The tapping mode is a method to achieve high resolution without inducing destructive friction forces. Also shown in the schematics of Fig. 19, the cantilever oscillates near its resonant frequency as it is scanned over the sample surface. The probe is brought closer to the sample surface until it begins to intermittently contact ('tap') on the surface. This contact with the sample causes the oscillation amplitude to be reduced. This change in oscillation enables the signal to reflect the surface topography accurately with no friction or damage incurred. Figure 20 further illustrates the tapping mode probe in free air and in contact with sample. We see that a laser interferometer can be used as a position sensor to detect the oscillation change [19, 20].

A good example of using AFM in CMP development is the study of surface roughness and W plug recess after W CMP using various slurries [17]. Table IV presents a comparison of the surface roughness. The measurements were taken at the center and edge of 200-mm wafers for different areas (large or small). Obviously, slurry A leads to the best post-W CMP smoothness for every measurement. This does not come as a surprise,

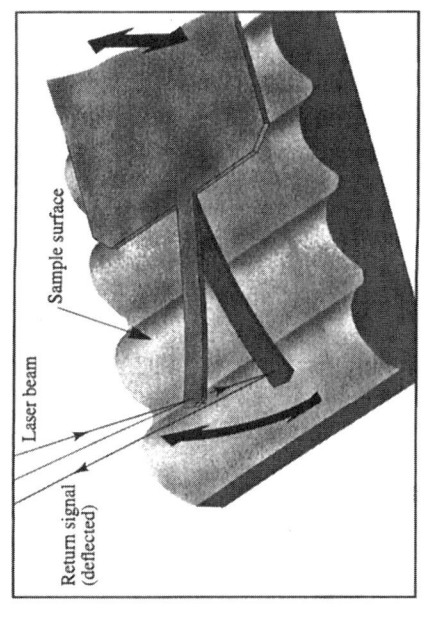

FIG. 20. Illustration of the actual AFM probe in tapping mode above and in contact with the surface. A laser beam is used to help amplify the change in oscillation. (Courtesy of Digital Instruments, Veeco Metrology Group, Santa Barbara, CA.)

TABLE IV

AFM SURFACE ROUGHNESS COMPARISON (RMS STANDS FOR ROOT MEAN SQUARE)

Position	Slurry Area (μm)	A Z range (nm)	A RMS (nm)	B Z range (nm)	B RMS (nm)	C with buff Z range (nm)	C with buff RMS (nm)
Center	10 × 10	3.1	0.2	120.6	19.9	55.9	1.1
Center	1 × 1	0.9	0.1	86.2	16.6	15.6	0.5
Edge	10 × 10	4.4	0.15	154.2	23.4	33	1.1
Edge	1 × 1	1.5	0.1	84.2	23.3	29.1	2.2

since slurries B and C use Al_2O_3 abrasives while slurry A does not. After the W is removed, the W CMP process actually polishes the TEOS oxide surface. If the abrasive is Al_2O_3, which is the one of the hardest materials known, we can expect that the surface will be rough and full of scratches. On the other hand, if the abrasive material is chosen to be one more similar to those for oxide CMP, as is the case with slurry A, the smoothness can be improved. It is also evident in Table IV that roughness can be ameliorated by using a post W CMP buff process.

The tungsten plug recess can be also easily measured using AFM. Table V presents an example of AFM measurements used to compare wafers after W CMP using various slurries. The measurements were taken at the centers and edges of 200-mm wafers, and from a dense array area or a sparse array area. The data indicate that although the recess was not significantly different among various slurries, slurry A tends to give slightly more recess. This may be due to the fact that the slurry A is H_2O_2 based and H_2O_2 dissolves tungsten (slurry B is KIO_3 based, and slurry C is $Fe(NO_3)_3$ based) [21].

TABLE V

AFM PLUG RECESS COMPARISON

Position	Slurry A (nm)	Slurry B (nm)	Slurry C with buff (nm)
Center, dense	24–36	5 ± 3	13–19
Center, isolated	50–80	10–25	24–32
Edge, dense	35–50	20–75	12–20
Edge, isolated	60–80	50–100	30–40

VIII. Four-Point Probe

Four-point probe is the most commonly used method to measure sheet resistance for metal films in silicon processing. Given known resistivities of each material, the film thickness can be obtained. Table VI lists the resistivities for the films often used in the back-end interconnect processes. The theory in four-point probe originally comes from Ohm's law:

$$R = \rho \times \frac{l}{A} = \rho \times \frac{l}{d \times w}$$

where R is the resistance, ρ is the resistivity, and l and A are the length and area, respectively, for the conducting materials (see Fig. 21 for the schematics). The area is that perpendicular to current flow, so equals the product of depth (d) and width (w). This equation can be rewritten as

$$R = R_s \times d \times \frac{l}{d \times w} = R_s \frac{1}{w}$$

$$R_s = \frac{\rho}{d}$$

$$d = \frac{\rho}{R_s}$$

where R_s is the sheet resistance. The sheet resistance has the units of ohms, but it is usually specified in ohms per square. This convention exists for convenience; when a distance unit l is divided by another distance unit w, a

TABLE VI

RESISTIVITIES FOR THE COMMONLY USED MATERIALS IN A METAL CMP PROCESS

Material	Resistivity $\mu\Omega$ cm	Reference
Al–0.5%Cu(PVD)	2.87	[22]
Ti (PVD)	56.85	[22]
TiN (PVD)	128.95	[22]
W (CVD)	10.2	[23]
Cu	1.8	[24]
Ta	13.1	[25]
TaN	~200	[26]
NbN	~50	[26]

PVD: physical vapor deposition.
CVD: chemical vapor deposition.

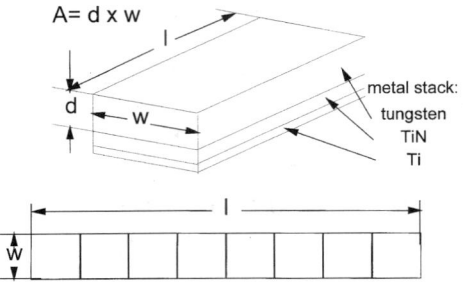

FIG. 21. Resistance of a metal block can be expressed as $R = \rho(l/A)$. The length divided by width is dimensionless, a square, or a multiple of squares.

dimensionless ratio (l/w), based on the concept of multiples of squares, results. As illustrated schematically in Fig. 21, this number of squares helps to stress that the value of a resistor can be given by the product of the number of equal square regions placed in a row times the sheet resistance, which is an intrinsic property of a film because of its assumed constant thickness [27].

A schematic for the four-point probe measurements is shown in Fig. 22. For an infinite sheet wafer (film on an unbounded wafer), the sheet resistance is

$$R_s = \frac{\pi}{\ln 2} \times \frac{V}{I} = 4.532 \times \frac{V}{I}$$

where V is the voltage, I is the current, and the $\pi/\ln 2$ or 4.532 is the shape correction factor for infinite dimension [28]. As the ratio between the sheet area (s) and probe distance (d) decreases, the correction factor decreases. Usually, the manufacturer of the tool provides these correction factors for various wafer sizes.

If a single metal film is deposited on an oxide, the sheet resistance measurement results can by easily interpreted and converted to the thickness. In practice, however, this is not usually the case. For example, in W CVD, the tungsten is not directly deposited on oxide due to high residual stress and unreliable adhesion. A titanium (Ti) layer must be first deposited as a glue layer. In addition, to prevent the fluorine in the CVD-precursor WF_6 from directly reacting with Ti (a strong catalytic reaction will occur), a barrier layer of titanium nitride (TiN) must be deposited on top of the Ti. As a result, we have a trilayer film of W on TiN on Ti on oxide, as shown schematically in Fig. 21. This poses some problems in accuracy in the four-point probe measurements. Based on the resistivities in Table VI, the

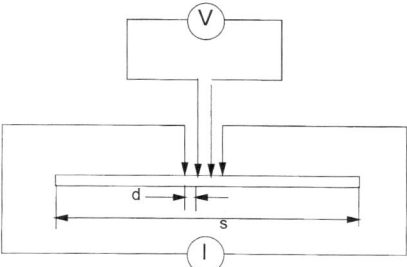

FIG. 22. Schematic of a four-point probe setup. The d/s value and its correction factor are typically provided by the equipment company.

thickness of the W film in the W–TiN–Ti metal stack must be larger than a certain thickness (also see Fig. 21); otherwise, the four-point probe measurements will incur a significant error. To be specific, assuming the thicknesses of Ti and TiN to be approximately 400 and 800 Å, respectively, the sheet resistance ratio between Ti, TiN, and W layers can be written as

$$R_{Ti}:R_{TiN}:R_W = \frac{56.85}{400}:\frac{128.95}{800}:\frac{10.2}{x} = 0.14:0.16:\frac{10.2}{x}$$

where 56.85, 128.95, and 10.2 are the resistivities of Ti, TiN, and W from Table VI. If x is, say, 100 Å, the resistance ratio between different layers will be 0.14:0.16:0.1, which are close to each other. In other words, current flux is shared significantly by each layer, rather than being confined to the W film only. On the other hand, if x equals 6000 Å, the resistance ratio will be 0.14:0.16:0.0017, where the resistance for the W layer is so much smaller than that of the Ti and TiN layers; that is, the conductance of the W film is so much larger. Thus the measurement is essentially that of the W layer alone, and the thickness derived will be accurate. Usually, there is no issue with the metal thickness measurement prior to W CMP. After the process, however, it may be necessary to take all three films into account.

REFERENCES

1. I. Ali, S. R. Roy, and G. Shinn, *Solid State Technol.* 37 (1994) 63.
2. S. E. Stokowski, *Proceedings of SPIE Integrated Circuit Metrology, Inspection, and Process Control IV*, vol. 1261, San Jose, CA, 1990, pp. 253–263.

3. *UV-1050 Thin Film Measurement System with StatTrax Version 7.0 User's Guide*, Prometrix (now KLA-Tencor, Santa Clara, CA) 1994.
4. W. L. Smith, K. Kruse, K. Holland, and R. Harwood, *Solid State Technol.* 39 (1996) 77–86.
5. F. Robert, STMicroelectronics, unpublished results, 1998.
6. J. Reynolds, A. L. Swecker, A. J. Strojwas, A. Levy, and B. Bell, *Solid State Technol.* 41. (1998) 39–44
7. A. E. Braun, *Semiconductor Int.* 21 (1998) 61–68.
8. P. Gise, *Solid State Technol.* 26 (1983) 163–156.
9. D. N. Schmidt, STMicroelectronics, internal publication, 1997.
10. Y. Zhang, R. Carpio, G. Schwartz, and A. Gonzalez, *Semiconductor Int.* Feb. (1997) 93.
11. Y. Zhang, P. Parikh, and H. Nguyen, *Proceedings of CMP-MIC (CMP Multilevel Interconnect Conference)*, Santa Clara, CA, 1996, p. 90.
12. Y. Zhang, P. Parikh, and B. Stephenson, *Proceedings of VMIC (VLSI Multilevel Interconnect Conference)*, Santa Clara, CA, 1996, p. 424.
13. *ADE Ultragage 9500 Operation Manual*, Rev. A, ADE Corporation, Westwood, MA, 1994.
14. B. D. Cullity, *Elements of X-ray Diffraction* (2nd edition, Addison-Wesley Publishing Co, London, 1978) chapter 1.
15. C. R. Brundle, C. Evans, and S. Wilson, editors, in *Encyclopedia of Materials Characterization* (Butterworth-Heinemann, Boston, 1992) chapter 5.
16. Charles Evans & Associates CEA Seminar Series, Charles Evans & Associates, Redwood City, CA, 1997.
17. S. H. Li, H. Banvillet, C. Augagneur, B. Miller, M. P. Nabot-Henaff, and K. Wooldridge, *Proceedings of CMP-MIC (Chemical Mechanical Polishing-Multilevel Interconnect Conference)*, Santa Clara, CA, 1998, p. 165.
18. Y. Martin, C, C. Williams, and H. K. Wickramasinghe, *J. Appl. Phys.*, 61 (1987) 4723.
19. C. B. Prater, and Y. E. Strausser, marketing materials AN2-5/94, Digital Instruments Veeco Metrology Group, Santa Barbara, CA, 1994.
20. *Dimension 5000 Scanning Probe Microscope Instruction Manual*, Digital Instruments, Veeco Metrology Group, Santa Barbara, CA, 1997.
21. E. A. Kneer, C. Raghunath, and S. Raghavan, *J. Electrochem. Soc.*, 143 (1996) 4095.
22. X. Breurec, STMicroelectronics, private communication, 1998.
23. M. McCafferty, STMicroelectronics, private communication, 1998.
24. C. H. Ting, *State-of-the-Art Seminar Visuals Booket of VMIC (VLSI Multilevel Interconnect Conference)*, Santa Clara, CA, 1998.
25. S. Wolf, Silicon *Processing for the VLSI Era* (Volume 2, *Process Integration*, Lattice Press, Sunset Beach, CA, 1990) p. 186.
26. S. M. Sze, *VLSI Technology* (2nd edition, McGraw-Hill Book Company, New York, 1988) p. 383.
27. R. S. Muller and T. I. Kamins, *Device Electronics for Integrated Circuits* (2nd edition, John Wiley and Sons, New York, 1977) p. 113.
28. S. M. Sze, *Physics of Semiconductor Devices* (2nd edition, John Wiley & Sons, New York, 1981) p. 31

CHAPTER 9

Applications and CMP-Related Process Problems

Shin Hwa Li, Visun Bucha, and Kyle Wooldridge

STMICROELECTRONICS
PHOENIX, ARIZONA

I. INTRODUCTION	245
II. OXIDE CMP WITHIN-WAFER NONUNIFORMITY (WIWNU)	246
1. Equipment and Consumables	246
2. Recipe Parameters	251
3. Effect of the End Effector	252
4. Deposit Less, Polish Less vs Deposit More, Polish More (Uniformity vs Planarity)	257
5. BPSG vs TEOS	261
III. POST-CMP OXIDE THICKNESS CONTROL	262
1. Lack of Endpoint Detection System	262
2. Wafer-to-Wafer Thickness Nonuniformity (WTWNU)	263
IV. DEFECTIVITY	265
1. Scratches	266
2. Other Defects	270
V. TUNGSTEN CMP PROBLEMS	273
1. Metal Contamination	273
2. Oxide Erosion	277
3. Polish Rate and Uniformity	277
VI. OTHER PROBLEMS	277
1. Missing Alignment Mark	278
2. Dishing	279
REFERENCES	280

I. Introduction

As the integrated circuit (IC) industry has chosen chemical mechanical planarization (CMP) as one of the indispensable processes in the generations of transistor gate lengths equal to or smaller than $0.35\,\mu\text{m}$, it is imperative that the CMP-related process problems be investigated and understood. Unlike processes in other areas, such as high-density-plasma etch (HDP-etch), electron-cyclotron-resonance chemical vapor deposition (ECR-CVD), and ion metal plasma physical vapor deposition (IMP-PVD), which are modified versions of existing technologies, the CMP process is of a new technology that has never been heavily used in the industry. Many problems either have not been reported or have no solutions yet. Nonetheless, the purpose of this chapter is to gather together this information and provide a better view of CMP-related problems in an IC manufacturing environment. Problem causes, and methods to prevent or reduce them from occurring, are discussed.

II. Oxide CMP Within-Wafer Nonuniformity (WIWNU)

Almost every CMP engineer's first problem to solve is the polish nonuniformity for oxide CMP. Since CMP is a newly adopted technology in the industry since the mid-1990s, there has been a terrible lack of experience and knowledge in this area. This is true not only for the engineers working for IC manufacturers, but also for the engineers of the CMP equipment and consumable vendors. As a result, when a CMP nonuniformity problem occurred, the engineers in manufacturing attributed this to equipment malfunction. The engineers of the equipment vendor suspected that the CMP consumables were not performing well, while those of the consumable vendors complained that the quality of incoming wafers was out of control. Actually, all these factors may play roles in the CMP nonuniformity problems, as discussed in the following.

1. Equipment and Consumables

If a manufacturing plant is in the beginning of a CMP learning curve, the nonuniformity problem can come from something as simple as an inexperienced technician not mounting the polish pad correctly on the platen, so that bubbles are trapped beneath the pad (see Fig. 1). The polish uniformity can be significantly degraded if the surface of a pad is not flat. In another instance, a wafer carrier is not rebuilt correctly, so that when a wafer is mounted, the wafer can not sit evenly, but is lopsided (see Fig. 2). Further, if a wafer carrier mounting film (also called an insert) is not cleaned thoroughly after polishing, the edge of the insert can be contaminated by slurry marks. The phenomenon of slurry marks is the result of the slurry touching the insert through the wafer notch during polishing, and the slurry

9 APPLICATIONS AND CMP-RELATED PROCESS PROBLEMS   247

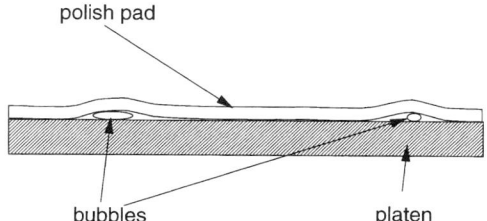

FIG. 1. Illustration of the loss of flatness of a CMP polish pad as air bubbles are entrapped beneath.

left on the insert not being cleaned afterward (see Fig. 3). A slurry mark-contaminated insert can cause CMP nonuniformity.

Beyond these problems, the choice of type of pad is very critical in reducing the CMP polish nonuniformity. A typical example is that in the beginning of the CMP development, there was no specially designed pad for oxide polishing in the back-end interconnect processes. CMP technology was originally used in lapping for silicon wafer polishing. The only purpose of lapping was to provide good polish uniformity, so soft pads were preferred to accomplish that goal. The typical soft pad, such as that with the commercial name of SubaIV, is made of polyurethane impregnated material [1]. As CMP has been adopted for back-end processes, the requirements have become stringent. CMP must provide not only good uniformity but also good capability for planarization. A kind of harder pad was designed to work with the soft pad. Its commercial name was the IC1000, and was made of a polyurethane material [1]. The harder pad (IC1000) was put on a soft pad (SubaIV) for CMP to fulfill the purposes of both uniformity and planarity. This kind of stacked pad was used and was very popular for a period of time, but it suffered from several disadvantages.

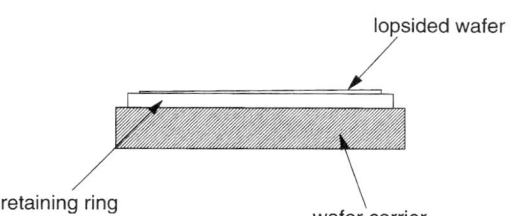

FIG. 2. Illustration of a lopsided wafer on a carrier due to the poor rebuilding of the wafer carrier.

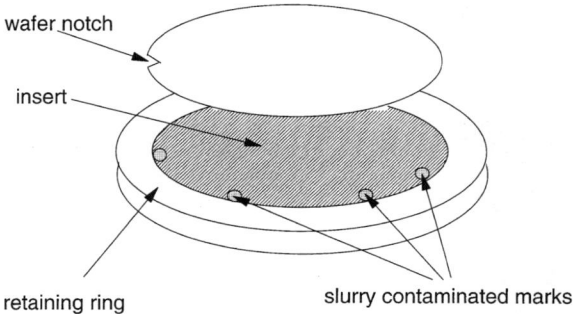

FIG. 3. Illustration of a contaminated insert due to the slurry dry-out through the wafer notch hole when not cleaned properly after polishing. Each mark was formed after each wafer was polished. These marks can accumulate to cause polish nonuniformity.

First, the top IC1000 pad was manually mounted on the bottom SubaIV pad. The performance of the combination of the pads varied with the skills of technicians who did the mounting. Second, the top pad was perforated, so that during polishing the slurry was allowed to go through the pad holes and stay between the top and bottom pads. The slurry could then react with the adhesive between the pads, causing breakdown of the adhesive, and leading to delamination of the top pad. Otherwise, the slurry could harden between two pads and form large particles so that the top pad would not be flat enough (the particles between pads behaving just like bubbles underneath the pad). In either case, the polish uniformity would be adversely affected (see Fig. 4 for an illustration).

Recently, machine-stacked IC1000–SubaIV pads have been available. Better yet, a special type of pad (commercial name: IC1400), consisting of a circularly grooved top pad plus foam-based bottom pad [1], machine-stacked), has been specifically designed for CMP in IC manufacturing. The bottom, foam-based pad can provide a better cushion to absorb uneven force, and the top, circularly grooved pad can better deliver slurry to the center of the wafer during polishing, hence increasing polishing rate [2]. More importantly, the top pad is not perforated, so that the slurry will not go between the top and bottom pads. Shown in Table I is a comparison between the IC1400 pad and the IC1000–SubaIV manually stacked pad from a design of experiments (DOE) using unpatterned wafers [3]. The IC1400 pad is termed pad A, and the IC1000–SubaIV combination of pads is termed pad B. The study was conducted on a single-head polisher, using a KOH-based slurry. From the results in Table I, we observe that the pad A outperformed pad B in terms of the polish uniformity (nonuniformity is

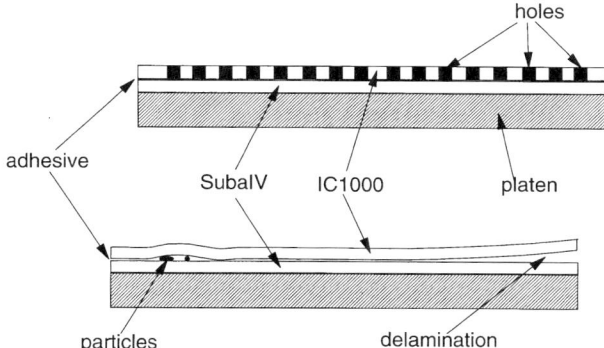

FIG. 4. Illustration of the failure modes of the combination of IC1000–SubaIV pads. The root cause is the slurry between the pads, which can form large particles or break down the adhesive.

less for pad A) for most of the experimental conditions, and its polish rate was also higher. Tables II and III demonstrate the split results for the patterned, production wafers for premetal dielectric (PMD) and interlevel dielectric (ILD) levels. Again, we see that the final thickness uniformity was improved for both levels using pad A. In addition, pad A sacrificed neither electrical results nor post-CMP planarity.

The disadvantages of pad A, as compared to pad B, are twofold: (1) the grooving imposing a pad-life limit, because once the grooves are worn out, the pad can no longer polish and (2) the difficulty in putting the whole pad successfully on a platen without entrapping bubbles; to accomplish this, more skilled technicians are required. Perforated pads do not have these problems.

Figure 5 shows a trend chart of polish nonuniformity. The data were collected from the daily qualification (three wafers per day) of a tool for a period of 1 month. The results indicate an unstable CMP process. As can be seen, the polish nonuniformity behavior was cyclic, gradually increasing and then suddenly dropping. These step-functional nonuniformity changes correspond closely to the polish pad changes. The reasons for these short-lived and run-away pads were probably the poor quality of the incoming pads; after this batch of pads was used up and a new batch of pads was introduced, the nonuniformity problem dramatically decreased.

Also apparent in Fig. 5 is that for the second half of the month, the tool lost consistency, with the nonuniformities of the three wafers tested each day varying significantly. In this period, the wafer-to-wafer problem was due to the previously mentioned effects: bubbles, delamination, poor pad choice, and so on.

TABLE I
THREE-LEVEL, FULL-FACTORIAL EXPERIMENT DESIGN, INTERACTION MODEL BETWEEN PAD A AND PAD B

Experiment no.	Down force (psi)	Platen speed (rpm)	Carrier speed (rpm)	Polishing rate (Å/min)			Nonuniformity (%)		
				Pad A	Pad B	A/B	Pad A	Pad B	A/B
1	5	20	25	1284	1310	0.98	4.6	8.92	0.52
2	9	20	25	2236	2150	1.04	4.3	8.25	0.52
3	5	36	25	2024	1952	1.04	7.4	8.08	0.91
4	9	36	25	3586	2485	1.44	4.9	4.47	1.1
5	5	20	39	1323	1350	0.98	6.7	12.5	0.54
6	9	20	39	2295	2195	1.05	5.7	6.85	0.83
7	5	36	39	2025	1924	1.05	7.4	7.49	0.98
8	9	36	39	3570	3109	1.15	4.6	7.17	0.64
9	7	28	32	2307	2185	1.06	4.2	8.43	0.5

TABLE II

SPLIT RESULTS FOR THE PMD LEVEL

	Wafers in split	Nonuniformity (%)		KLA particles	Intrametal leak	n, p contact	Planarity step height
		Mean	1 sigma				
Pad A	12	1.01a	0.28b	0.72c	<50 pA	Normal	Noise
Pad B	11	a	b	c	<50 pA	Normal	Noise

TABLE III

SPLIT RESULTS FOR THE ILD LEVEL

	Wafers in split	Nonuniformity (%)		KLA particles	Intermetal leak	Via resistance	Planarity step height
		Mean	1 sigma				
Pad A	12	0.78d	0.24e	0.84f	<50 pA	Normal	Noise
Pad B	9	d	e	f	<50 pA	Normal	Noise

2. RECIPE PARAMETERS

Typically, there are three principal parameters in a CMP recipe, namely, the down force, the platen rotation speed, and the carrier rotation speed. If we put the results of the DOE study from Table I into contour plots, it is easier to see the influences of each factor on the polish nonuniformity. Figures 6–8 show the relationships between recipe parameters and the polish rate and nonuniformity. We see, from Fig. 6, that the polish rate and nonuniformity can be simultaneously improved by using a larger down force. However, if the down force is set too high, scratches on a wafer may result. Decreasing the platen speed and the carrier speed decreases nonuniformity, as we can see from Figs. 6 and 7. However, decrease of the platen speed also decreases the polish rate, as shown in Fig. 6.

Another variable in CMP recipes is the back pressure. Usually, if the nonuniformity problem is identified to be due to a center-slow–edge-fast process, back pressure can be used to push the back of a wafer and accelerate the center polish rate. Thus, the uniformity can be improved accordingly.

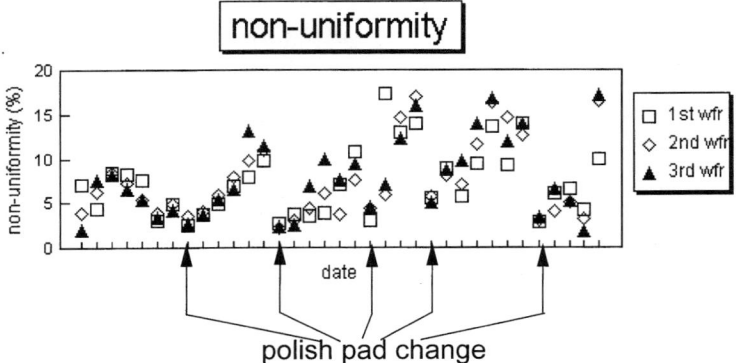

FIG. 5. Trend chart showing nonuniformity in daily qualification of a CMP tool for a whole month. Three wafers were tested per day.

3. Effect of the End Effector

An end effector, also called a pad dressor or a pad conditioner, is used to condition the polish pad to retrieve polish rate. If this is not done, the surface of a pad can become glazed and the pad austerity lost (see Fig. 9). The austerity of a pad is required in CMP; otherwise, hydroplaning will occur, which means that contact between a pad and a wafer surface no longer exists.

An end effector consists of diamond grit or similar silicon carbide materials. These extremely hard materials can scrape off the topmost layer of a pad during conditioning; if properly deployed, an end effector can help flatten a polish pad and improve polish uniformity. If not, the surface can be roughened and the nonuniformity worsened.

No matter how good the consumables, recipes, and equipment are in CMP, there is an intrinsic nonuniformity problem, in which a wafer is always polished more at its edge and less at its center. This is due to the fact that the relative linear velocity between a rotating wafer and a rotating platen is larger for positions at the edge than those at the center (an edge position always has a larger diameter than a center position). Hence a polish profile can be generated on a polish pad. The areas contacting the wafer center are less polished and the areas contacting the wafer edge are more polished (see Fig. 10). Once this kind of particular profile is formed on a polish pad, the normal polish uniformity is lost. In other words, the intrinsic nonuniformity problem can trigger an extrinsic problem on the polish pad. The only way to counter this is to use an end effector. The end effector can

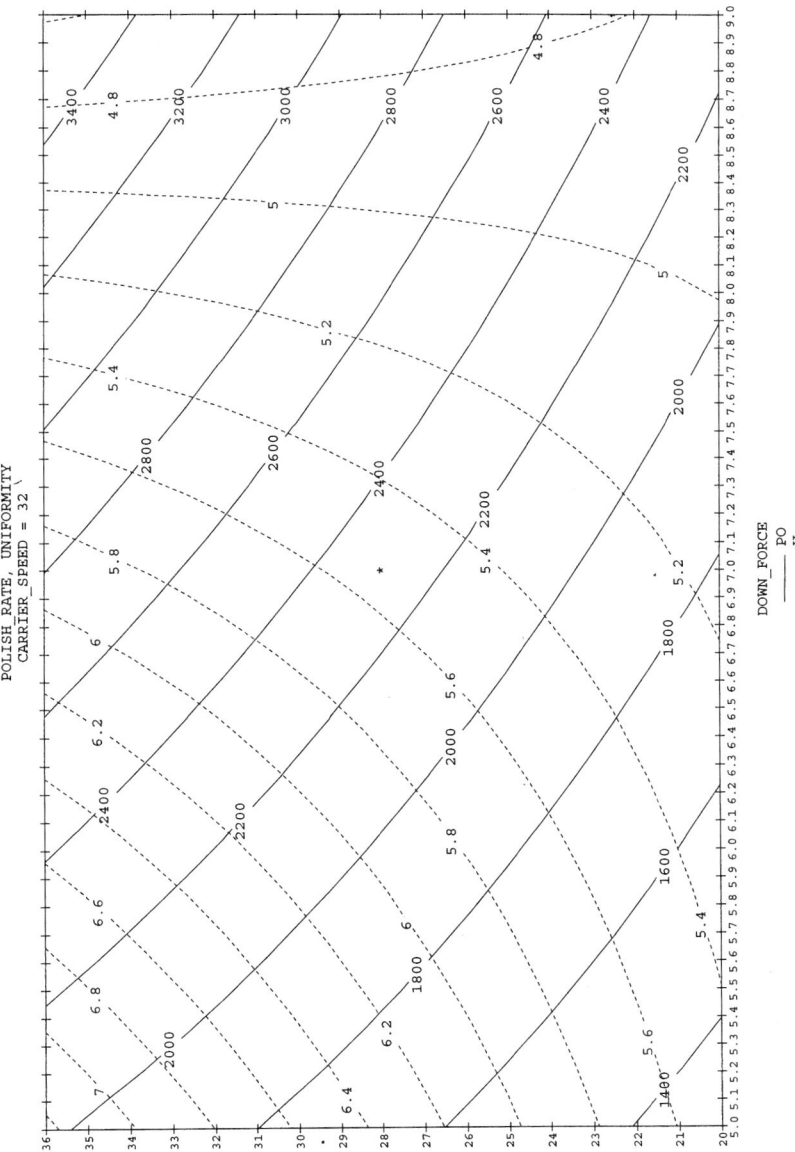

FIG. 6. Contour plot of the CMP down force vs the platen speed. The output responses are polish nonuniformity and rate.

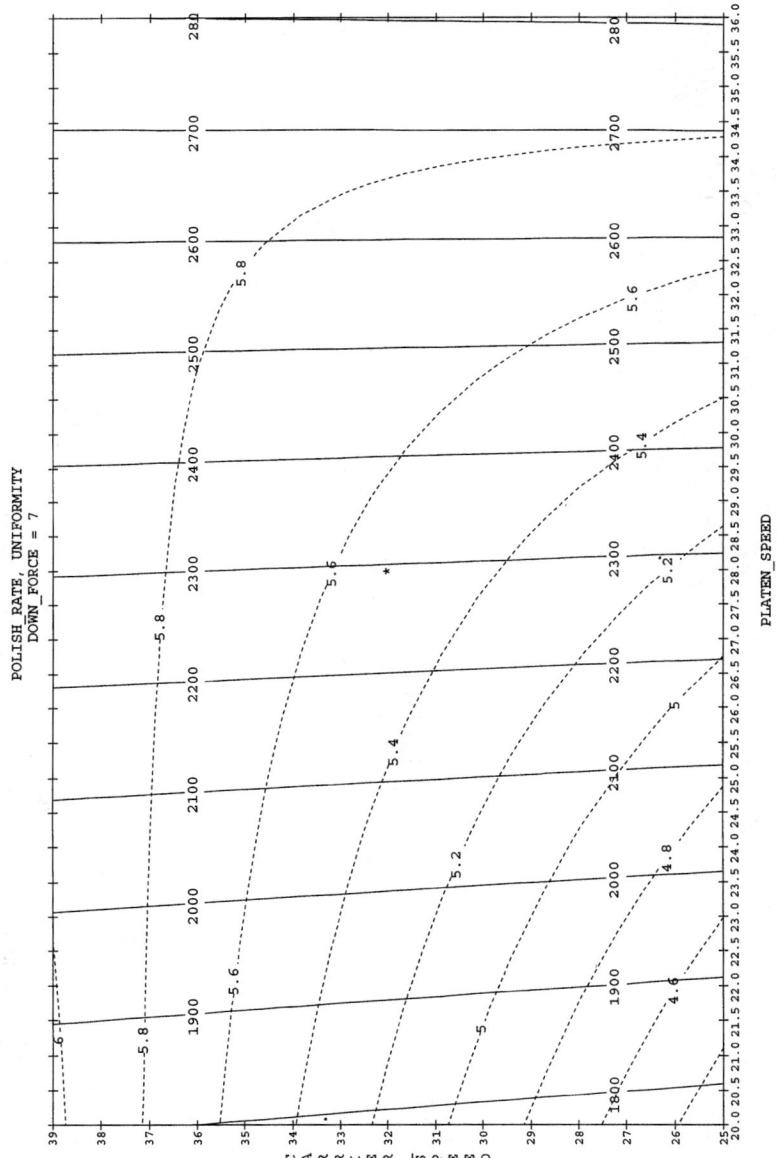

FIG. 7. Contour plot of the CMP down force vs the carrier speed. The output responses are polish nonuniformity and rate.

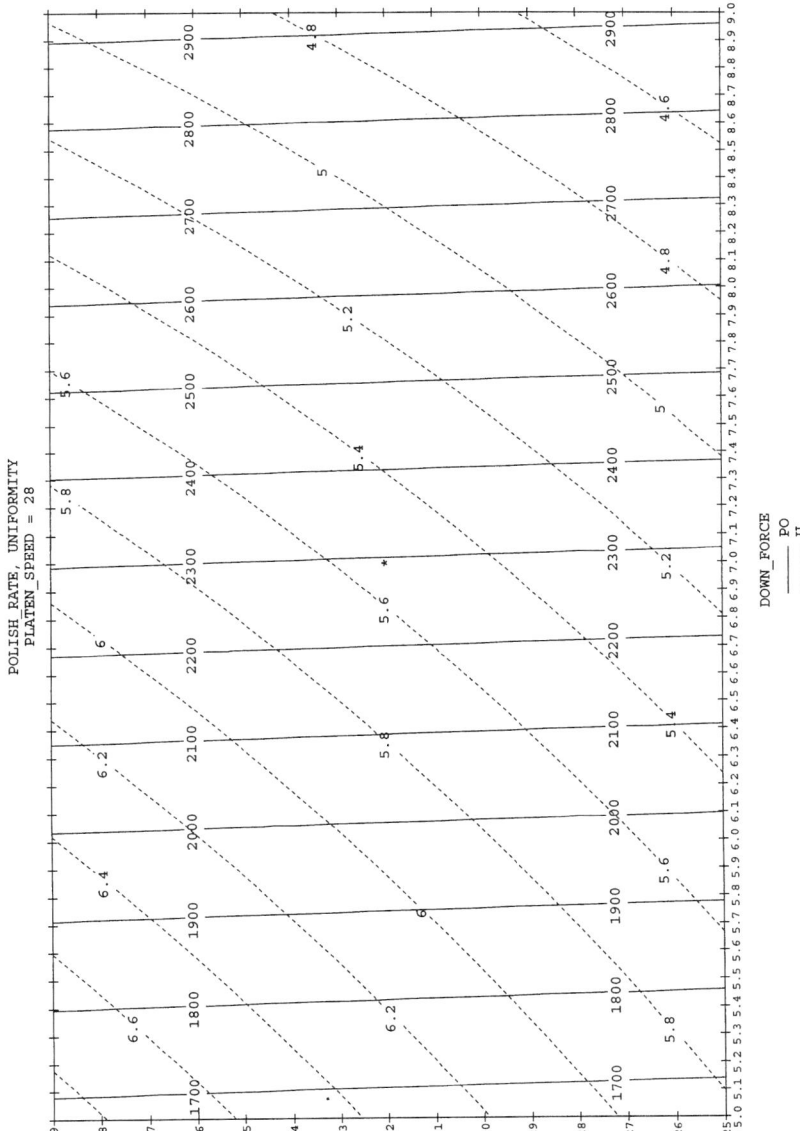

FIG. 8. Contour plot of the CMP platen speed vs the carrier speed. The output responses are polish nonuniformity and rate.

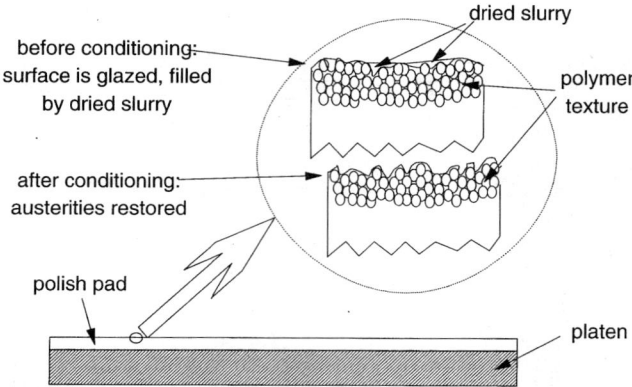

FIG. 9. Schematics of a microscopic view of a polish pad before and after conditioning. Before conditioning, the pad is glazed and filled with dried slurry; after conditioning, the top layer of the pad is removed and the austerity is restored.

intentionally condition more (longer time) at some places and less (shorter time) at others. Most CMP polishers provide an end effector conditioning profile program that enables user setup. The program can make the effector move to different positions of a pad and condition each place for a different duration. In addition, conditioning can be set up *in situ* or *ex situ* during polishing.

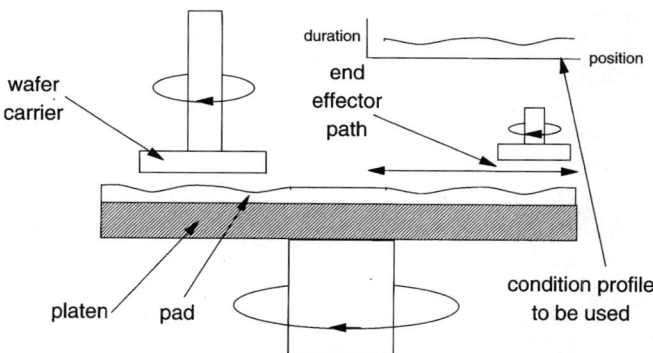

FIG. 10. Illustration of the polish pad surface profile created by the wafer carrier due to the polish (wear) rate being higher at the wafer edge than at the wafer center. An end effector can be used to compensate the profile if the duration vs position profile is set up correctly.

If back pressure is used in CMP, the wafer can be polished more at its center. The resultant pad profile is flat. In this case, the end effector conditioning profile would also be set flat. Otherwise, a mismatch between the polishing profile and the end effector conditioning profile would cause more problems.

Another issue with the end effector is its down force during conditioning. This down force must be as low as possible, as long as the polish rate remains stable. If the down force is set too high, the resultant high wear rate shortens the pad life. Once the grooves on the pad are worn out, the pad can no longer deliver slurry.

4. Deposit Less, Polish Less vs Deposit More, Polish More (Uniformity vs Planarity)

The primary purpose of using CMP in back-end interconnect processes is to planarize the surface. Hence, the depth of focus of existing photolithography tools can be extended into sub-0.35-μm technologies [1]. A question arises, however, as to how much sacrificial thickness is required for polishing away to planarize the surface. Intuitively, the more thickness polished, the better planarity achieved; however, at the same time, the across-wafer final thickness nonuniformity becomes worse. This concept of the relationship between the thickness polished and the remaining step height can be understood from Fig. 11. Planarity is a local, microscopic term, which is defined as

$$\text{Planarity} = \left(1 - \frac{\text{Final step height}}{\text{Initial step height}}\right) \times 100\%$$

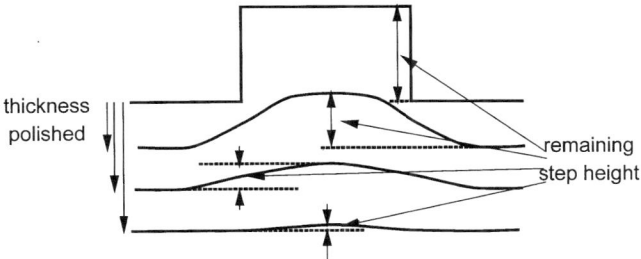

FIG. 11. Illustration of the relationship between the thickness polished and the remaining step height. The more thickness removed, the less step height left (better planarity).

Planarity can vary from one pattern feature to another. On the other hand, nonuniformity is a global, macroscopic term, which is defined as

$$\text{Nonuniformity} = \frac{\text{Standard deviation}}{\text{Mean}} \times 100\%$$

where

$$\text{Mean} = \frac{\sum y}{N}$$

and

$$\text{Standard deviation} = \sqrt{\frac{\sum (y - \text{mean})^2}{N}}$$

where y is a single measured value and N is the total number of the measurements performed across a wafer [4]. Nonuniformity indicates the thickness variation across a wafer. Because there is no direct mathematical relationship between planarity and nonuniformity, it is possible that the planarity is 100% for some features, while the final thickness nonuniformity is as bad as or worse than 20%, as explained later.

The nonuniformity in CMP is sometimes confusing because there are three different thickness specifications: pre- and post-thickness and thickness delta. To describe the CMP tool performance, and for the daily tool qualification, usually the nonuniformity of delta thickness is used. However, for quality control of the manufacturing of integrated circuits, the nonuniformity for the final thickness is used. The relation between the delta and the final thickness nonuniformities is explained as in the Fig. 12.

The figure assumes that a perfectly flat incoming wafer with oxide thickness of 20,000 Å is being polished, 15,000 Å (mean of 49 measurement sites on the wafer) remains after polishing, and that 5000 Å is polished. Since the CMP tool used is not perfect (due to the pad, equipment, and human factors, as already mentioned), nonuniformity exists. If the standard deviation of the final thickness measurements is 250 Å, then the final thickness nonuniformity is $250 \text{ Å}/15,000 \text{ Å} = 1.67\%$. However, the delta thickness nonuniformity is $250 \text{ Å}/5000 \text{ Å} = 5\%$. This 5% nonuniformity indicates the tool polishing capability. If a better pad is used, or a better adjustment has been made on the tool, this nonuniformity of 5% can go lower. On the other hand, given a fixed tool with a nonuniformity of 5%, if more thickness is polished, more standard deviation will be generated. This can be understood

9 APPLICATIONS AND CMP-RELATED PROCESS PROBLEMS

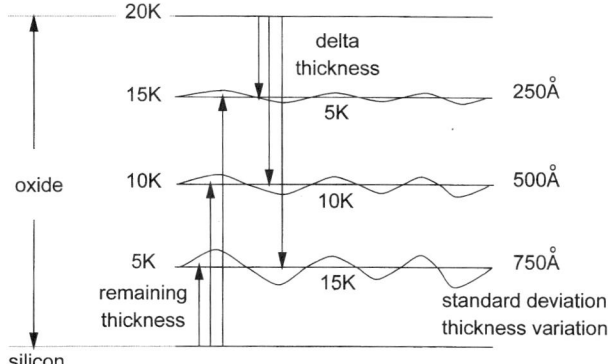

FIG. 12. Illustration of the relationship between the delta and the final thicknesses across a wafer. Thickness variation is caused by the CMP process. The more thickness polished, the more thickness variation (higher standard deviation), and the higher the nonuniformity.

if we polish 10,000 Å rather than 5000 Å. The standard deviation on the final thickness will be 5% × 10,000 Å (delta thickness) = 500 Å rather than 250 Å. This causes the final thickness nonuniformity change from 250 Å/15,000 Å = 1.67% to 500 Å/10,000 Å (final thickness) = 5%, since not only the standard deviation changes but also the final mean thickness. Similarly, if we use a tool with 5% nonuniformity to polish 15,000 Å and let 5000 Å remain, the nonuniformity for the final thickness will be 5% × 15000 Å/5000 Å = 15%.

This derivation tells us that given known CMP tool capability and final thickness of the integrated circuit design, we must minimize the polished (delta) thickness to get better uniformity. On the other hand, we need to know the relation between planarity change and the polished thickness. Grillaert et al. reported that the planarity change in CMP is strongly dependent on the pattern densities [5]. Figure 13 shows the empirical relations between step height and thickness polished for various levels (PMD, ILD1, ILD2, and ILD3) and different features (lines of 5 μm width and a box of 20 × 20 μm²) for positions at center and edge of each wafer. As shown in PMD and ILD plots, there is no major difference in the planarity change between the lines and the box at the center and edge. In the ILD3 and ILD4 plots, however, we see that a box feature is more difficult to planarize than a line feature, and a feature at a central position is more difficult to planarize than that at an edge position. The so-called "last spot" feature in the ILD3 plot indicates a trench feature between two large boxes at the center of a wafer. Comparing the plots at PMD and ILD1

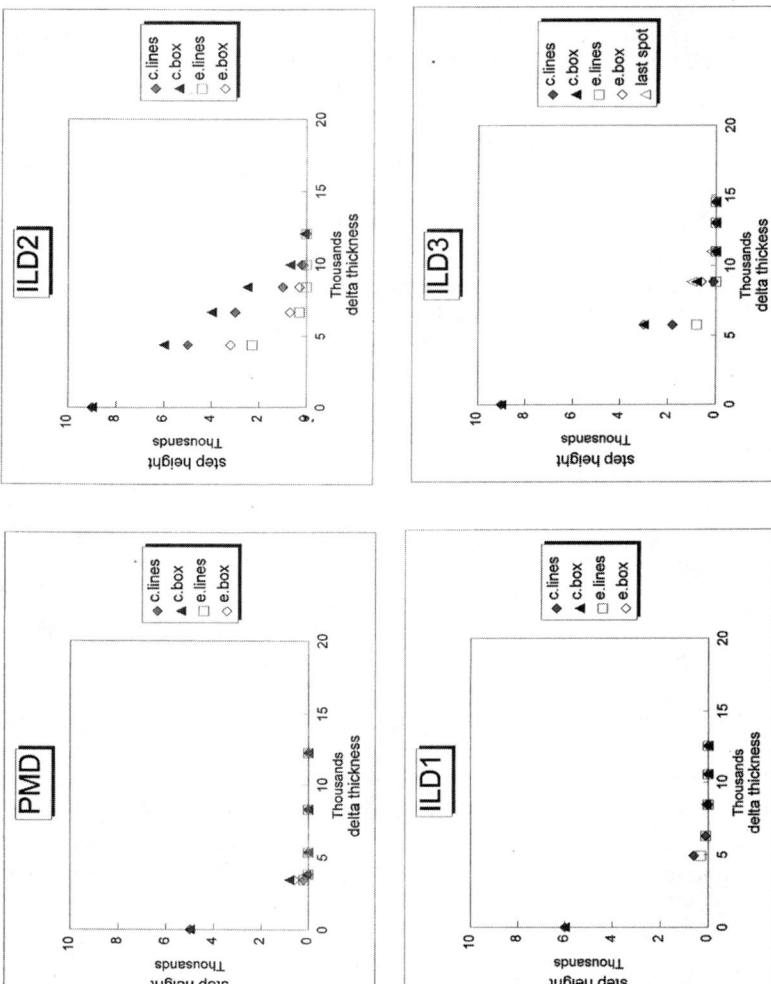

FIG. 13. Empirical relations between step height and thickness polished for various levels (PMD, ILD1, ILD2, and ILD3) and different features (lines 5 μm wide and a box of 2 × 20 μm²) for positions at the center and edge of each wafer. The so-called "last spot" feature in the ILD3 plot indicates a trench feature between two large boxes at the center of a wafer.

levels to the plots at the ILD2 and ILD3 levels, we might draw the conclusion that the patterns at ILD2 and ILD3 are more difficult to planarize than those at the PMD and ILD1 levels. In fact, a more important, invisible factor is whether the polish pad is being conditioned properly and/or is being used near to the end of its lifetime. In this particular case, we determined that the cause of the difficulty in planarization at ILD2 and ILD3 was a pad problem. The lesson learned here is that if the polish pad is in a good condition, the dependence of the pattern density vs planarity is minimized. As the pad loses its capability to planarize, however, the dependence of planarity on pattern density increases.

If the step height of a certain pattern feature can not be planarized, the nonplanarity problem may cause more serious outcomes, such as the metal stringers, which lead to short circuits [6]. Sometimes, poor planarity is not caused by the smaller delta thickness polished. If the end effector for the pad conditioning does not function normally and/or the pad austerity cannot be restored effectively after conditioning, deposit more–polish more may not truly resolve the poor planarity problems.

5. BPSG vs TEOS

Borophosphosilicate glass (BPSG) has been traditionally used in PMD in metal-oxide semiconductor (MOS) integrated circuits. The addition of phosphorus to the glass makes the layer an excellent getter of Na or other mobile ions; otherwise, contamination by mobile ions can cause instabilities in V_T (transistor threshold voltage) of the MOS devices. The addition of boron to the glass reduces the viscous flow temperatures, enabling the deposited layer to be flowed and planarized. As CMP has gained more importance in Si processing, however, these advantages have to be reconsidered: (1) if CMP is a perfect planarization process, we no longer need B in the glass to help reflow or planarize and (2) B and P dopants in the glass are not distributed perfectly uniformly across the wafer and through the depth of the film. This will cause thickness nonuniformity after CMP [7].

A study was done to compare the results of the CMP nonuniformities for polishing BPSG and TEOS (tetra-ethyl-ortho-silicate) [8]. A production lot was split between a BPSG only and BPSG–TEOS stacks in PMD. The input parameters and output responses of the experiments are listed in Table IV. The P concentration was fixed at 6% and BPSG thickness varied (splits 1–5: BPSG–TEOS stacks; split 6: standard BPSG only). The output responses included post-CMP thickness uniformity and yield. The results show improvement of the BPSG–TEOS in PMD in post-CMP thickness uniformity and yield. For the BPSG–TEOS stacks, only TEOS was

TABLE IV

INPUT AND OUTPUT RESPONSES FOR THE EXPERIMENTS CONDUCTED

Split	BPSG thk, A	PETEOS thk, A	Total ox thk, post-CMP	Delta thk	CMP (nu%)	2K redep.	Yield (%)*
1	2K	12K	13K	6K	2.5	2K	$a - 3$
2	4K	10K	13K	6K	3.22	2K	$a + 14.1$
3	6K	8K	13K	6K	3.33	2K	$a + 14.4$
4	4K	12K	15K	8K	2.75	n/a	$a + 10.4$
5	6K	10K	15K	8K	2.82	n/a	$a + 17.2$
6	18K	n/a	13K	10K	7.74	2K	a

*The test vehicle was based on 0.35-μm CMOS technology.

polished. This suggests that unless necessary for other purposes, it is better to replace BPSG with TEOS to improve uniformity.

III. Post-CMP Oxide Thickness Control

In addition to a tight distribution of the thickness variation within a wafer, the average of a group of individual thicknesses must also be targeted within a certain range. Statistically, the control of the WIWNU is the control of standard deviation of individual thicknesses, and the control of final thickness post CMP is the control of the mean. The variation of the mean from wafer to wafer is called wafer-to-wafer nonuniformity (WTWNU). All the thicknesses mentioned in this section are actually the means of many individual thickness measurements in each wafer. Control is not easy, for reasons discussed in the following.

1. LACK OF ENDPOINT DETECTION SYSTEM

For oxide CMP at PMD and ILD levels, a pilot wafer must be run to estimate the optimal polish time, prior to processing a whole lot. This not only wastes precious production time but also adds a human factor into the process control, a major obstacle that hinders CMP from becoming a true mass-production tool in Si processing. In addition, due to the fact that the polish rate may vary from wafer to wafer in CMP, the final remaining PMD or ILD thickness of each wafer varies. As a result, either underpolished wafers need to be repolished, or overpolished wafers need to be redeposited.

Many endpoint detection systems, based on mechanisms, such as those based on reflected optical light [9], spindle motor current [10], pad temperature [11, 12], have been used to resolve this problem, with limited success. Some systems may work with blank wafers or wafers with relatively low pattern density (at the STI level, for example), but for the PMD or ILD levels no useful results have been reported. The presence of a pattern at the PMD or ILD levels adds a great deal of complexity to the signals. Currently, use of an endpoint detection system to control the final post-CMP thickness is still a fertile topic for research and development.

2. WAFER-TO-WAFER THICKNESS NONUNIFORMITY

The wafer-to-wafer thickness nonuniformity (WTWNU) in oxide CMP can be attributed primarily to variation in the CMP polish rate, and secondarily to the thickness variation in incoming wafers.

a. WTWNU Due to Variation in CMP Polish Rate

The variation in CMP polish rate has several symptoms: (1) erratic, (2) gradually decreasing, and (3) gradually increasing. Figure 14 is a companion figure to Fig. 5 of the polish rate in a trend chart of daily qualification of a tool for a unstable CMP process. As we can see, the polish rate was up and down. Seemingly, the polish rate is cyclic, corresponding to the pad life, but, in fact, it is more inclined to be erratic. This behavior was later determined

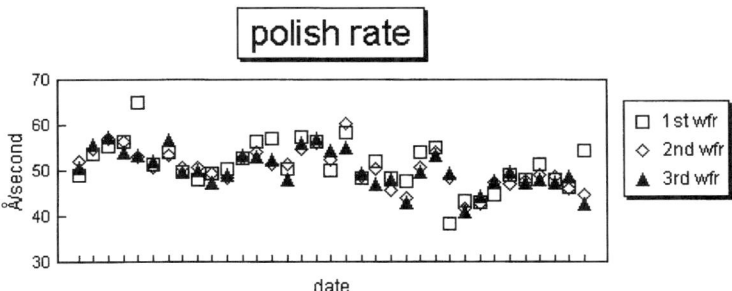

FIG. 14. Trend chart of polish rate from a daily qualification of a CMP tool for a whole month. Three wafers were tested per day.

to be due to the poor pad quality. After changing to another batch of pads, the problem went away.

If the pad quality is stable, a pad can usually run 400 to 600 wafers, with 7-lb down force in the recipe and 3–4 min of the polish time per wafer. However, based on the current CMP pad technology, we cannot predict when a pad will fail, or at which wafer. Even when the majority of a batch of pads is good, one or two pads may still have a short life. If a pad fails in the middle of polishing a lot, the polish rate will drop gradually, and the lot will have wafer-to-wafer final thickness variation. On the other hand, if a pad is not broken in enough prior to use, the polish rate will gradually increase.

The so-called break-in is a preparation step to stabilize the polish rate. Basically, a certain number of dummy wafers must be run. If a poor quality pad is used, the pad may require more wafers to break in, but even so may still run with unstable polish rates.

b. WTWNU Due to Thickness Variation in Incoming Wafers

Assuming the CMP polish rate is constant, and the polish time is fixed, the post-CMP wafer-to-wafer thickness variation (WTWNU) will reflect the same thickness variation pattern as that of the incoming wafers, but a far worse magnitude of variation. The reason is that in dielectric film deposition, the control is typically within $\pm 10\%$. For example, for a target of 8000 Å, the upper and lower limits will be 8800 and 7200 Å. When the CMP is used, however, the film must be deposited with additional thickness to be sacrificed in polishing. In other words, the pre-CMP thickness target must be changed to 15,000 Å, for example, whereby a 10% variation becomes ± 1500 Å. Assuming the CMP polishing is perfect, removing 7000 down to 8000 Å, and not introducing any additional wafer-to-wafer thickness variation, the variation in the incoming wafers, still 1500 Å, will make post-CMP results look terrible. The post-CMP results will have the target thickness of 8000 Å, but a wafer-to-wafer thickness variation of 1500 Å, which means that the WTWNU is 1500 Å/8000 Å = 18.7%.

Figure 15 records an example of this relationship between incoming thickness variation and post-CMP thickness variation. The incoming thickness variation is due to the fact that the TEOS film was deposited in two different plasma-enhanced chemical vapor deposition (PECVD) chambers, which are not calibrated identically [13]. As can be seen in Fig. 15, the pre-CMP wafer-to-wafer thickness variation pattern has been perfectly preserved after CMP.

Another example is a situation in which the PECVD chamber runs a

9 APPLICATIONS AND CMP-RELATED PROCESS PROBLEMS 265

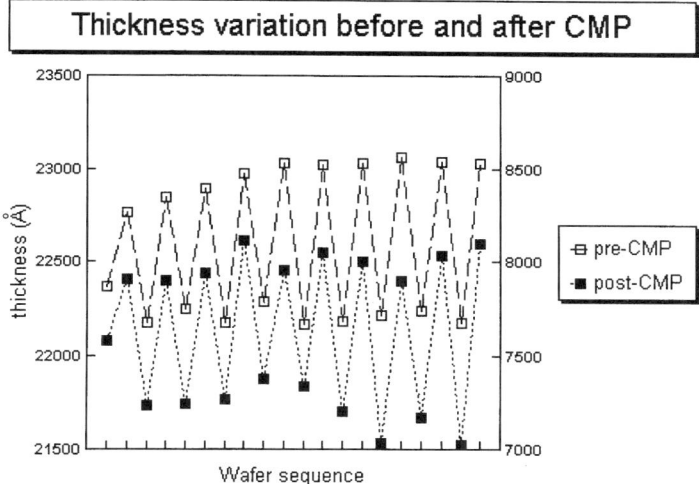

FIG. 15. The relationship between incoming wafer-to-wafer thickness variation and the post-CMP thickness variation. The pre-CMP scale is on the left axis, and the post-CMP scale on the right axis.

self-cleaning process once per two or three wafers, causing the wafer deposited just after the cleaning process to have low deposition rates [13], as illustrated in Fig. 16. Such a thickness variation in the incoming wafers will cause post-CMP thickness variation.

IV. Defectivity

IC manufacturing using sub-0.35-μm CMOS (complimentary metal-oxide semiconductor) technologies requires an extremely clean environment, sufficient to produce fewer than 50 0.2-μm-size defects per wafer for an oxide monitor wafer, and fewer than 7 defects (of any size) per production wafer. The defect density should be less than $0.02/cm^2$. A CMP process, however, uses slurries for polishing, and a slurry is composed of 15–30% of silica particles (abrasives) [14]. These particles are intentionally pressed into the dielectric films to remove film materials. It is highly probable that if something goes wrong with the slurry and/or if the post-CMP cleaning is not effective, the CMP process will produce defects, as discussed next.

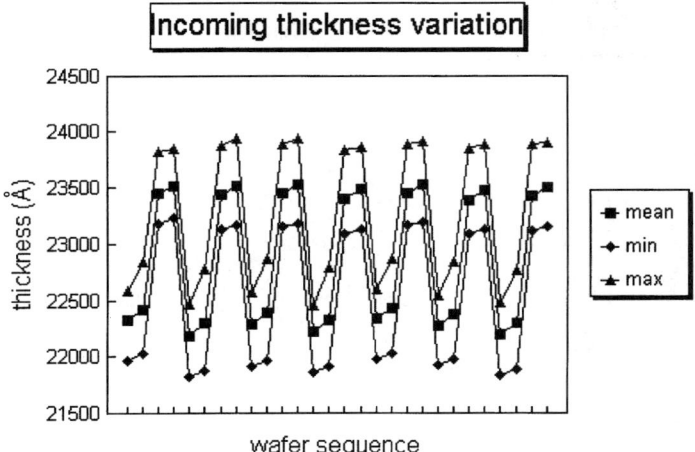

Fig. 16. Deposited thickness as a function of wafer sequence in a PECVD chamber. The variation in thickness is due to the fact that the wafer deposited immediately after a clean process will have low deposition rates (max: maximum value; min: minimum value measured).

1. Scratches

The most common defects in CMP are scratches. If a slurry has a tight distribution of particle sizes, the post-CMP scratches should be minimized. Normal post-CMP scratches are shallow and can be removed by the post-CMP buffing and/or cleaning processes. Occasionally, however, the particle size can be out of control due to agglomeration, which results from poor temperature or moisture control in storage. Because slurry is a thermodynamically unstable colloidal dispersion, anything affecting the kinetics leads to a phase change [15]. Once these large-size particles (a few microns in diameter, as compared to a mean value of 500–1000 Å for the normal abrasives) are formed and delivered to the polishing pad, abnormally deep scratches can be produced on the wafer. Typical CMP scratches are shown in Figs. 17a–17c; these include scanning electron microscopy (SEM) pictures with different magnifications (Figs. 17a and 17b) and a cross-sectional view (Fig. 17c). As we can see, a via hole was affected by the scratch, undoubtedly affecting functionality of that particular via.

To prevent large slurry particles from scratching wafers, filters can be used on the slurry lines. The drawbacks of filters in the slurry lines, however, are that (1) the slurry flow rate may decrease as the filter is being clogged and (2) the CMP polish rate can decrease as the filter approaches its end of life.

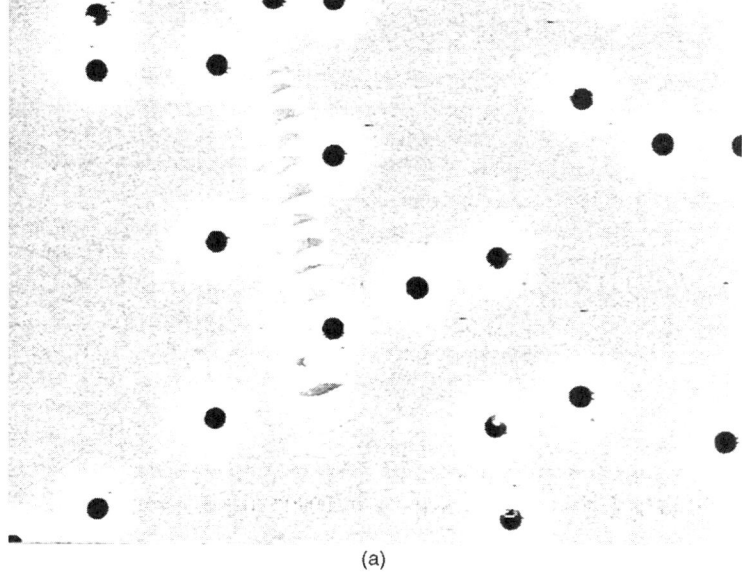

(a)

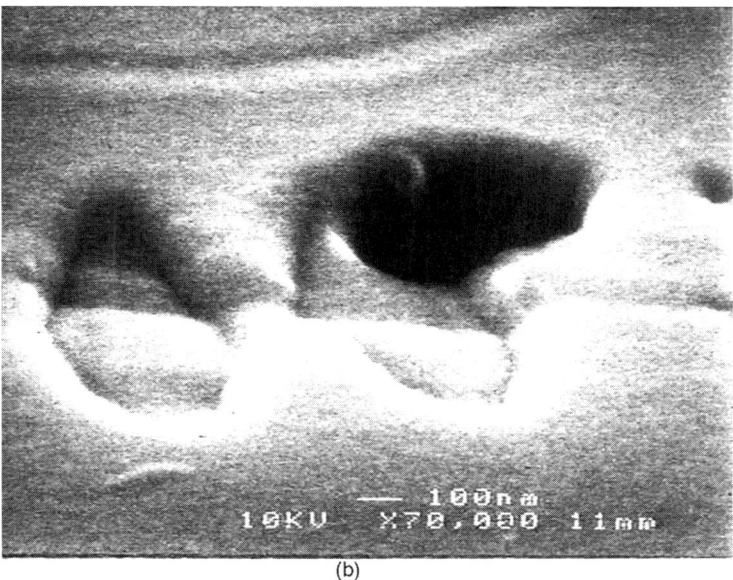

(b)

FIG. 17. Typical CMP scratches. (a) and (b) are SEM micrographs with different magnifications, and (c) is a cross-sectional view.

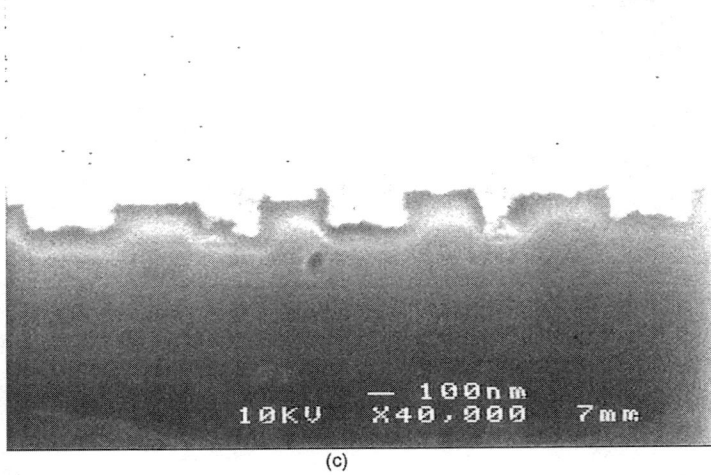

(c)

Fig. 17. Continued.

A choice of a filter with the correct membrane pore size can prevent the shift in polish rate. The pore size chosen must be significantly larger than the slurry particle sizes. For example, if the mean of the slurry particle size is 1000 Å, the pore size should be a few microns, so that no normal sized particles will be filtered out. Frequent filter changes can prevent filter clogging.

Scratches can also be caused by particles by another means. Since in CMP the slurry (particles included) stays on the polish pad, and the polish pad material is removed by the end effector, the end product of this complicated slurry–pad material reaction is a kind of substance with unknown characteristics and sizes. These substances can be the cause of the scratches. Figure 18 shows such substances collected by an end effector [16]. If the end effector is not properly cleaned, these particles can become hardened and cause severe scratches. To prevent this, a high-pressure water spray nozzle or a rotating brush can be used.

Another cause of scratches is the drop out of diamond grit from the end effector. In contrast to the aforementioned microscratches, which may be 1 μm wide and a few microns long, these will cause macroscratches and destruction of entire wafers. Since the diamond grit is nickel-bonded on the end effector, the bonding can be vulnerable if the slurry is chemically corrosive.

9 APPLICATIONS AND CMP-RELATED PROCESS PROBLEMS

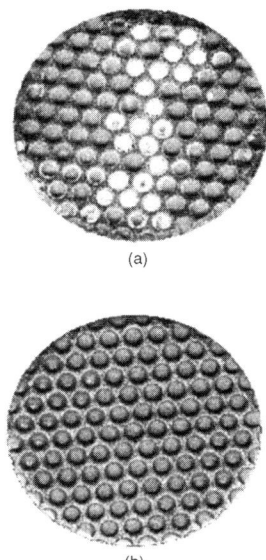

FIG. 18. (a) End effector collecting substances that can cause scratches. (b) If a high-pressure nozzle or a clean brush is used, the end effector is clean [16].

The signature of this kind of scratch depends on the polishing mechanisms of the polishers. Figure 19a shows the scratches caused by the diamond grit from a rotating polisher, and Fig. 19b shows those from an orbiting polisher. A new technique has been developed to epitaxially grow a thin diamond film to bond the diamond grit [17], which can greatly reduce the risk of the fall off of the diamond materials.

Scratches are also common in W CMP, as discussed later.

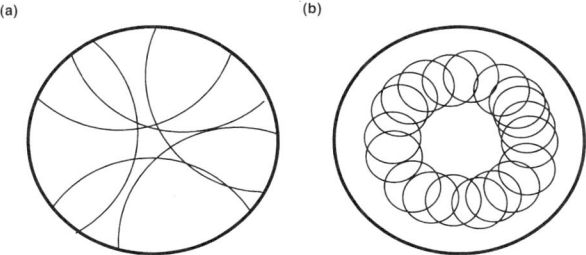

FIG. 19. Scratches on a wafer caused by fall-off of diamond grit from an end effector for (a) a rotating polisher and (b) an orbiting polisher.

2. OTHER DEFECTS

Figures 20a and 20b are SEM micrographs of dielectric cracks from a top view and from a cross-sectional view. The root cause was determined to be poor stress control in the TEOS film, not the CMP process. However, this problem may not have been as serious without CMP (i.e., CMP may exacerbate the problem). For CMP, the dielectric film must be highly compressive. If a film is tensile or not compressive enough, the overwhelming down force of CMP will impart significant shear stress, enough to break down the film. This is particularly true for the areas above the metal lines at ILD levels, where the surface has an abrupt step height in topography. A crack can start from the concave corner of the step and travel to the corner of a metal line, as shown in Fig. 20c.

Figures 21a and 21b are SEM micrographs of metal line corrosion defects from the top view and from the cross-sectional view. The cause was

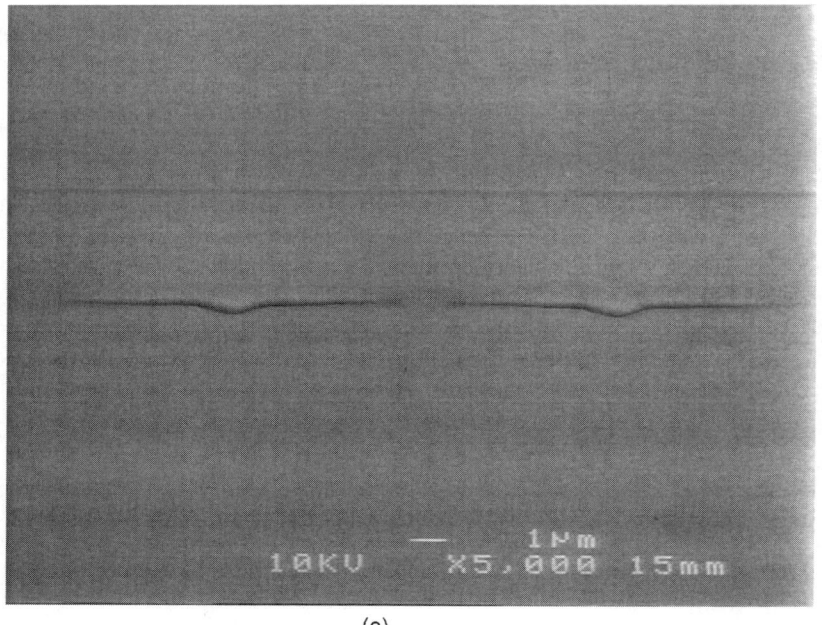

(a)

FIG. 20. SEM micrographs of dielectric cracks from (a) a top view and (b) a cross-sectional view. The crack will run from the corner of the step height to the corner of a metal line, as illustrated in (c).

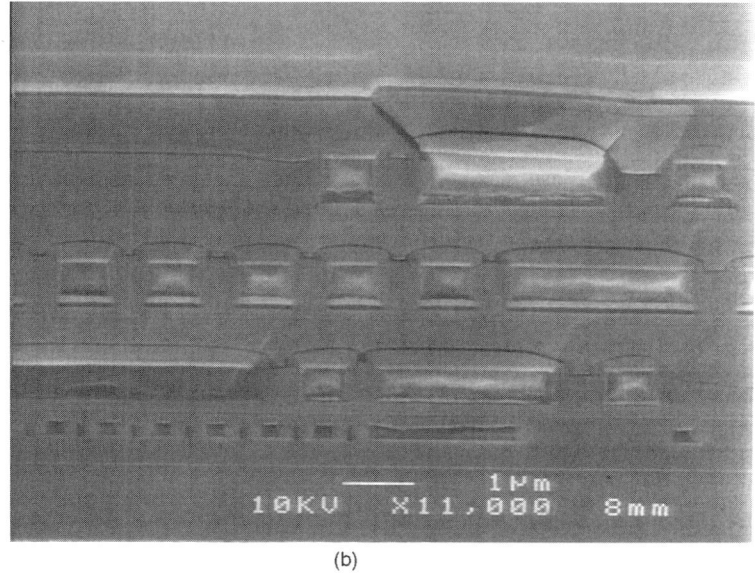

(b)

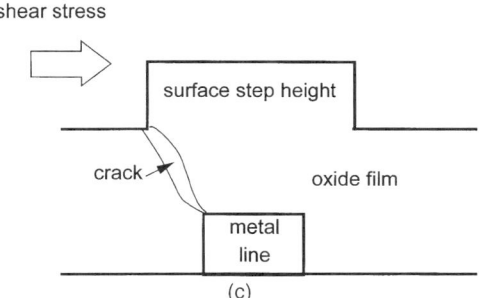

(c)

FIG. 20. Continued.

determined to be microcracks in the dielectric films. The metal lines were corroded because strong acids, such as the HF (hydrogen fluoride) or HCl (hydrogen chloride), were used in the post-CMP cleaning. These chemicals passed through the cracks and reacted with the metal materials. Once this happens, a whole piece of metal will disappear. The post-CMP cleaning chemistry must be adjusted to be just strong enough to remove post-CMP particles, but not strong enough to cause metal corrosion.

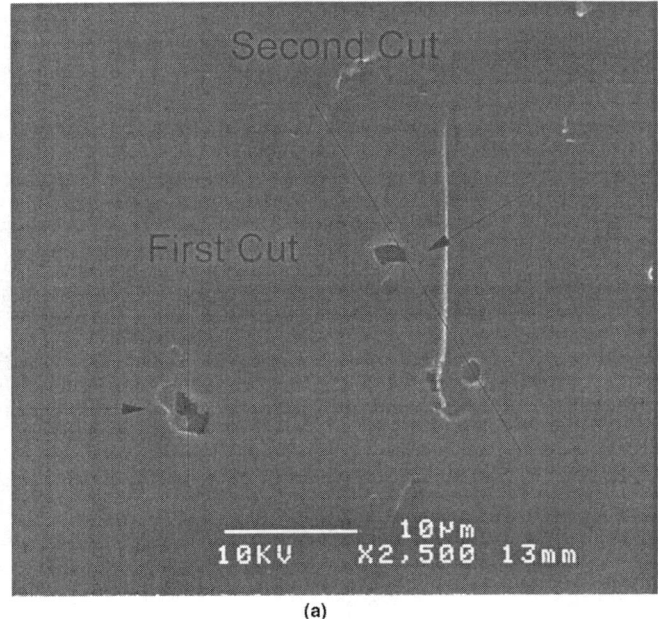

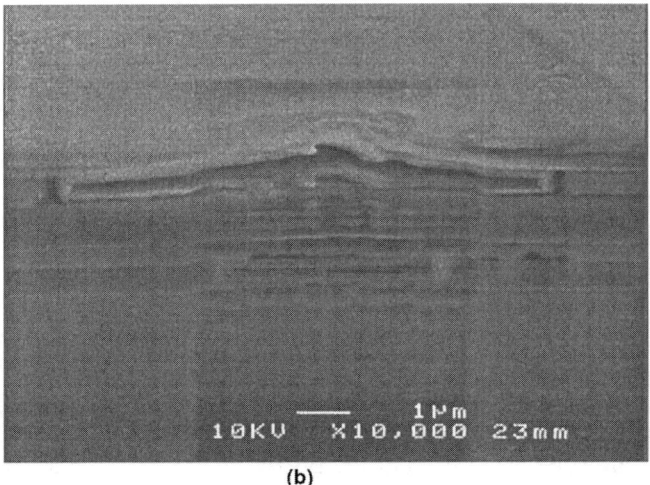

Fig. 21. SEM micrographs for metal line corrosion defects post-CMP cleaning from (a) a top view and (b) a cross-sectional view.

V. Tungsten CMP Problems

As the industry continues to move through generations of devices with geometries smaller and smaller, tungsten (W) CMP becomes preferred to tungsten etch-back (WEB) for plug and damascene applications. This is largely because WEB usually leaves W residuals that will cause intralevel metal line shorts [18]. Overetch is required to reduce residuals, at the cost of a recessed W plug, which in turn may cause metal voids [19]. In contrast to such a three-dimensional, chemical process, W CMP utilizes a two-dimensional, mechanical polishing mechanism. W, TiN (titanium nitride), and Ti (titanium) on the W–TiN–Ti film stack are removed sequentially. Hence, despite the fact that sometimes some areas may be polished incompletely, it is unlikely that W remains. Residuals of TiN or Ti are not as harmful as W because in the subsequent etch metal line definition process, TiN and Ti can be easily etched away by the BCl_3 aluminum etch chemistry, while W remains [20].

While some of the WEB drawbacks can be eliminated by W CMP, W CMP itself generates new problems. Because W CMP uses alumina (Al_2O_3, one of the hardest materials known) abrasive in a $Fe(NO_3)_3$-based slurry to polish, defectivity (scratches) and Fe contamination become issues. For the former, oxide buff can be used to reduce defects. However, this may lead to an oxide erosion problem, as discussed in a later section.

1. Metal Contamination

While the metal (primarily Fe, iron) contamination of W CMP does not directly lead to any functionality or yield loss in the manufacturing of products of the current 0.35-μm or larger technologies, it is questionable that the Fe contamination can be acceptable for the products with device geometries 0.25 μm and smaller. As the speed of the transistor increases to a certain crossover point, the speed of the integrated circuit as a whole becomes predominately dictated by the back-end processes [21].

It is always interesting to know the fate of the Fe species remaining on the surface. For the W plug application, after W CMP, the subsequent relevant processes are the deposition, patterning and etch of the Ti–Al (aluminum)–TiN stack. The Fe residuals on the surface will not cause any problem for the metal deposition because the entire surface is covered by metals. However, a problem may occur in the metal etch process because the BCl_3 plasma chemistry can not etch away Fe. The reaction products between Fe and BCl_3 are FeCl complexes, which are nonvolatile [22]. They will stay in an etch chamber, unable to be pumped out. Therefore, the fate of the Fe species is likely similar to that of W residuals in WEB, that is, ending up as a potential bridging source between metal lines. For the W damascene application, there is no metal etch process at all, so clearly Fe residuals on the surface will not be removed.

A study regarding the Fe contamination has been previously conducted [23]. Different chemistries, cleaning tools, and temperatures, and with and without oxide buff, were used as the input parameters, and the output responses were Fe contamination determined by TXRF (total x-ray fluorescence), plug integrity by visual check, and the contact or via resistance determined by electrical measurements. The results are summarized in Table V. Table V is organized in such a way that in each row there are several types of wafers that were run using the same experimental condition. Since each wafer has its own purpose, we could view all output responses at the same time for each experimental condition. For the baseline process (experimental condition 0), using scrubber and NH_4OH, we see that despite the compatibility of the existing process to the W plug (based on the visual check results and reasonable electrical data), the metal contamination (from TXRF) was unacceptable.

Experimental conditions 1–4 record our effort to identify the sources of the problem. Comparing the baseline and experiment 1, we see that the buffing process is not the true cause of the metal contamination. From experiment 2 we observe that SC1–HF combination can dramatically reduce the problem. Comparing the baseline and experiments 3 and 4, we find that NH_4OH had very little effect on the metal contamination. In experiments 5 to 8, we used citric acid and SC1 solutions, and found that citric acid can remove metal ions more effectively than SC1. However, neither chemical can remove Fe surface atoms completely. On the other hand, although both SC1 and citric acid splits yielded good via or contact resistance, visual check revealed that SC1 caused W plug keyholes. The data also indicate that more metal ions remained with our buffing process than without buffing.

Experiments 9–12 explore the use of an SAT (spray acid tool, which rotates the wafers during cleaning) with SC1–HF with various concentrations. We found out that none of the experimental conditions yielded acceptable electrical data. Both SC1 and HF corroded W plugs. Experiments 13 and 14 use SAT and novel chemicals; that is, TMAH ($N(CH_3)_4OH$, tetramethyl ammonium hydroxide), and/or NH_4F (ammonium fluoride). These results were the best in terms of both the Fe contamination and the W plug compatibility.

All the data in Table V are displayed in Fig. 22 to show the effects more clearly. The Fe contamination and the via or contact resistance are plotted

TABLE V
Experimental Conditions vs Output Responses for Fe Contamination in W CMP

Experimental conditions	Buff	Tool*	Chemistry**	Wfr Type[†]	TXRF[‡]	Visual	Rs[§]
Baseline	Yes	Scrub.	NH_4OH	1	25,000		
				2		Good	1.23v
				2		Good	4.8pl
				3		Good	6.7n
							5.9p
1	No	Scrub.	NH_4OH	1	24,000		
2	No	SAT	5:1SC1–HF	1	8		
3	Yes	SRD	H_2O	1	29,000		
4	No	SRD	H_2O	1	19,000		
5	Yes	Scrub.	10:1SC1	1	26,000		
				2		Keyhole	1.23v
6	No	Scrub.	10:1SC1	1	5,200		
				2		Keyhole	1.26v
7	Yes	Scrub.	Citric acid	1	500		
				2		Good	1.25v
8	No	Scrub.	Citric acid	1	300		
				2		Good	1.36v
9	Yes	SAT	5:1SC1–HF	2		Open	Open
				2		Open	Open
10	Yes	SAT	5:1SC1	2		Poor	Open
				2		Open	21.5pl
11	Yes	SAT	10:1SC1–HF	2		Poor	Open
				2		Open	Open
12	Yes	SAT	10:1SC1	2		Poor	Open
				2		Open	22.9pl
13	Yes	SAT	TMAH	1	50		
				3		Good	6.8n
							6.1p
14	Yes	SAT	NH_4F	1	20		
				3		Good	7.2n
							6.3p
15	Yes	SAT	TMAH–NH_4F	1	60		
				3		Good	7.1n
							6.1p

*The one-sided scrubber, spin-rinse-dry (SRD), or spray acid tool (SAT).

**Chemical; ratio; temperature:
  NH_4OH; 0.7% in DI; 25°C.
  SC1; x:1:1/4(DI:H_2O_2:NH_4OH); 75°C.
  HF (49%); 500:1 (DI:HF); 25°C.
  Citric acid; 20:1 (DI:citric acid); 25°C.
  NH_4F (40%); 6:1 (DI:NH_4F) with FC93 surfactant; 60°C.

[†]Type Pre-W CMP condition; output response; metrology
  (1) blanket TEOS or Ti–TiN/W stack; metal contamination; total x-ray fluorescence.
  (2) patterned short loop with W plug; plug integrity; visual, electrical (via, poly).
  (3) patterned production wafer; plug integrity; visual, electrical (n, p cont.).

[‡]The TXRF unit is $E10/cm^2$.

[§]The unit for the contact–via resistance is ohms/unit; n: n-contact; p: p-contact; v: via; pl: poly contact.

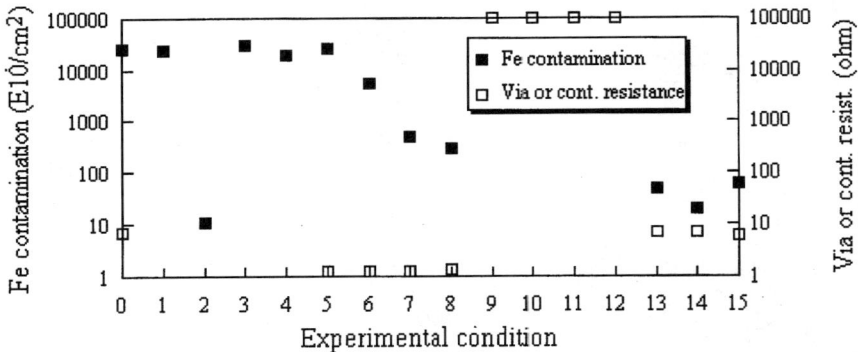

FIG. 22. Fe contamination and via or contact resistance are plotted against the experimental conditions listed in Table V. Experimental conditions 13–15 give acceptable results for both Fe contamination and plug integrity (electrical).

against the experimental conditions. It is clear that to get both acceptable Fe contamination and via or contact resistance, only experimental conditions 13–15 are feasible.

We observe that the Fe metal contamination problem is related to the post-W CMP cleaning chemistry used. We have found that the effectiveness of the chemistry to remove Fe contamination is in the order of

$$SC1/HF > NH_4F > N(CH_3)_4OH > \text{citric acid} > SC1 > NH_4OH$$

However, the chemical compatibility with the W plug, based on the visual and electrical checks, show that SC1 and HF are corrosive to W plug. This reduces the above sequence to

$$NH_4F > N(CH_3)_4OH > \text{citric acid} > NH_4OH$$

If we combine these results of Fe contamination, W-plug compatibility, or the electrical results, we conclude that NH_4F and $N(CH_3)_4OH$ are promising solutions in W CMP cleaning.

Many companies expended great efforts to prepare the W CMP slurry in nonferric solutions. Good candidates include H_5IO_6-based and H_2O_2-based slurries [24].

2. Oxide Erosion

Global oxide erosion can be caused by two factors: the unintentional erosion due to poor selectivity between metal (for W, TiN, and Ti) and oxide CMP and the intentional erosion due to the additional step of oxide buffing for the purpose of eliminating defects (primarily scratches) generated by the W CMP process. Nowadays, we rarely see that the erosion is caused by poor selectivity. Usually, as long as the selectivity between W and oxide is greater than 10, its impact on oxide erosion is insignificant. This is easily accomplished, since most W slurries are acidic and oxide must be polished in a basic slurry. The global oxide erosion observed currently is most likely the result of oxide buffing process. Although global oxide erosion is not detrimental to integrated circuit production, it is troublesome for (1) ILD thickness control and (2) defectivity monitoring, because oxide erosion may cause surface color variation, which impairs defectivity visibility.

Local oxide erosion is associated with dishing in damascene process, as discussed in the following.

3. Polish Rate and Uniformity

Instability in polish rate and nonuniformity in W CMP are also daily problems, as they are in oxide CMP. They are attributable to the quantity of consumables, tool maintenance, and the like; however, their impact on products is not as direct as in oxide CMP, because the W CMP process is self-limited by the metal–oxide selectivity (selectivity of tungsten to oxide greater than 10 is required). The polish time in W CMP is intentionally set up longer than necessary to overpolish by 25%. As a result, even if the polish rate is unstable, as long as the W, TiN, and Ti layers can be polished completely, there is no concern with overpolishing. However, if the polish rate is too low, 25% overpolish is not enough, and metal residuals may stay on the surface, eventually causing short circuits.

The nonuniformity problem in W CMP is also not detrimental. Again, as long as the W CMP can stop at oxide film, it does not matter if the metal at some areas is removed sooner or later.

VI. Other Problems

Some problems that are not classifiable into the preceding categories are considered in this section. These problems may occur occasionally, or they may appear only for certain applications.

1. MISSING ALIGNMENT MARK

A planarized surface is advantageous for photolithography in that the exposure of a pattern from a mask to the photoresist on a wafer can be better focused and less blurred; that is, little depth of focus latitude is required. On the other hand, it also has the disadvantage that, if the surface is too flat, the pattern underneath the planarized film cannot be aligned with the current mask. Two common examples of this problem are alignments after W CMP and STI CMP processes. The case of W CMP is illustrated in Fig. 23. On the left side of the figure are W plugs, and on the right side is a photolithographic alignment mark. Since the plug size is usually less than 1 μm in diameter, the plug can be completely filled by W. However, the alignment mark has a width of 5 to 10 μm, so that after the W deposition, its trench is not filled. On the alignment mark, the surface is intentionally left nonplanarized, so that after the metal deposition, at photolithography, the photostepper can discern the alignment mark due to the nonplanarity. Several things can happen to make this fail. If the W deposited thickness is large enough, then after W CMP, the surface at the alignment can be planarized. Then, after the metal deposition, since the metal film is opaque, the alignment mark is too obscure to be seen by the stepper. Even if nonplanarity remains but the step abruptness has been reduced, after the metal deposition, the alignment mark image can be lost in the noise

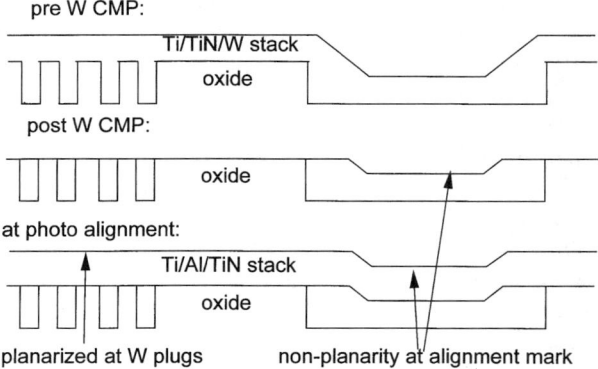

FIG. 23. Illustration of the nonplanarity at the photoalignment mark required for photolithography. This nonplanarity can be lost if the W deposited thickness prior to W CMP is too large or if nonplanarity exists but is insufficient due to the step abruptness having been smeared out. For the former, the surface at the alignment mark will be planarized after CMP; for the latter, after metal deposition, the image of the alignment mark can be obscured by noise due to the grainy metal surface.

generated by the grainy metal surface, and the alignment will fail the subsequent overlay inspection.

The same is true for the STI process, because after the CMP and the nitride strip, polysilicon is deposited. Polysilicon is also an opaque material, and if the nonplanarity is not present or reduced, the stepper cannot locate the alignment mark to align the polysilicon pattern to the STI pattern.

2. DISHING

Basically, dishing is defined as the percentage of the maximum recess to the nominal depth of the filling material in a plug (zero-dimensional feature), in a trench (one-dimensional feature), or in a well (two-dimensional feature). The severity of dishing is dependent on the material and the size of the structure. For the STI process, the material in a trench (or a well) is oxide, which is refractory, and dishing is hardly a problem even though the area of the trench or well is several hundreds of microns square. Similarly, for the W CMP with the plug application, since the size of the plug is small (less than 1 μm), there is also no dishing problem. For W CMP in a damascene application, however, dishing is significant [25]. Figure 24 shows

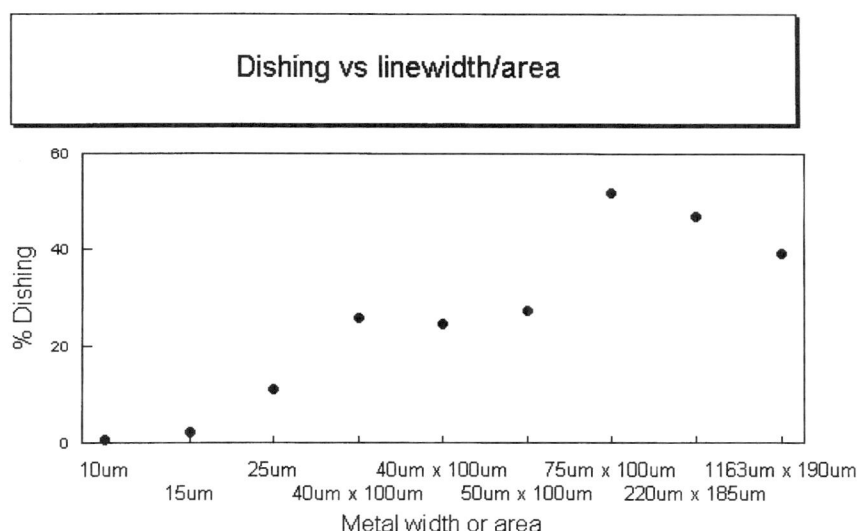

FIG. 24. Metal linewidth and area dependence of dishing for W CMP. Note: the x axis is not continuous, but represents various pattern features.

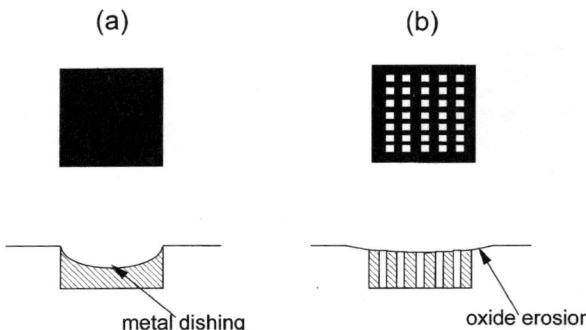

FIG. 25. Probe pad structure (a) without and (b) with oxide studs in the two-dimensional wells. The latter can be used to reduce dishing. However, local oxide erosion may be still a problem.

a metal linewidth or area dependence of the dishing. As we can see, as the size increases, more dishing is present. For two-dimensional features, we observe that amount of dishing is inversely related to perimeter; for instance, the dishing of a square-shaped pad is worse than that of a rectangle-shaped pad if their areas are the same. To reduce dishing, several things can be done. First, the down force of W CMP can be reduced. This will reduce the polish rate also, but higher platen speed can be used to compensate the rate loss. Second, a harder polish pad can be used rather than a soft pad. A harder pad works better, because it provides a better conformity, so that the pad will not push into the metal regions. When a harder pad (IC1000-type, for example) is used, however, more pad conditioning is required.

Dishing can be also reduced by adding oxide studs in the two-dimensional wells (see Fig. 25). Since the two-dimensional well structures are usually probe pads for electrical measurements, the addition of oxide studs will not affect the pad function as long as the test probe can still touch metal materials. Although this method can reduce dishing, local oxide erosion may still exist. The mechanism of local oxide erosion due to dishing is still unclear at this point of time [26].

REFERENCES

1. J. M. Steigerwald, S. P. Murarka, and R. J. Gutmann, *Chemical Mechanical Planarization of Microelectronic Materials* (Wiley & Sons, New York, 1997), pp. 66–84, 22–25.
2. M. Weling, C. Drill, W. Parmantie, and G. Fawley, *Proceedings of CMP-MIC (Chemical Mechanical Polishing Multi-Interconnect Conference)*, Santa Clara, CA, Feb. 1996, p. 40.

3. S. H. Li, J. Ling, and C. Spinner, *Proceedings of Advanced Metallization and Interconnect Systems for ULSI Applications,* 1996, Boston, MA (eds. R. Havemann, J. Schmitz, H. Komiyama, and K. Tsubouchi, MRS Society, Pittsburgh, PA, 1997), p. 567.
4. G. E. P. Box, W. G. Hunter, and J. S. Hunter, *Statistics for Experimenters* (Wiley & Sons, New York, 1978), p. 39.
5. J. Grillaert, H. Meynen, J. Waeterloosk, B. Coenegrachts, and L. Van den Hove, *Proceedings of Advanced Metallization and Interconnect Systems for ULSI Applications,* 1996, Boston, MA (eds. R. Havemann, J. Schmitz, H. Komiyama, and K. Tsubouchi, MRS Society, Pittsburgh, PA, 1997), p. 525
6. S. Wolf, *Silicon Processing for the VLSI Era*, vol. 2 (Lattice Press, Sunset Beach, CA, 1990), p. 201.
7. W. J. Schaffer and H. W. Fry, *Semiconductor Int.,* July (1996) p. 205.
8. S. H. Li, Y. Mansoori, A. Tissier, S. Saunders, B. Miller, and K. Wooldridge, *Proceedings of VLSI Multilevel Interconnection Conference* (VMIC), Santa Clara, CA, 1998, p. 556.
9. P. Holzapfel and J. Schlueter, *Proceedings of CMP-MIC (Chemical Mechanical Polishing Multi-Interconnect Conference),* 1997, Santa Clara, CA, p. 44.
10. H. E. Litvak and H. M. Tzeng, *Semiconductor Int.,* July 1996, p. 259.
11. H. W. Chiou, L. J. Chen, and H. C. Chen, *Proceedings of CMP-MIC (Chemical Mechanical Polishing Multi-Interconnect Conference),* 1997, Santa Clara, CA, p. 131.
12. R. L. Lane and G. Mlynar, *Proceedings of CMP-MIC (Chemical Mechanical Polishing Multi-Interconnect Conference),* 1997, Santa Clara, CA, p. 139.
13. S. L. Toh and M. Q. Lu, STMicroelectronics, private communication, 1998.
14. Description of SS12 slurry, Microelectronics Materials Division, Cabot Corporation, 1995.
15. S. H. Maron and J. B. Lando, *Fundamentals of Physical Chemistry* (Macmillan Publishing Co., New York, 1974), p. 776.
16. F. Robert, STMicroelectronics, private communication, 1998.
17. A. Inamdar, M. A. Fury, D. Towery, A. B. Stubbmann, and J. W. Zimmer, *Proceedings of CMP-MIC (Chemical Mechanical Polishing Multi-Interconnect Conference),* 1998, Santa Clara, CA, p. 169.
18. T. L. Myers, M. A. Fury, and W. C. Krusell, *Solid State Technology,* Oct. (1995), p. 59.
19. F. Cazzaniga, L. Riva, and G. Queirolo, *Advanced Metallization and Interconnect Systems for ULSI Application* in 1996 (eds. R. Havemann, J. Schmitz, H. Komiyama, and K. Tsubouchi, MRS Society, Pittsburgh, PA, 1997), p. 135.
20. C. Yi, W. C. Y. Tu, K. Tsai, S. Hsieh, and H. C. Chen, *Proceedings of CMP-MIC (Chemical Mechanical Polishing Multi-Interconnect Conference),* Santa Clara, CA, 1997, p. 107.
21. M. T. Bohr, *Advanced Metallization and Interconnect Systems for ULSI Application* in 1996 (eds. R. Havemann, J. Schmitz, H. Komiyama, and K. Tsubouchi, MRS Society, Pittsburgh, PA, 1997), p. 3.
22. F. A. Cotton and G. Wilkinson, *Advanced Inorganic Chemistry* (5th ed., Wiley & Sons, New York, 1988), p. 718.
23. S. H. Li, J. Ling, B. Miller, C. Spinner, and M. Jolley, Conference of Advanced Metallization and Interconnect Systems for ULSI Application in 1997, San Diego, CA, 1997.
24. S. H. Li, H. Banvillet, C. Augagneur, B. Miller, M.-P. Nabot-Henaff, and K. Wooldridge, *Proceedings of CMP-MIC (Chemical-Mechanical Polish for ULSI Multilevel Interconnection Conference),* Santa Clara, CA, 1998, p. 165.
25. V. Chaku, STMicroelectronics, private communication, 1998.
26. N. Elbel, B. Neureither, B. Ebersberger, and P. Lahnor, *J. Electrochem. Soc.,* 145 (1998) 1659.

Index

A

Abrasives
   agglomeration, 143–145
   alumina, 140
   for metal slurries, 142–143
   milling, 145, 147
   for oxide slurries, 140, 143
   silica, 140
Acid media, slurry removal in, 205–206
Agitation, 74–75
Alignment marks, missing, 278–279
Alkaline media, slurry removal in, 204
Alpha phase alumina, 144
Alumina, 140, 144
Aluminum, 188
Alum process, 144
Asperity contact model, 162–165
Atomic force microscopy (AFM), 236–240

B

Backpressure devices, 67–68
BPSG versus TEOS, 261–262
Bulk chemical distribution (BCD) systems, 30

C

Chemical cleaning agents, 84
Chemical mechanical planarization (CMP)
   applications, 139
   cost of, 7–8
   current work on, 3
   future of, 3–4
   history and development of, 2, 6
   major suppliers of, 7

Chemical shock, 61–62
Chip level effect, 100
Contact wear models, 128–130
Copper, 188
   cleanup, 212
Copper (Cu) damascene, 3, 5, 99
   equipment requirements and, 38–40
   modeling, 130–132
Corrosion effects, 186–192

D

Damaged-layer
   effects, 186
   removal, 208–210
Daytanks, 71–74
   agitation, 74–75
   replenishment, 59–61
Dead-head piping systems, 63–64
Deep trench capacitors, 5
Defectivity
   monitoring, 226–229
   other types of, 270–271
   scratches, 266–269
Density, calculation of effective pattern, 114–116
Density weighting function, 110–112
Dielectric CMP, pads for, 170–177
Die-level/scale models, 104, 166–167
   layout density dependence, 105–106
   planarization length and response function, 108–113
   planarization length determination, 113–118
   shallow trench isolation, 118–119
   thickness evolution and, 106–108

Digital image comparison, 228–229
Dishing, 279–280
Distributed polish head, 23
Down force, 251, 257
Dry-wafer-out (DWO) process, 30
Dynamic random access memories (DRAMs), 8

E

Edge exclusion region, 16–19
Electrochemical corrosion, 186–190
Electrostatic forces, 194
Electrostatic readhesion, preventing, 197–202
End effector effects, 252–257
Endpoint detection (EPD), 37
Equipment design, 8
    conclusions, 41–42
    copper polishing and, 38–40
    first-generation polishers, 10–11
    integration of, 30–38
    pad conditioning, 25–30
    platens, 24–25
    second-generation polishers, 11–13
    third-generation polishers, 13–16
    300–mm tools, 40–41
    throughput improvements, 9–10
    wafer carriers, 16–24

F

Facilitization
    agitation, 74–75
    backpressure devices, 67–68
    daytank replenishment, 59–61
    filtration, 78–82
    metrology, 76–78
    mixing slurry constituents, 61–62
    piping systems, 62–65
    pressure and flow consistency, 66–67
    slurry blending technology, 54–55
    slurry consumption ramp, 68–69
    slurry dispense engines, 53–54
    slurry distribution system, evolution of, 52–53
    slurry distribution system, maintenance of, 83–84
    slurry distribution system, overview of, 48–49, 50
    slurry distribution system, redundancy of, 69
    slurry handling, 49, 51–52
    slurry measuring techniques, 55–59
    slurry room location, 66
    slurry settling, 65–66
    storage tanks/daytanks, 71–74
    valve boxes, 69–71
    waste disposal, 84–87
Fast Fourier transform (FFT) techniques, 116
Feature-scale models, 100–104
Filtration, 78–82
First-generation polishers, 10–11
Floating carriers, 21–23
Flow meter times time measuring, 57
Four-point probe, 241–243

G

Gamma phase alumina, 144
Gimbaled carriers, 20–21
Global circulation loop filtration, 81

H

Hydrodynamic effects, 164–165
Hydrostatic pressure, use of, 22–23

I

In-line metrology, 35–37
In situ metrology, 37
Inter-level-dielectric (ILD) planarization, 5, 139
    die-level modeling of, 104–125
Inter metal dielectric (IMD), 184
IPEC Precision, 37

K

Kinetic aspect, 189–190

L

Laser-induced fluorescence, 165
Laser scanning, 228
Layout density dependence, 105–106
Linear polisher and pressure control, 23–24
Load balancing, 12
Local tungsten interconnects, 5
Loading effect, 203

M

Masks, layout, 113–114
Measuring techniques, 55–59
Megasonic cleaning, 34–35
Metallic contamination
  effects, 185–186
  removal, 206–208
  tungsten, 273–276
Metal polishing
  models for, 125–132
  pads for, 177–178
Metal slurries
  abrasives for, 143–144
  solution for, 150–151
Metering pumps times strokes measuring, 57
Metrology
  atomic force microscopy, 236–240
  defectivity monitoring, 226–229
  four-point probe, 241–243
  in-line, 35–37
  in situ, 37
  methods, 76–78
  noncontact capacitive measurement, 229–232
  reflectometry, 216–226
  stylus profilometry, 236
  total x-ray fluorescence, 233–235
Milling, 145, 147
Mixing slurry constituents, 61–62
Models
  applications, 124–125
  asperity contact, 162–165
  die-level/scale, 104–118, 166–167
  feature-scale, 100–104
  for metal polishing, 125–132
  patterned, 98–104
  shallow trench isolation, 118–119
  for step height reduction, 120–124
  wafer-scale, 90–98, 167–169
Multiwafer polishing systems, 12

N

Noncontact capacitive measurement, 229–230
  bow and warp, 232
  flatness, 231–232
Nova, 37

O

OnTrak, 23, 34
Optimal weighting function, 113
Orbital polishers, 14–16
  slurry delivery, 25
Order skipping, 218
Oxide erosion, 277
Oxide slurries
  abrasives for, 140, 143
  comparison of, 151
  solution for, 149–150

P

Pad(s)
  classes of, 156, 157
  deformation, 110–112
  development of, 156
  for dielectric CMP, 170–177
  function of, 155
  manufacturing processes, 156–162
  material property effects on polishing process 169–170
  for metal CMP, 177–178
  shaping, 28, 30
  wafer nonuniformity and, 97
Pad conditioning, 25–30
  acoustic energy for, 28
  concurrent versus sequential, 26–27
  diamond-coated disk for, 27–28
  high-pressure deionized water for, 28
Pad feed (web) polishers, 16
Parallel piping systems, 64
Particle effects, 184–185

Particle-removal mechanisms, 193–202
Patterned wafer modeling, 98–104
PFA Teflon®, 64
Photoassisted corrosion, 190–192
pH shock, 61–62
Pinch valves, 67–68
Piping systems, 62
   materials for, 64–65
   types of, 63–64
Planarization length
   determination, 113–118
   response function and, 108–113
Platen polishers
   hard, 24
   multiwafer per, 12
   slurry delivery, 25
   temperature control, 24–25
Point-of-use (POU)
   filtration, 81
   mixing, 58–59
Post-CMP cleaning (PCMPC), 30
   corrosion effects, 186–192
   damaged-layer effects, 186
   damaged-layer removal, 208–210
   double-sided brush scrubbers, 32–34
   examples of, 210–212
   megasonic cleaning, 34–35
   metallic contamination effects, 185–186
   metallic contamination removal, 206–208
   particle effects, 184–185
   requirements, 184–186
   rules for, 31
   slurry removal, 193–206
Pourbaix diagrams, 187
Premetal dielectrics (PMD), 139
Pressure and flow consistency, 66–67
   backpressure devices, 67–68
Preston equation/coefficient, 91, 92, 93, 95, 105

R

Reflectometry, 216–217
   advantages, 218
   integration issues, 225–226
   measurement of patterned wafers, 224
   measurement patterns and points, 220–224
   substrate modeling, 218–220

S

Scattering, 220
Scratches, 266–269
Scrubber cleaning, 202–203
Second-generation polishers, 11–13
Sequential linear polishers, 13
Sequential rotational systems, 12–13
Series/parallel piping systems, 64
Series piping systems, 64
Shallow trench isolation (STI), 3, 5, 99, 118–119, 139, 184, 211
Silica
   applications, 140
   colloidal, 140, 143, 151
   fumed, 140, 143, 151
Silicon oxide CMP, 211
Slurries
   abrasives, 140–147
   agitation, 74–75
   backpressure devices, 67–68
   blending technology, 54–55
   comparisons of, 151–154
   costs and, 37–38
   daytank replenishment, 59–61
   delivery, 25
   dispense engines, 53–54
   flow and wafer-scale models, 96–97
   handling, 49, 51–52
   measuring techniques, 55–59
   mixing constituents, 61–62
   pressure and flow consistency, 66–67
   ready-to-use, 55
   reprocessing, 37–38, 40, 86
   settling, 65–66
Slurry distribution system
   bulk, 53
   chemical cleaning agents for, 84
   consumption ramp, 68–69
   evolution of, 52–53
   filtration, 78–82
   maintenance, 83–84
   manual, 52–53
   overview of, 48–49, 50
   piping systems, 62–65
   redundancy, 69
   room location, 66
   storage tanks/daytanks, 71–74
   valve boxes, 69–71
   waste disposal, 84–86

INDEX

Slurry mix ratio parameter endpoint measuring, 57–58
Slurry removal
  in acid media, 205–206
  in alkaline media, 204
  forces involved, 193–195
  particle-removal mechanisms, 193–202
  prevention of electrostatic readhesion, 197–202
  scrubber cleaning, 202–203
  separation from substrate, 195–196
  wet cleaning, 204–206
Slurry solution, 146–151
  for metal slurries, 149
  for oxide slurries, 146–148
Spray acid tool (SAT), 274
Step height reduction, 100
  models for, 120–124
Storage tanks. See Daytanks
Stress, effects of, 19–20
Stylus profilometry, 236
Substrate modeling, 218–220
Suppliers
  of CMP equipment, 7
  marketing CMP waste treatment, 86
  for slurry dispense systems, 54
Supply drum filtration, 80

T

Tantalum, 188
Temperature control, 24–25
TEOS, BPSG versus, 261–262
Thermodynamic aspect, 187–189
Third-generation polishers, 13–16
300-mm equipment, 40–41
Titanium, 188
Total x-ray fluorescence (TXRF), 233–235, 274
Tungsten, 188
  cleanup, 211–212
  metal contamination, 273–276
  oxide erosion, 277
  polish rate and uniformity, 277
Tungsten etch-back (WEB), 273
Tungsten modeling, 126–128
  contact wear, 128–130
Tungsten slurries, comparison of, 151–154

V

Valve boxes, 69–71
van der Waals forces, 193–195
Volumetric mixing, 56

W

Wafer carriers
  edge exclusion region, 16–19
  floating, 21–23
  gimbaled, 20–21
  importance of, 16
  linear polisher and pressure control, 23–24
  requirements of, 17–19
  stress, effects of, 19–20
Wafer handling efficiency, 9
Wafer-scale models, 167–169
  empirical approaches to, 97
  macroscopic-bulk polish, 90–92
  pad-related nonuniformity, 97
  pressure variation, 95
  slurry flow, 96–97
  sources of nonuniformity, 92–95, 168
  status of, 97–98
Wafer-to-wafer nonunifority (WTWNU)
  due to thickness variation in incoming wafers, 264–265
  due to variation in polish rate, 263–264
  endpoint detection system, lack of, 262–263
Waste disposal, 84–86
Water consumption, 85
Weighting function
  density, 110–112
  optimal, 113
Weight scale measuring, 56–57
Wet cleaning, 204–206
Within-wafer nonuniformity (WIWNU), 20
  backpressure, 251
  BPSG versus TEOS, 261–262
  down force, 251, 257
  end effector effects, 252–257
  equipment and consumables, 246–251
  uniformity versus planarity, 257–261

Contents of Volumes in This Series

Volume 1 Physics of III–V Compounds

C. *Hilsum*, Some Key Features of III–V Compounds
F. *Bassani*, Methods of Band Calculations Applicable to III–V Compounds
E. O. *Kane*, The k-p Method
V. L. *Bonch-Bruevich*, Effect of Heavy Doping on the Semiconductor Band Structure
D. *Long*, Energy Band Structures of Mixed Crystals of III–V Compounds
L. M. *Roth and P. N. Argyres*, Magnetic Quantum Effects
S. M. *Puri and T. H. Geballe*, Thermomagnetic Effects in the Quantum Region
W. M. *Becker*, Band Characteristics near Principal Minima from Magnetoresistance
E. H. *Putley*, Freeze-Out Effects, Hot Electron Effects, and Submillimeter Photoconductivity in InSb
H. *Weiss*, Magnetoresistance
B. *Ancker-Johnson*, Plasma in Semiconductors and Semimetals

Volume 2 Physics of III–V Compounds

M. G. *Holland*, Thermal Conductivity
S. I. *Novkova*, Thermal Expansion
U. *Piesbergen*, Heat Capacity and Debye Temperatures
G. *Giesecke*, Lattice Constants
J. R. *Drabble*, Elastic Properties
A. U. *Mac Rae and G. W. Gobeli*, Low Energy Electron Diffraction Studies
R. *Lee Mieher*, Nuclear Magnetic Resonance
B. *Goldstein*, Electron Paramagnetic Resonance
T. S. *Moss*, Photoconduction in III–V Compounds
E. *Antoncik and J. Tauc*, Quantum Efficiency of the Internal Photoelectric Effect in InSb
G. W. *Gobeli and I. G. Allen*, Photoelectric Threshold and Work Function
P. S. *Pershan*, Nonlinear Optics in III–V Compounds
M. *Gershenzon*, Radiative Recombination in the III–V Compounds
F. *Stern*, Stimulated Emission in Semiconductors

Volume 3   Optical of Properties III–V Compounds

M. Hass, Lattice Reflection
W. G. Spitzer, Multiphonon Lattice Absorption
D. L. Stierwalt and R. F. Potter, Emittance Studies
H. R. Philipp and H. Ehrenveich, Ultraviolet Optical Properties
M. Cardona, Optical Absorption above the Fundamental Edge
E. J. Johnson, Absorption near the Fundamental Edge
J. O. Dimmock, Introduction to the Theory of Exciton States in Semiconductors
B. Lax and J. G. Mavroides, Interband Magnetooptical Effects
H. Y. Fan, Effects of Free Carries on Optical Properties
E. D. Palik and G. B. Wright, Free-Carrier Magnetooptical Effects
R. H. Bube, Photoelectronic Analysis
B. O. Seraphin and H. E. Bennett, Optical Constants

Volume 4   Physics of III–V Compounds

N. A. Goryunova, A. S. Borschevskii, and D. N. Tretiakov, Hardness
N. N. Sirota, Heats of Formation and Temperatures and Heats of Fusion of Compounds $A^{III}B^{V}$
D. L. Kendall, Diffusion
A. G. Chynoweth, Charge Multiplication Phenomena
R. W. Keyes, The Effects of Hydrostatic Pressure on the Properties of III–V Semiconductors
L. W. Aukerman, Radiation Effects
N. A. Goryunova, F. P. Kesamanly, and D. N. Nasledov, Phenomena in Solid Solutions
R. T. Bate, Electrical Properties of Nonuniform Crystals

Volume 5   Infrared Detectors

H. Levinstein, Characterization of Infrared Detectors
P. W. Kruse, Indium Antimonide Photoconductive and Photoelectromagnetic Detectors
M. B. Prince, Narrowband Self-Filtering Detectors
I. Melngalis and T. C. Harman, Single-Crystal Lead-Tin Chalcogenides
D. Long and J. L. Schmidt, Mercury-Cadmium Telluride and Closely Related Alloys
E. H. Putley, The Pyroelectric Detector
N. B. Stevens, Radiation Thermopiles
R. J. Keyes and T. M. Quist, Low Level Coherent and Incoherent Detection in the Infrared
M. C. Teich, Coherent Detection in the Infrared
F. R. Arams, E. W. Sard, B. J. Peyton, and F. P. Pace, Infrared Heterodyne Detection with Gigahertz IF Response
H. S. Sommers, Jr., Macrowave-Based Photoconductive Detector
R. Sehr and R. Zuleeg, Imaging and Display

Volume 6   Injection Phenomena

M. A. Lampert and R. B. Schilling, Current Injection in Solids: The Regional Approximation Method
R. Williams, Injection by Internal Photoemission
A. M. Barnett, Current Filament Formation

R. Baron and J. W. Mayer, Double Injection in Semiconductors
W. Ruppel, The Photoconductor-Metal Contact

Volume 7   Application and Devices
Part A

J. A. Copeland and S. Knight, Applications Utilizing Bulk Negative Resistance
F. A. Padovani, The Voltage-Current Characteristics of Metal-Semiconductor Contacts
P. L. Hower, W. W. Hooper, B. R. Cairns, R. D. Fairman, and D. A. Tremere, The GaAs Field-Effect Transistor
M. H. White, MOS Transistors
G. R. Antell, Gallium Arsenide Transistors
T. L. Tansley, Heterojunction Properties

Part B

T. Misawa, IMPATT Diodes
H. C. Okean, Tunnel Diodes
R. B. Campbell and Hung-Chi Chang, Silicon Junction Carbide Devices
R. E. Enstrom, H. Kressel, and L. Krassner, High-Temperature Power Rectifiers of $GaAs_{1-x}P_x$

Volume 8   Transport and Optical Phenomena

R. J. Stirn, Band Structure and Galvanomagnetic Effects in III–V Compounds with Indirect Band Gaps
R. W. Ure, Jr., Thermoelectric Effects in III–V Compounds
H. Piller, Faraday Rotation
H. Barry Bebb and E. W. Williams, Photoluminescence I: Theory
E. W. Williams and H. Barry Bebb, Photoluminescence II: Gallium Arsenide

Volume 9   Modulation Techniques

B. O. Seraphin, Electroreflectance
R. L. Aggarwal, Modulated Interband Magnetooptics
D. F. Blossey and Paul Handler, Electroabsorption
B. Batz, Thermal and Wavelength Modulation Spectroscopy
I. Balslev, Piezopptical Effects
D. E. Aspnes and N. Bottka, Electric-Field Effects on the Dielectric Function of Semiconductors and Insulators

Volume 10   Transport Phenomena

R. L. Rhode, Low-Field Electron Transport
J. D. Wiley, Mobility of Holes in III–V Compounds
C. M. Wolfe and G. E. Stillman, Apparent Mobility Enhancement in Inhomogeneous Crystals
R. L. Petersen, The Magnetophonon Effect

Volume 11 Solar Cells

H. J. Hovel, Introduction; Carrier Collection, Spectral Response, and Photocurrent; Solar Cell Electrical Characteristics; Efficiency; Thickness; Other Solar Cell Devices; Radiation Effects; Temperature and Intensity; Solar Cell Technology

Volume 12 Infrared Detectors (II)

W. L. Eiseman, J. D. Merriam, and R. F. Potter, Operational Characteristics of Infrared Photodetectors
P. R. Bratt, Impurity Germanium and Silicon Infrared Detectors
E. H. Putley, InSb Submillimeter Photoconductive Detectors
G. E. Stillman, C. M. Wolfe, and J. O. Dimmock, Far-Infrared Photoconductivity in High Purity GaAs
G. E. Stillman and C. M. Wolfe, Avalanche Photodiodes
P. L. Richards, The Josephson Junction as a Detector of Microwave and Far-Infrared Radiation
E. H. Putley, The Pyroelectric Detector — An Update

Volume 13 Cadmium Telluride

K. Zanio, Materials Preparations; Physics; Defects; Applications

Volume 14 Lasers, Junctions, Transport

N. Holonyak, Jr. and M. H. Lee, Photopumped III–V Semiconductor Lasers
H. Kressel and J. K. Butler, Heterojunction Laser Diodes
A Van der Ziel, Space-Charge-Limited Solid-State Diodes
P. J. Price, Monte Carlo Calculation of Electron Transport in Solids

Volume 15 Contacts, Junctions, Emitters

B. L. Sharma, Ohmic Contacts to III–V Compounds Semiconductors
A. Nussbaum, The Theory of Semiconducting Junctions
J. S. Escher, NEA Semiconductor Photoemitters

Volume 16 Defects, (HgCd)Se, (HgCd)Te

H. Kressel, The Effect of Crystal Defects on Optoelectronic Devices
C. R. Whitsett, J. G. Broerman, and C. J. Summers, Crystal Growth and Properties of $Hg_{1-x}Cd_xSe$ alloys
M. H. Weiler, Magnetooptical Properties of $Hg_{1-x}Cd_x Te$ Alloys
P. W. Kruse and J. G. Ready, Nonlinear Optical Effects in $Hg_{1-x}Cd_x Te$

Volume 17 CW Processing of Silicon and Other Semiconductors

J. F. Gibbons, Beam Processing of Silicon
A. Lietoila, R. B. Gold, J. F. Gibbons, and L. A. Christel, Temperature Distributions and Solid Phase Reaction Rates Produced by Scanning CW Beams

A. Leitoila and J. F. Gibbons, Applications of CW Beam Processing to Ion Implanted Crystalline Silicon
N. M. Johnson, Electronic Defects in CW Transient Thermal Processed Silicon
K. F. Lee, T. J. Stultz, and J. F. Gibbons, Beam Recrystallized Polycrystalline Silicon: Properties, Applications, and Techniques
T. Shibata, A. Wakita, T. W. Sigmon, and J. F. Gibbons, Metal-Silicon Reactions and Silicide
Y. I. Nissim and J. F. Gibbons, CW Beam Processing of Gallium Arsenide

Volume 18 Mercury Cadmium Telluride

P. W. Kruse, The Emergence of $(Hg_{1-x}Cd_x)Te$ as a Modern Infrared Sensitive Material
H. E. Hirsch, S. C. Liang, and A. G. White, Preparation of High-Purity Cadmium, Mercury, and Tellurium
W. F. H. Micklethwaite, The Crystal Growth of Cadmium Mercury Telluride
P. E. Petersen, Auger Recombination in Mercury Cadmium Telluride
R. M. Broudy and V. J. Mazurczyck, (HgCd)Te Photoconductive Detectors
M. B. Reine, A. K. Soad, and T. J. Tredwell, Photovoltaic Infrared Detectors
M. A. Kinch, Metal-Insulator-Semiconductor Infrared Detectors

Volume 19 Deep Levels, GaAs, Alloys, Photochemistry

G. F. Neumark and K. Kosai, Deep Levels in Wide Band-Gap III–V Semiconductors
D. C. Look, The Electrical and Photoelectronic Properties of Semi-Insulating GaAs
R. F. Brebrick, Ching-Hua Su, and Pok-Kai Liao, Associated Solution Model for Ga-In-Sb and Hg-Cd-Te
Y. Ya. Gurevich and Y. V. Pleskon, Photoelectrochemistry of Semiconductors

Volume 20 Semi-Insulating GaAs

R. N. Thomas, H. M. Hobgood, G. W. Eldridge, D. L. Barrett, T. T. Braggins, L. B. Ta, and S. K. Wang, High-Purity LEC Growth and Direct Implantation of GaAs for Monolithic Microwave Circuits
C. A. Stolte, Ion Implantation and Materials for GaAs Integrated Circuits
C. G. Kirkpatrick, R. T. Chen, D. E. Holmes, P. M. Asbeck, K. R. Elliott, R. D. Fairman, and J. R. Oliver, LEC GaAs for Integrated Circuit Applications
J. S. Blakemore and S. Rahimi, Models for Mid-Gap Centers in Gallium Arsenide

Volume 21 Hydrogenated Amorphous Silicon
Part A

J. I. Pankove, Introduction
M. Hirose, Glow Discharge; Chemical Vapor Deposition
Y. Uchida, di Glow Discharge
T. D. Moustakas, Sputtering
I. Yamada, Ionized-Cluster Beam Deposition
B. A. Scott, Homogeneous Chemical Vapor Deposition

F. J. Kampas, Chemical Reactions in Plasma Deposition
P. A. Longeway, Plasma Kinetics
H. A. Weakliem, Diagnostics of Silane Glow Discharges Using Probes and Mass Spectroscopy
L. Gluttman, Relation between the Atomic and the Electronic Structures
A. Chenevas-Paule, Experiment Determination of Structure
S. Minomura, Pressure Effects on the Local Atomic Structure
D. Adler, Defects and Density of Localized States

Part B

J. I. Pankove, Introduction
G. D. Cody, The Optical Absorption Edge of a-Si:H
N. M. Amer and W. B. Jackson, Optical Properties of Defect States in a-Si:H
P. J. Zanzucchi, The Vibrational Spectra of a-Si:H
Y. Hamakawa, Electroreflectance and Electroabsorption
J. S. Lannin, Raman Scattering of Amorphous Si, Ge, and Their Alloys
R. A. Street, Luminescence in a-Si:H
R. S. Crandall, Photoconductivity
J. Tauc, Time-Resolved Spectroscopy of Electronic Relaxation Processes
P. E. Vanier, IR-Induced Quenching and Enhancement of Photoconductivity and Photoluminescence
H. Schade, Irradiation-Induced Metastable Effects
L. Ley, Photoelectron Emission Studies

Part C

J. I. Pankove, Introduction
J. D. Cohen, Density of States from Junction Measurements in Hydrogenated Amorphous Silicon
P. C. Taylor, Magnetic Resonance Measurements in a-Si:H
K. Morigaki, Optically Detected Magnetic Resonance
J. Dresner, Carrier Mobility in a-Si:H
T. Tiedje, Information about band-Tail States from Time-of-Flight Experiments
A. R. Moore, Diffusion Length in Undoped a-Si:H
W. Beyer and J. Overhof, Doping Effects in a-Si:H
H. Fritzche, Electronic Properties of Surfaces in a-Si:H
C. R. Wronski, The Staebler-Wronski Effect
R. J. Nemanich, Schottky Barriers on a-Si:H
B. Abeles and T. Tiedje, Amorphous Semiconductor Superlattices

Part D

J. I. Pankove, Introduction
D. E. Carlson, Solar Cells
G. A. Swartz, Closed-Form Solution of I–V Characteristic for a a-Si:H Solar Cells
I. Shimizu, Electrophotography
S. Ishioka, Image Pickup Tubes

P. G. LeComber and W. E. Spear, The Development of the a-Si:H Field-Effect Transistor and Its Possible Applications
D. G. Ast, a-Si:H FET-Addressed LCD Panel
S. Kaneko, Solid-State Image Sensor
M. Matsumura, Charge-Coupled Devices
M. A. Bosch, Optical Recording
A. D'Amico and G. Fortunato, Ambient Sensors
H. Kukimoto, Amorphous Light-Emitting Devices
R. J. Phelan, Jr., Fast Detectors and Modulators
J. I. Pankove, Hybrid Structures
P. G. LeComber, A. E. Owen, W. E. Spear, J. Hajto, and W. K. Choi, Electronic Switching in Amorphous Silicon Junction Devices

Volume 22 Lightwave Communications Technology
Part A

K. Nakajima, The Liquid-Phase Epitaxial Growth of InGaAsP
W. T. Tsang, Molecular Beam Epitaxy for III–V Compound Semiconductors
G. B. Stringfellow, Organometallic Vapor-Phase Epitaxial Growth of III–V Semiconductors
G. Beuchet, Halide and Chloride Transport Vapor-Phase Deposition of InGaAsP and GaAs
M. Razeghi, Low-Pressure Metallo-Organic Chemical Vapor Deposition of $Ga_xIn_{1-x}As P_{1-y}$ Alloys
P. M. Petroff, Defects in III–V Compound Semiconductors

Part B

J. P. van der Ziel, Mode Locking of Semiconductor Lasers
K. Y. Lau and A. Yariv, High-Frequency Current Modulation of Semiconductor Injection Lasers
C. H. Henry, Special Properties of Semiconductor Lasers
Y. Suematsu, K. Kishino, S. Arai, and F. Koyama, Dynamic Single-Mode Semiconductor Lasers with a Distributed Reflector
W. T. Tsang, The Cleaved-Coupled-Cavity (C^3) Laser

Part C

R. J. Nelson and N. K. Dutta, Review of InGaAsP InP Laser Structures and Comparison of Their Performance
N. Chinone and M. Nakamura, Mode-Stabilized Semiconductor Lasers for 0.7–0.8- and 1.1–1.6-μm Regions
Y. Horikoshi, Semiconductor Lasers with Wavelengths Exceeding 2 μm
B. A. Dean and M. Dixon, The Functional Reliability of Semiconductor Lasers as Optical Transmitters
R. H. Saul, T. P. Lee, and C. A. Burus, Light-Emitting Device Design
C. L. Zipfel, Light-Emitting Diode-Reliability
T. P. Lee and T. Li, LED-Based Multimode Lightwave Systems
K. Ogawa, Semiconductor Noise-Mode Partition Noise

Part D

F. *Capasso,* The Physics of Avalanche Photodiodes
T. P. *Pearsall and M. A. Pollack,* Compound Semiconductor Photodiodes
T. *Kaneda,* Silicon and Germanium Avalanche Photodiodes
S. R. *Forrest,* Sensitivity of Avalanche Photodetector Receivers for High-Bit-Rate Long-Wavelength Optical Communication Systems
J. C. *Campbell,* Phototransistors for Lightwave Communications

Part E

S. *Wang,* Principles and Characteristics of Integrable Active and Passive Optical Devices
S. *Margalit and A. Yariv,* Integrated Electronic and Photonic Devices
T. *Mukai, Y. Yamamoto, and T. Kimura,* Optical Amplification by Semiconductor Lasers

Volume 23 Pulsed Laser Processing of Semiconductors

R. F. *Wood, C. W. White, and R. T. Young,* Laser Processing of Semiconductors: An Overview
C. W. *White,* Segregation, Solute Trapping, and Supersaturated Alloys
G. E. *Jellison, Jr.,* Optical and Electrical Properties of Pulsed Laser-Annealed Silicon
R. F. *Wood and G. E. Jellison, Jr.,* Melting Model of Pulsed Laser Processing
R. F. *Wood and F. W. Young, Jr.,* Nonequilibrium Solidification Following Pulsed Laser Melting
D. H. *Lowndes and G. E. Jellison, Jr.,* Time-Resolved Measurement During Pulsed Laser Irradiation of Silicon
D. M. *Zebner,* Surface Studies of Pulsed Laser Irradiated Semiconductors
D. H. *Lowndes,* Pulsed Beam Processing of Gallium Arsenide
R. B. *James,* Pulsed CO_2 Laser Annealing of Semiconductors
R. T. *Young and R. F. Wood,* Applications of Pulsed Laser Processing

Volume 24 Applications of Multiquantum Wells, Selective Doping, and Superlattices

C. *Weisbuch,* Fundamental Properties of III–V Semiconductor Two-Dimensional Quantized Structures: The Basis for Optical and Electronic Device Applications
H. *Morkoc and H. Unlu,* Factors Affecting the Performance of (Al, Ga)As/GaAs and (Al, Ga)As/InGaAs Modulation-Doped Field-Effect Transistors: Microwave and Digital Applications
N. T. *Linh,* Two-Dimensional Electron Gas FETs: Microwave Applications
M. *Abe et al.,* Ultra-High-Speed HEMT Integrated Circuits
D. S. *Chemla, D. A. B. Miller, and P. W. Smith,* Nonlinear Optical Properties of Multiple Quantum Well Structures for Optical Signal Processing
F. *Capasso,* Graded-Gap and Superlattice Devices by Band-Gap Engineering
W. T. *Tsang,* Quantum Confinement Heterostructure Semiconductor Lasers
G. C. *Osbourn et al.,* Principles and Applications of Semiconductor Strained-Layer Superlattices

Volume 25 Diluted Magnetic Semiconductors

W. Giriat and J. K. Furdyna, Crystal Structure, Composition, and Materials Preparation of Diluted Magnetic Semiconductors

W. M. Becker, Band Structure and Optical Properties of Wide-Gap $A^{II}_{1-x}Mn_xB_{IV}$ Alloys at Zero Magnetic Field

S. Oseroff and P. H. Keesom, Magnetic Properties: Macroscopic Studies

T. Giebultowicz and T. M. Holden, Neutron Scattering Studies of the Magnetic Structure and Dynamics of Diluted Magnetic Semiconductors

J. Kossut, Band Structure and Quantum Transport Phenomena in Narrow-Gap Diluted Magnetic Semiconductors

C. Riquaux, Magnetooptical Properties of Large-Gap Diluted Magnetic Semiconductors

J. A. Gaj, Magnetooptical Properties of Large-Gap Diluted Magnetic Semiconductors

J. Mycielski, Shallow Acceptors in Diluted Magnetic Semiconductors: Splitting, Boil-off, Giant Negative Magnetoresistance

A. K. Ramadas and R. Rodriquez, Raman Scattering in Diluted Magnetic Semiconductors

P. A. Wolff, Theory of Bound Magnetic Polarons in Semimagnetic Semiconductors

Volume 26 III–V Compound Semiconductors and Semiconductor Properties of Superionic Materials

Z. Yuanxi, III–V Compounds

H. V. Winston, A. T. Hunter, H. Kimura, and R. E. Lee, InAs-Alloyed GaAs Substrates for Direct Implantation

P. K. Bhattacharya and S. Dhar, Deep Levels in III–V Compound Semiconductors Grown by MBE

Y. Ya. Gurevich and A. K. Ivanov-Shits, Semiconductor Properties of Supersonic Materials

Volume 27 High Conducting Quasi-One-Dimensional Organic Crystals

E. M. Conwell, Introduction to Highly Conducting Quasi-One-Dimensional Organic Crystals

I. A. Howard, A Reference Guide to the Conducting Quasi-One-Dimensional Organic Molecular Crystals

J. P. Pouquet, Structural Instabilities

E. M. Conwell, Transport Properties

C. S. Jacobsen, Optical Properties

J. C. Scott, Magnetic Properties

L. Zuppiroli, Irradiation Effects: Perfect Crystals and Real Crystals

Volume 28 Measurement of High-Speed Signals in Solid State Devices

J. Frey and D. Ioannou, Materials and Devices for High-Speed and Optoelectronic Applications

H. Schumacher and E. Strid, Electronic Wafer Probing Techniques

D. H. Auston, Picosecond Photoconductivity: High-Speed Measurements of Devices and Materials

J. A. Valdmanis, Electro-Optic Measurement Techniques for Picosecond Materials, Devices, and Integrated Circuits.

J. M. Wiesenfeld and R. K. Jain, Direct Optical Probing of Integrated Circuits and High-Speed Devices

G. Plows, Electron-Beam Probing

A. M. Weiner and R. B. Marcus, Photoemissive Probing

Volume 29 Very High Speed Integrated Circuits: Gallium Arsenide LSI

M. Kuzuhara and T. Nazaki, Active Layer Formation by Ion Implantation
H. Hasimoto, Focused Ion Beam Implantation Technology
T. Nozaki and A. Higashisaka, Device Fabrication Process Technology
M. Ino and T. Takada, GaAs LSI Circuit Design
M. Hirayama, M. Ohmori, and K. Yamasaki, GaAs LSI Fabrication and Performance

Volume 30 Very High Speed Integrated Circuits: Heterostructure

H. Watanabe, T. Mizutani, and A. Usui, Fundamentals of Epitaxial Growth and Atomic Layer Epitaxy
S. Hiyamizu, Characteristics of Two-Dimensional Electron Gas in III–V Compound Heterostructures Grown by MBE
T. Nakanisi, Metalorganic Vapor Phase Epitaxy for High-Quality Active Layers
T. Nimura, High Electron Mobility Transistor and LSI Applications
T. Sugeta and T. Ishibashi, Hetero-Bipolar Transistor and LSI Application
H. Matsueda, T. Tanaka, and M. Nakamura, Optoelectronic Integrated Circuits

Volume 31 Indium Phosphide: Crystal Growth and Characterization

J. P. Farges, Growth of Discoloration-free InP
M. J. McCollum and G. E. Stillman, High Purity InP Grown by Hydride Vapor Phase Epitaxy
T. Inada and T. Fukuda, Direct Synthesis and Growth of Indium Phosphide by the Liquid Phosphorous Encapsulated Czochralski Method
O. Oda, K. Katagiri, K. Shinohara, S. Katsura, Y. Takahashi, K. Kainosho, K. Kohiro, and R. Hirano, InP Crystal Growth, Substrate Preparation and Evaluation
K. Tada, M. Tatsumi, M. Morioka, T. Araki, and T. Kawase, InP Substrates: Production and Quality Control
M. Razeghi, LP-MOCVD Growth, Characterization, and Application of InP Material
T. A. Kennedy and P. J. Lin-Chung, Stoichiometric Defects in InP

Volme 32 Strained-Layer Superlattices: Physics

T. P. Pearsall, Strained-Layer Superlattices
F. H. Pollack, Effects of Homogeneous Strain on the Electronic and Vibrational Levels in Semiconductors
J. Y. Marzin, J. M. Gerárd, P. Voisin, and J. A. Brum, Optical Studies of Strained III–V Heterolayers
R. People and S. A. Jackson, Structurally Induced States from Strain and Confinement
M. Jaros, Microscopic Phenomena in Ordered Superlattices

Volume 33 Strained-Layer Superlattices: Materials Science and Technology

R. Hull and J. C. Bean, Principles and Concepts of Strained-Layer Epitaxy
W. J. Schaff, P. J. Tasker, M. C. Foisy, and L. F. Eastman, Device Applications of Strained-Layer Epitaxy

S. T. Picraux, B. L. Doyle, and J. Y. Tsao, Structure and Characterization of Strained-Layer Superlattices
E. Kasper and F. Schäffer, Group IV Compounds
D. L. Martin, Molecular Beam Epitaxy of IV–VI Compounds Heterojunction
R. L. Gunshor, L. A. Kolodziejski, A. V. Nurmikko, and N. Otsuka, Molecular Beam Epitaxy of II–VI Semiconductor Microstructures

Volume 34 Hydrogen in Semiconductors

J. I. Pankove and N. M. Johnson, Introduction to Hydrogen in Semiconductors
C. H. Seager, Hydrogenation Methods
J. I. Pankove, Hydrogenation of Defects in Crystalline Silicon
J. W. Corbett, P. Deák, U. V. Desnica, and S. J. Pearton, Hydrogen Passivation of Damage Centers in Semiconductors
S. J. Pearton, Neutralization of Deep Levels in Silicon
J. I. Pankove, Neutralization of Shallow Acceptors in Silicon
N. M. Johnson, Neutralization of Donor Dopants and Formation of Hydrogen-Induced Defects in n-Type Silicon
M. Stavola and S. J. Pearton, Vibrational Spectroscopy of Hydrogen-Related Defects in Silicon
A. D. Marwick, Hydrogen in Semiconductors: Ion Beam Techniques
C. Herring and N. M. Johnson, Hydrogen Migration and Solubility in Silicon
E. E. Haller, Hydrogen-Related Phenomena in Crystalline Germanium
J. Kakalios, Hydrogen Diffusion in Amorphous Silicon
J. Chevalier, B. Clerjaud, and B. Pajot, Neutralization of Defects and Dopants in III–V Semiconductors
G. G. DeLeo and W. B. Fowler, Computational Studies of Hydrogen-Containing Complexes in Semiconductors
R. F. Kiefl and T. L. Estle, Muonium in Semiconductors
C. G. Van de Walle, Theory of Isolated Interstitial Hydrogen and Muonium in Crystalline Semiconductors

Volume 35 Nanostructured Systems

M. Reed, Introduction
H. van Houten, C. W. J. Beenakker, and B. J. van Wees, Quantum Point Contacts
G. Timp, When Does a Wire Become an Electron Waveguide?
M. Büttiker, The Quantum Hall Effects in Open Conductors
W. Hansen, J. P. Kotthaus, and U. Merkt, Electrons in Laterally Periodic Nanostructures

Volume 36 The Spectroscopy of Semiconductors

D. Heiman, Spectroscopy of Semiconductors at Low Temperatures and High Magnetic Fields
A. V. Nurmikko, Transient Spectroscopy by Ultrashort Laser Pulse Techniques
A. K. Ramdas and S. Rodriguez, Piezospectroscopy of Semiconductors
O. J. Glembocki and B. V. Shanabrook, Photoreflectance Spectroscopy of Microstructures
D. G. Seiler, C. L. Littler, and M. H. Wiler, One- and Two-Photon Magneto-Optical Spectroscopy of InSb and $Hg_{1-x}Cd_xTe$

Volume 37 The Mechanical Properties of Semiconductors

A.-B. Chen, A. Sher and W. T. Yost, Elastic Constants and Related Properties of Semiconductor Compounds and Their Alloys
D. R. Clarke, Fracture of Silicon and Other Semiconductors
H. Siethoff, The Plasticity of Elemental and Compound Semiconductors
S. Guruswamy, K. T. Faber and J. P. Hirth, Mechanical Behavior of Compound Semiconductors
S. Mahajan, Deformation Behavior of Compound Semiconductors
J. P. Hirth, Injection of Dislocations into Strained Multilayer Structures
D. Kendall, C. B. Fleddermann, and K. J. Malloy, Critical Technologies for the Micromachining of Silicon
I. Matsuba and K. Mokuya, Processing and Semiconductor Thermoelastic Behavior

Volume 38 Imperfections in III/V Materials

U. Scherz and M. Scheffler, Density-Functional Theory of sp-Bonded Defects in III/V Semiconductors
M. Kaminska and E. R. Weber, El2 Defect in GaAs
D. C. Look, Defects Relevant for Compensation in Semi-Insulating GaAs
R. C. Newman, Local Vibrational Mode Spectroscopy of Defects in III/V Compounds
A. M. Hennel, Transition Metals in III/V Compounds
K. J. Malloy and K. Khachaturyan, DX and Related Defects in Semiconductors
V. Swaminathan and A. S. Jordan, Dislocations in III/V Compounds
K. W. Nauka, Deep Level Defects in the Epitaxial III/V Materials

Volume 39 Minority Carriers in III–V Semiconductors: Physics and Applications

N. K. Dutta, Radiative Transitions in GaAs and Other III–V Compounds
R. K. Ahrenkiel, Minority-Carrier Lifetime in III–V Semiconductors
T. Furuta, High Field Minority Electron Transport in p-GaAs
M. S. Lundstrom, Minority-Carrier Transport in III–V Semiconductors
R. A. Abram, Effects of Heavy Doping and High Excitation on the Band Structure of GaAs
D. Yevick and W. Bardyszewski, An Introduction to Non-Equilibrium Many-Body Analyses of Optical Processes in III–V Semiconductors

Volume 40 Epitaxial Microstructures

E. F. Schubert, Delta-Doping of Semiconductors: Electronic, Optical, and Structural Properties of Materials and Devices
A. Gossard, M. Sundaram, and P. Hopkins, Wide Graded Potential Wells
P. Petroff, Direct Growth of Nanometer-Size Quantum Wire Superlattices
E. Kapon, Lateral Patterning of Quantum Well Heterostructures by Growth of Nonplanar Substrates
H. Temkin, D. Gershoni, and M. Panish, Optical Properties of $Ga_{1-x}In_xAs/InP$ Quantum Wells

Volume 41 High Speed Heterostructure Devices

F. Capasso, F. Beltram, S. Sen, A. Pahlevi, and A. Y. Cho, Quantum Electron Devices: Physics and Applications
P. Solomon, D. J. Frank, S. L. Wright, and F. Canora, GaAs-Gate Semiconductor–Insulator–Semiconductor FET
M. H. Hashemi and U. K. Mishra, Unipolar InP-Based Transistors
R. Kiehl, Complementary Heterostructure FET Integrated Circuits
T. Ishibashi, GaAs-Based and InP-Based Heterostructure Bipolar Transistors
H. C. Liu and T. C. L. G. Sollner, High-Frequency-Tunneling Devices
H. Ohnishi, T. More, M. Takatsu, K. Imamura, and N. Yokoyama, Resonant-Tunneling Hot-Electron Transistors and Circuits

Volume 42 Oxygen in Silicon

F. Shimura, Introduction to Oxygen in Silicon
W. Lin, The Incorporation of Oxygen into Silicon Crystals
T. J. Schaffner and D. K. Schroder, Characterization Techniques for Oxygen in Silicon
W. M. Bullis, Oxygen Concentration Measurement
S. M. Hu, Intrinsic Point Defects in Silicon
B. Pajot, Some Atomic Configurations of Oxygen
J. Michel and L. C. Kimerling, Electical Properties of Oxygen in Silicon
R. C. Newman and R. Jones, Diffusion of Oxygen in Silicon
T. Y. Tan and W. J. Taylor, Mechanisms of Oxygen Precipitation: Some Quantitative Aspects
M. Schrems, Simulation of Oxygen Precipitation
K. Simino and I. Yonenaga, Oxygen Effect on Mechanical Properties
W. Bergholz, Grown-in and Process-Induced Effects
F. Shimura, Intrinsic/Internal Gettering
H. Tsuya, Oxygen Effect on Electronic Device Performance

Volume 43 Semiconductors for Room Temperature Nuclear Detector Applications

R. B. James and T. E. Schlesinger, Introduction and Overview
L. S. Darken and C. E. Cox, High-Purity Germanium Detectors
A. Burger, D. Nason, L. Van den Berg, and M. Schieber, Growth of Mercuric Iodide
X. J. Bao, T. E. Schlesinger, and R. B. James, Electrical Properties of Mercuric Iodide
X. J. Bao, R. B. James, and T. E. Schlesinger, Optical Properties of Red Mercuric Iodide
M. Hage-Ali and P. Siffert, Growth Methods of CdTe Nuclear Detector Materials
M. Hage-Ali and P Siffert, Characterization of CdTe Nuclear Detector Materials
M. Hage-Ali and P. Siffert, CdTe Nuclear Detectors and Applications
R. B. James, T. E. Schlesinger, J. Lund, and M. Schieber, $Cd_{1-x}Zn_xTe$ Spectrometers for Gamma and X-Ray Applications
D. S. McGregor, J. E. Kammeraad, Gallium Arsenide Radiation Detectors and Spectrometers
J. C. Lund, F. Olschner, and A. Burger, Lead Iodide
M. R. Squillante, and K. S. Shah, Other Materials: Status and Prospects
V. M. Gerrish, Characterization and Quantification of Detector Performance
J. S. Iwanczyk and B. E. Patt, Electronics for X-ray and Gamma Ray Spectrometers
M. Schieber, R. B. James, and T. E. Schlesinger, Summary and Remaining Issues for Room Temperature Radiation Spectrometers

Volume 44 II–IV Blue/Green Light Emitters: Device Physics and Epitaxial Growth

J. Han and R. L. Gunshor, MBE Growth and Electrical Properties of Wide Bandgap ZnSe-based II–VI Semiconductors
S. Fujita and S. Fujita, Growth and Characterization of ZnSe-based II–VI Semiconductors by MOVPE
E. Ho and L. A. Kolodziejski, Gaseous Source UHV Epitaxy Technologies for Wide Bandgap II–VI Semiconductors
C. G. Van de Walle, Doping of Wide-Band-Gap II–VI Compounds—Theory
R. Cingolani, Optical Properties of Excitons in ZnSe-Based Quantum Well Heterostructures
A. Ishibashi and A. V. Nurmikko, II–VI Diode Lasers: A Current View of Device Performance and Issues
S. Guha and J. Petruzello, Defects and Degradation in Wide-Gap II–VI-based Structures and Light Emitting Devices

Volume 45 Effect of Disorder and Defects in Ion-Implanted Semiconductors: Electrical and Physiochemical Characterization

H. Ryssel, Ion Implantation into Semiconductors: Historical Perspectives
You-Nian Wang and Teng-Cai Ma, Electronic Stopping Power for Energetic Ions in Solids
S. T. Nakagawa, Solid Effect on the Electronic Stopping of Crystalline Target and Application to Range Estimation
G. Müller, S. Kalbitzer and G. N. Greaves, Ion Beams in Amorphous Semiconductor Research
J. Boussey-Said, Sheet and Spreading Resistance Analysis of Ion Implanted and Annealed Semiconductors
M. L. Polignano and G. Queirolo, Studies of the Stripping Hall Effect in Ion-Implanted Silicon
J. Stoemenos, Transmission Electron Microscopy Analyses
R. Nipoti and M. Servidori, Rutherford Backscattering Studies of Ion Implanted Semiconductors
P. Zaumseil, X-ray Diffraction Techniques

Volume 46 Effect of Disorder and Defects in Ion-Implanted Semiconductors: Optical and Photothermal Characterization

M. Fried, T. Lohner and J. Gyulai, Ellipsometric Analysis
A. Seas and C. Christofides, Transmission and Reflection Spectroscopy on Ion Implanted Semiconductors
A. Othonos and C. Christofides, Photoluminescence and Raman Scattering of Ion Implanted Semiconductors. Influence of Annealing
C. Christofides, Photomodulated Thermoreflectance Investigation of Implanted Wafers. Annealing Kinetics of Defects
U. Zammit, Photothermal Deflection Spectroscopy Characterization of Ion-Implanted and Annealed Silicon Films
A. Mandelis, A. Budiman and M. Vargas, Photothermal Deep-Level Transient Spectroscopy of Impurities and Defects in Semiconductors
R. Kalish and S. Charbonneau, Ion Implantation into Quantum-Well Structures
A. M. Myasnikov and N. N. Gerasimenko, Ion Implantation and Thermal Annealing of III-V Compound Semiconducting Systems: Some Problems of III-V Narrow Gap Semiconductors

Volume 47   Uncooled Infrared Imaging Arrays and Systems

R. G. Buser and M. P. Tompsett, Historical Overview
P. W. Kruse, Principles of Uncooled Infrared Focal Plane Arrays
R. A. Wood, Monolithic Silicon Microbolometer Arrays
C. M. Hanson, Hybrid Pyroelectric-Ferroelectric Bolometer Arrays
D. L. Polla and J. R. Choi, Monolithic Pyroelectric Bolometer Arrays
N. Teranishi, Thermoelectric Uncooled Infrared Focal Plane Arrays
M. F. Tompsett, Pyroelectric Vidicon
T. W. Kenny, Tunneling Infrared Sensors
J. R. Vig, R. L. Filler and Y. Kim, Application of Quartz Microresonators to Uncooled Infrared Imaging Arrays
P. W. Kruse, Application of Uncooled Monolithic Thermoelectric Linear Arrays to Imaging Radiometers

Volume 48   High Brightness Light Emitting Diodes

G. B. Stringfellow, Materials Issues in High-Brightness Light-Emitting Diodes
M. G. Craford, Overview of Device issues in High-Brightness Light-Emitting Diodes
F. M. Steranka, AlGaAs Red Light Emitting Diodes
C. H. Chen, S. A. Stockman, M. J. Peanasky, and C. P. Kuo, OMVPE Growth of AlGaInP for High Efficiency Visible Light-Emitting Diodes
F. A. Kish and R. M. Fletcher, AlGaInP Light-Emitting Diodes
M. W. Hodapp, Applications for High Brightness Light-Emitting Diodes
I. Akasaki and H. Amano, Organometallic Vapor Epitaxy of GaN for High Brightness Blue Light Emitting Diodes
S. Nakamura, Group III-V Nitride Based Ultraviolet-Blue-Green-Yellow Light-Emitting Diodes and Laser Diodes

Volume 49   Light Emission in Silicon: from Physics to Devices

D. J. Lockwood, Light Emission in Silicon
G. Abstreiter, Band Gaps and Light Emission in Si/SiGe Atomic Layer Structures
T. G. Brown and D. G. Hall, Radiative Isoelectronic Impurities in Silicon and Silicon-Germanium Alloys and Superlattices
J. Michel, L. V. C. Assali, M. T. Morse, and L. C. Kimerling, Erbium in Silicon
Y. Kanemitsu, Silicon and Germanium Nanoparticles
P. M. Fauchet, Porous Silicon: Photoluminescence and Electroluminescent Devices
C. Delerue, G. Allan, and M. Lannoo, Theory of Radiative and Nonradiative Processes in Silicon Nanocrystallites
L. Brus, Silicon Polymers and Nanocrystals

Volume 50   Gallium Nitride (GaN)

J. I. Pankove and T. D. Moustakas, Introduction
S. P. DenBaars and S. Keller, Metalorganic Chemical Vapor Deposition (MOCVD) of Group III Nitrides
W. A. Bryden and T. J. Kistenmacher, Growth of Group III-A Nitrides by Reactive Sputtering
N. Newman, Thermochemistry of III-N Semiconductors
S. J. Pearton and R. J. Shul, Etching of III Nitrides

S. M. Bedair, Indium-based Nitride Compounds
A. Trampert, O. Brandt, and K. H. Ploog, Crystal Structure of Group III Nitrides
H. Morkoc, F. Hamdani, and A. Salvador, Electronic and Optical Properties of III–V Nitride based Quantum Wells and Superlattices
K. Doverspike and J. I. Pankove, Doping in the III-Nitrides
T. Suski and P. Perlin, High Pressure Studies of Defects and Impurities in Gallium Nitride
B. Monemar, Optical Properties of GaN
W. R. L. Lambrecht, Band Structure of the Group III Nitrides
N. E. Christensen and P. Perlin, Phonons and Phase Transitions in GaN
S. Nakamura, Applications of LEDs and LDs
I. Akasaki and H. Amano, Lasers
J. A. Cooper, Jr., Nonvolatile Random Access Memories in Wide Bandgap Semiconductors

Volume 51A Identification of Defects in Semiconductors

G. D. Watkins, EPR and ENDOR Studies of Defects in Semiconductors
J.-M. Spaeth, Magneto-Optical and Electrical Detection of Paramagnetic Resonance in Semiconductors
T. A. Kennedy and E. R. Glaser, Magnetic Resonance of Epitaxial Layers Detected by Photoluminescence
K. H. Chow, B. Hitti, and R. F. Kiefl, μSR on Muonium in Semiconductors and Its Relation to Hydrogen
K. Saarinen, P. Hautojärvi, and C. Corbel, Positron Annihilation Spectroscopy of Defects in Semiconductors
R. Jones and P. R. Briddon, The Ab Initio Cluster Method and the Dynamics of Defects in Semiconductors

Volume 51B Identification of Defects in Semiconductors

G. Davies, Optical Measurements of Point Defects
P. M. Mooney, Defect Identification Using Capacitance Spectroscopy
M. Stavola, Vibrational Spectroscopy of Light Element Impurities in Semiconductors
P. Schwander, W. D. Rau, C. Kisielowski, M. Gribelyuk, and A. Ourmazd, Defect Processes in Semiconductors Studied at the Atomic Level by Transmission Electron Microscopy
N. D. Jager and E. R. Weber, Scanning Tunneling Microscopy of Defects in Semiconductors

Volume 52 SiC Materials and Devices

K. Järrendahl and R. F. Davis, Materials Properties and Characterization of SiC
V. A. Dmitriev and M. G. Spencer, SiC Fabrication Technology: Growth and Doping
V. Saxena and A. J. Steckl, Building Blocks for SiC Devices: Ohmic Contacts, Schottky Contacts, and p-n Junctions
M. S. Shur, SiC Transistors
C. D. Brandt, R. C. Clarke, R. R. Siergiej, J. B. Casady, A. W. Morse, S. Sriram, and A. K. Agarwal, SiC for Applications in High-Power Electronics
R. J. Trew, SiC Microwave Devices

J. Edmond, H. Kong, G. Negley, M. Leonard, K. Doverspike, W. Weeks, A. Suvorov, D. Waltz, and C. Carter, Jr., SiC-Based UV Photodiodes and Light-Emitting Diodes

H. Morkoç, Beyond Silicon Carbide! III–V Nitride-Based Heterostructures and Devices

Volume 53 Cumulative Subject and Author Index Including Tables of Contents for Volume 1–50

Volume 54 High Pressure in Semiconductor Physics I

W. Paul, High Pressure in Semiconductor Physics: A Historical Overview

N. E. Christensen, Electronic Structure Calculations for Semiconductors under Pressure

R. J. Neimes and M. I. McMahon, Structural Transitions in the Group IV, III-V and II-VI Semiconductors Under Pressure

A. R. Goni and K. Syassen, Optical Properties of Semiconductors Under Pressure

P. Trautman, M. Baj, and J. M. Baranowski, Hydrostatic Pressure and Uniaxial Stress in Investigations of the EL2 Defect in GaAs

M. Li and P. Y. Yu, High-Pressure Study of DX Centers Using Capacitance Techniques

T. Suski, Spatial Correlations of Impurity Charges in Doped Semiconductors

N. Kuroda, Pressure Effects on the Electronic Properties of Diluted Magnetic Semiconductors

Volume 55 High Pressure in Semiconductor Physics II

D. K. Maude and J. C. Portal, Parallel Transport in Low-Dimensional Semiconductor Structures

P. C. Klipstein, Tunneling Under Pressure: High-Pressure Studies of Vertical Transport in Semiconductor Heterostructures

E. Anastassakis and M. Cardona, Phonons, Strains, and Pressure in Semiconductors

F. H. Pollak, Effects of External Uniaxial Stress on the Optical Properties of Semiconductors and Semiconductor Microstructures

A. R. Adams, M. Silver, and J. Allam, Semiconductor Optoelectronic Devices

S. Porowski and I. Grzegory, The Application of High Nitrogen Pressure in the Physics and Technology of III-N Compounds

M. Yousuf, Diamond Anvil Cells in High Pressure Studies of Semiconductors

Volume 56 Germanium Silicon: Physics and Materials

J. C. Bean, Growth Techniques and Procedures

D. E. Savage, F. Liu, V. Zielasek, and M. G. Lagally, Fundamental Crystal Growth Mechanisms

R. Hull, Misfit Strain Accommodation in SiGe Heterostructures

M. J. Shaw and M. Jaros, Fundamental Physics of Strained Layer GeSi: Quo Vadis?

F. Cerdeira, Optical Properties

S. A. Ringel and P. N. Grillot, Electronic Properties and Deep Levels in Germanium-Silicon

J. C. Campbell, Optoelectronics in Silicon and Germanium Silicon

K. Eberl, K. Brunner, and O. G. Schmidt, $Si_{1-y}C_y$ and $Si_{1-x-y}Ge_xC_y$ Alloy Layers

Volume 57 Gallium Nitride (GaN) II

R. J. Molnar, Hydride Vapor Phase Epitaxial Growth of III-V Nitrides
T. D. Moustakas, Growth of III-V Nitrides by Molecular Beam Epitaxy
Z. Liliental-Weber, Defects in Bulk GaN and Homoepitaxial Layers
C. G. Van de Walle and N. M. Johnson, Hydrogen in III-V Nitrides
W. Götz and N. M. Johnson, Characterization of Dopants and Deep Level Defects in Gallium Nitride
B. Gil, Stress Effects on Optical Properties
C. Kisielowski, Strain in GaN Thin Films and Heterostructures
J. A. Miragliotta and D. K. Wickenden, Nonlinear Optical Properties of Gallium Nitride
B. K. Meyer, Magnetic Resonance Investigations on Group III-Nitrides
M. S. Shur and M. Asif Khan, GaN and AlGaN Ultraviolet Detectors
C. H. Qiu, J. I. Pankove, and C. Rossington, III-V Nitride-Based X-ray Detectors

Volume 58 Nonlinear Optics in Semiconductors I

A. Kost, Resonant Optical Nonlinearities in Semiconductors
E. Garmire, Optical Nonlinearities in Semiconductors Enhanced by Carrier Transport
D. S. Chemla, Ultrafast Transient Nonlinear Optical Processes in Semiconductors
M. Sheik-Bahae and E. W. Van Stryland, Optical Nonlinearities in the Transparency Region of Bulk Semiconductors
J. E. Millerd, M. Ziari, and A. Partovi, Photorefractivity in Semiconductors

Volume 59 Nonlinear Optics in Semiconductors II

J. B. Khurgin, Second Order Nonlinearities and Optical Rectification
K. L. Hall, E. R. Thoen, and E. P. Ippen, Nonlinearities in Active Media
E. Hanamura, Optical Responses of Quantum Wires/Dots and Microcavities
U. Keller, Semiconductor Nonlinearities for Solid-State Laser Modelocking and Q-Switching
A. Miller, Transient Grating Studies of Carrier Diffusion and Mobility in Semiconductors

Volume 60 Self-Assembled InGaAs/GaAs Quantum Dots

Mitsuru Sugawara, Theoretical Bases of the Optical Properties of Semiconductor Quantum Nano-Structures
Yoshiaki Nakata, Yoshihiro Sugiyama, and Mitsuru Sugawara, Molecular Beam Epitaxial Growth of Self-Assembled InAs/GaAs Quantum Dots
Kohki Mukai, Mitsuru Sugawara, Mitsuru Egawa, and Nobuyuki Ohtsuka, Metalorganic Vapor Phase Epitaxial Growth of Self-Assembled InGaAs/GaAs Quantum Dots Emitting at 1.3 μm
Kohki Mukai and Mitsuru Sugawara, Optical Characterization of Quantum Dots
Kohki Mukai and Mitsuru Sugawara, The Photon Bottleneck Effect in Quantum Dots
Hajime Shoji, Self-Assembled Quantum Dot Lasers
Hiroshi Ishikawa, Applications of Quantum Dot to Optical Devices
Mitsuru Sugawara, Kohki Mukai, Hiroshi Ishikawa, Koji Otsubo, and Yoshiaki Nakata, The Latest News

Volume 61 Hydrogen in Semiconductors II

Norbert H. Nickel, Introduction to Hydrogen in Semiconductors II
Noble M. Johnson and Chris G. Van de Walle, Isolated Monatomic Hydrogen in Silicon
Yurij V. Gorelkinskii, Electron Paramagnetic Resonance Studies of Hydrogen and Hydrogen-Related Defects in Crystalline Silicon
Norbert H. Nickel, Hydrogen in Polycrystalline Silicon
Wolfhard Beyer, Hydrogen Phenomena in Hydrogenated Amorphous Silicon
Chris G. Van de Walle, Hydrogen Interactions with Polycrystalline and Amorphous Silicon — Theory
Karen M. McNamara Rutledge, Hydrogen in Polycrystalline CVD Diamond
Roger L. Lichti, Dynamics of Muonium Diffusion, Site Changes and Charge-State Transitions
Matthew D. McCluskey and Eugene E. Haller, Hydrogen in III-V and II-VI Semiconductors
S. J. Pearton and J. W. Lee, The Properties of Hydrogen in GaN and Related Alloys
Jörg Neugebauer and Chris G. Van de Walle, Theory of Hydrogen in GaN

Volume 62 Intersubband Transitions in Quantum Wells: Physics and Device Applications I

Manfred Helm, The Basic Physics of Intersubband Transitions
Jerome Faist, Carlo Sirtori, Federico Capasso, Loren N. Pfeiffer, Ken W. West, Deborah L. Sivco, and Alfred Y. Cho, Quantum Interference Effects in Intersubband Transitions
H. C. Liu, Quantum Well Infrared Photodetector Physics and Novel Devices
S. D. Gunapala and S. V. Bandara, Quantum Well Infrared Photodetector (QWIP) Focal Plane Arrays

ISBN 0-12-752172-0